CONSERVING THE SACRED
For Biodiversity Management

Editors
P.S. Ramakrishnan
K.G. Saxena
U.M. Chandrashekara

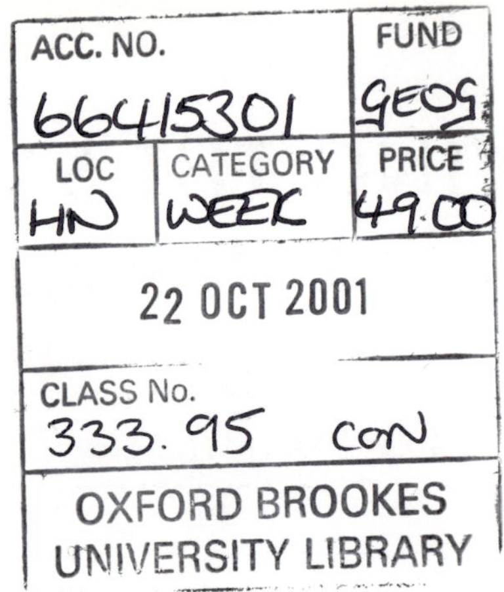

SCIENCE PUBLISHERS, INC.
Post Office Box 699
Enfield, New Hampshire 03748
United States of America

Internet site: *http://www.scipub.net*

sales@scipub.net (marketing department)
editor@scipub.net (editorial department)
info@scipub.net (for all other enquiries)

ISBN 1-57808-036-3

The designations employed and the presentation of the material in this publication do not imply the expression of any opinion whatsoever on the part of the publishers concerning the legal status of any country or territory, or of its authorities, or concerning the frontiers of any country or territory.

The authors are responsible for the choice and the presentation of the facts contained in this book and for the opinions expressed therein, which are not necessarily those of Unesco and do not commit the organization.

Printed in India

Foreword

Improving scientific understanding of natural and social processes relating to man's interactions with his environment, providing information useful to decision-making on resource use, promoting the conservation of genetic diversity as an integral part of land management, enjoining the efforts of scientists, policy-makers and local people in problem-solving ventures, mobilizing resources for field activities, strengthening of regional co-operative frameworks, these are some of the generic characteristics of UNESCO's Man and Biosphere (MAB) Programme.

Programmes on sacred sites and cultural values are considered important under UNESCO's MAB Programme, as they facilitate better understanding of these self-organized systems of traditional resource management. UNESCO, with financial support from the Dutch Embassy in New Delhi, implemented a one year study during 1997 on the 'Role of Sacred Groves in Conservation and Management of Biological Diversity in India' with partners such as the Kerala Forest Research Institute, Peechi; Madurai Kamraj University, Madurai; G.B. Pant Institute for Himalayan Environment and Development, Almora; World Wildlife Fund, New Delhi; Centre for Sustainable Environment and Heritage, New Delhi, with academic co-ordination from the School of Environmental Sciences, Jawaharlal Nehru University, New Delhi. This study examined, analyzed and documented the ecological, religious, socio-cultural and economic aspects of sacred groves, particularly, in the Western Ghats region (Kerala and Tamil Nadu), Himalayas (Garhwal and Himachal Pradesh) and Rajasthan, with the view to improving the scientific knowledge of sacred groves and for developing participatory approaches for conservation and management of their biodiversity.

This study was a prelude to a UNESCO Regional Workshop on the Role of Sacred Groves in Conservation and Management of Biological Diversity, hosted by the Kerala Forest Research Institute, Peechi, from 8-11 December 1997, and co-sponsored by the State Committee on Science, Technology and Environment in Kerala. Interesting insights were obtained — several sacred sites continue to be protected-although to a much lower degree in the present time. This can be attributed to (a) persistence-continuation of traditional practices, (b) revival-resumption of earlier practices of traditional protection-this is mainly dependent on local communities perceiving tangible losses of benefits on destruction/liquidation of sacred groves, and (c) emergence-institution of protection on a site without an earlier tradition. Following serious depletion and mismanagement of living resources by the modern industrial societies, there is now interest in revival of such self-organized systems of conservation. Their future prospect would largely depend on availability of appropriate tangible benefits to communities, and on effective community organization, and devolution of authority to them. It is reasonable to expect that these self-organized systems of conservation would continue to play an important role in times to come. We hope to carry forward, the work done so far, through a larger sub-regional research project in South and Central Asia, which would look in greater depth at issues of management and conservation of traditional systems.

The present volume 'Conserving the Sacred: For Biodiversity Management contains a wealth of information. It aims at synthesizing the current state-of-art on sacred groves, and includes presentations made during this workshop including findings of the research study. We hope that this publication will be of interest to national government, international, bilateral and multilateral donor agencies, scientists researchers, managers, NGOs and Students concerned with approaches to sustainable use of biodiversity and its management.

It is a pleasure for UNESCO to thank the Dutch Embassy in New Delhi, for the financial support which made it possible to undertake this very interesting study on sacred groves and to bring out this publication. Our sincere thanks are also due to the tireless efforts and enthusiasm of all our collaborators in this study, and to all those who prepared the synthesis paper. We would also like to thank Dr. K.S.S. Nair, Director, Kerala Forest Research Institute, Peechi, for hosting the Regional Workshop and particularly, Dr. Sankar and Dr. Chandrashekara for their valuable support and tireless efforts. We would like to record our gratitute and sincere thanks to Prof. P.S. Ramakrishnan, School of Environmental Sciences, Jwaharlal Nehru University, New Delhi, for providing academic co-ordination to both the research study and the workshop, and for his enthusiasm, diligence and perseverence in seeing the present volume into print with his very able and dedicated colleague, Dr. K.G. Saxena, who also helped us in several ways. Finally, my sincere thanks to my colleague, Ms. Sudha Mehndiratta, for conceptualizing and co-ordinating this programme.

Prof. Moegiadi
Director, UNESCO, New Delhi

Preface

Many traditional societies all over the world have often maintained a portion of the natural ecosystem type as a 'sacred grove', for cultural/religious reasons. Looking at the functional attributes of the 'sacred groves' maintained by the Khasi tribe of Meghalaya in north-eastern India, as part of a major multidisciplinary programme mounted during the early 70s, towards a comprehensive analysis of shifting agriculture in the context of forest ecosytem functioning and landuse and land cover dynamics in the region, it was realized that the available information on them are largely descriptive. More recently, a programme for a comprehensive study on such groves and their value for biodiversity conservation and management was conceptualized in close consultation with Ms. Sudha Mehndiratta of UNESCO, New Delhi. What emerged during the discussions resulted in a critical evaluation of the 'sacred', sponsored by UNESCO, from a variety of perspectives: (a) ecological, social, anthropological and cultural attributes of the sacred groves; (b) consideration of the spatial dimensions of the 'sacred', leading to species to landscape level analysis; (c) the ecological/social interphase of conserving the 'sacred', determining ecosystem/landscape level functional attributes; (d) the whole issue of managing the 'sacred' in the contemporary context of declining natural resources, land degradation and rehabilitation ecology; and (e) management related policy implications.

The research work was taken forward through a UNESCO-sponsored 1-year research project on 'Study on the Role of Sacred Groves in Conservation and Management of Biological Diversity in India' 1996-97, funded by the Dutch Government. This project had five case study sites in the Western Ghat regions of Kerala and Tamil Nadu, the Garhwal and Himachal region of the Himalaya, and in the arid zone region of Rajasthan.

Working within and outside the sacred grove zones, the ecological significance of the concept of 'sacred species' in the contemporary context of biodiversity management got further sharpened. Ecologically important keystone species were often shown to be socially selected, by rural communities, this having implication for ensuring people's participation in ecosystem rehabilitation and management on the basis of a value system with which local communities could identify themselves. This approach has added a new dimension to 'conserving the sacred' for biodiversity.

On the other hand, the opportunity that I had to ecologically evaluate the 'Demojong' sacred landscape of the Tibetan Buddists of the Sikkim region in India — a landscape, which cover a whole variety of natural and human-managed ecosystem types, right from the alpine zone reaching up to the sub-tropical humid rainforest zone, with a whole variety of traditional human societies well integrated into the landscape-gave an orienation towards a landscape approach to this study. The Rajasthan case study is an example, where the sacred groves maintained by the Bishnoi community was recognised as part of an integrated and interacting 'sacred landscape.' The 'sacred grove' forests and other natural systems along with the wildlife maintained by the Bishnois became part of a cluster of village ecosystems, contributing to many landscape functions.

Thus, this volume sponsored by UNESCO, New Delhi, emphasizes upon conserving the 'sacred', in all its spatial dimensions ranging from species to landscape levels. Starting with conceptual issues, the volume covers historical, social, cultural and ecogical perspectives in Part 1. As part of the historical development of the concept of the 'sacred' the last paper in this section raises some ethical issues, emphasizing the thin dividing line between what is often distinguished, unfortunately, as the 'profane', contrasted against the 'sacred'. Considering the sacred groves in India in detail, in view of the large body of information available from here, Part II also looks at the groves from the south Asian region, and those from other parts of the world. Part III leads on to biodiversity conservation issues in the context of ecosystem/landscape functions. Management strategies and policy considerations are also discussed in this Section.

The Volume aims at synthesizing the current state of knowledge on the subject through a series of commissioned articles from international experts and through case studies from India and from the South Asian region. In depth dicussions during a UNESCO, New Delhi sponsored workshop, entitle 'Role of Sacared Groves in Conservation and Management of Biological Diversity', at the Kerala Forest Research Institute, Peechi, India, December 8-11, 1997, formed the basis for a revision towards a greater focus on management and policy related issues. The emphasis here is therefore towards: (a) a better understanding of ecosystem functions which would contribute towards better management of the 'sacred' for biodiversity, (b) defining the concept of the 'sacred' to include spatial dimensions extending from species to landscape levels, the concept of the sacred grove falling in between, (c) a better understanding of the concept of the 'sacred' for rehabilitation of degraded landscapes, (d) designing short-term strategy for sustainable livelihood of local communities and place the same in the context of a long-term strategy for sustainable development of the landscape, and (e) evaluating management possibilities in the context of a policy framework or the absence of it. On this last point, a consensus emerged towards future research efforts emphasising more upon management and policy implications of conserving the 'sacred' in the contemporary context of rapid land degradation and desertification and declining biodiversity.

As academic coordinator of this activity, it is my previlege to record my sincere thanks to Ms. Sudha Mehndiratta, Programme Officer of UNESCO, New Delhi for her initiative and vision and to Professor Moegiadi, Director of UNESCO for his encouragement which immensely contributed to its successful completion. The financial support given by the Dutch Government through UNESCO for the 1-year research study, was liberally supplemented by the UNESCO itself, through support towards the organization of the concluding international workshop and for the publication of this Volume. The support received from the Division of Ecological Sciences, UNESCO, Paris, that provided by the State Committee on Science, Technology and Environment, Govt. of Kerala, and that given by the Kerala Forest Research Institute for hosting the December; 1997 workshop and the efforts put in by S. Sankar are acknowledged. This unique synthesis Volume would not have been possible without rigorous editorial support provided by the two co-editors, namely, K.G. Saxena and U.M. Chandrashekara, and that received from K.K. Sen, S.M. Semwal and Kirtika Singh, the enthusiastic support from all the collaborators in the 1-year research project which preceded the workshop, and of those who painstakingly prepared the synthesis papers.

This activity is a contribution to the **Special Target Area of Research, Element 10 on 'Human Dimensions', of the 'DIVERSITAS'**, an International Programme of Biodiversity Science.

P.S. Ramakrishnan

Contents

Conserving the Sacred: For Biodiversity Management
Conclusions and Recommendations

EXECUTIVE SUMMARY

P.S. Ramakrishnan

THE CONCEPT

Many traditional societies all over the world value a large number of plant species from the wild for a variety of reasons-for food, fibre, shelter or medicine. Partly, perhaps, arising out of this close human-forest linkage and partly because of anemistic belief system of the forest dwelling traditional societies (Boojh and Ramakrishnan, 1983; Kiewtam and Ramakrishnan, 1989), the protected refugia of the natural ecosystem in a given region has existed as 'sacred groves' in many societies all over the world. These groves are small patches of native vegetation types traditionally protected and managed by the local communities, though a wide range of management practices are possible. The 'sacred groves' may range in size from less than a hectare to up to a few km^2 (Box 1).

• On the one extreme, planted trees are often considered as 'sacred groves' as in many religious temple lands of India and elsewhere; it obviously is a more recent development. This, in a sense, is based on the concept of 'sacred species'. *Ficus* of the Fig family, is one such species which is culturally valued across Asia and Africa (Ramakrishnan; Khaneghah; Michaloud and Dury, among others, this Volume-see Appendix). There are many others, locally valued cultural entities too, such as *Quercus,* the Oaks in the Central Himalaya (Ramakrishnan; Sinha and Maikhuri; this Volume). The concept of the 'sacred species' could be recognised as a social evolution from the concept of 'sacred groves', through a process of condensation.

• Elaboration of the concept of the 'sacred grove' representating a given ecosystem types leads into a set of interacting ecosystem types, both natural (terrestrial, wetland and/or freshwater systems) and human-made

Box 1. Distribution of sacred groves in India (from Ramakrishnan, 1996)

In India sacred groves occur under a variety of ecological situations. They have evolved under resource-rich situations such as in Meghalaya in north-eastern India, western Ghat region in southern India or in the Bastar region in the state of Madhya Pradesh in central India. In that sense, it contradicts the view point that forest conservation measures always follow from some perception of resource scarcity, because all the above examples are from resource-rich situations, unlike the situation elsewhere, as in arid regions of Rajasthan or elsewhere in East Africa (Rodgers, 1994). Religion and culture are the over riding considerations in the Indian context.

- According to a report by the Centre for Earth System Studies in Kerala, at present there are 240 groves in the Western region of Kerala, in southern India. The largest of these (20 ha) is the famous Iringole Kavu. Many of them are linked to temples premises. These 'kavus' contain all the rare and endangered species of the western Ghats and rare medicinal plants of the region.
- Elsewhere in the Maharashtra region of the Western Ghats further north, Gadgil and Vartak (1974) report an equal number of groves locally called 'deorais', many of them being degraded by humans due to a general decline in the value system, as elsewhere in the country.
- Known as 'Sarnas', the groves of Chhotanagpur region of Bihar in northern India was established around 600 BC by the 'Munda' tribe, as an abode of their Godly spirits. Because removal of trees could be done against sacrificial offering of animals and with permission of the village priest, many of the groves are degraded (Anonymous, 1994).
- Variously called 'Vanis', 'Kenkris', 'Oraans' or 'Shamlet Deha', the groves of the arid region of Rajasthan in northwestern India act as refugia of biodiversity for the people of the desert (Anonymous, 1994).
- Then there is the story of a sect, 'Bishnois', founded about 500 years ago in the Rajasthan desert that led to absolute protection not only to the 'khejadi tree (*Prosopis cinerarea*), a multipurpose legume tree valued by the villagers, but also promote plant and animal biodiversity within their village ecosystem boundary. These trees are valued by the local people for pods for food, leaves for fodder and manure and branches as construction material. It is said that some 350 years ago, many Bishnois even laid down their lives when the Prince of Jodhpur tried to fell Khejadi trees for his lime kilns.
- In Meghalaya in the north-eastern hill region, many sacred groves are still well protected, inspite of a rapid decline in the traditional value system, with the advent of Christianity (Boojh and Ramakrishnan, 1983; Khiewtam and Ramakrishnan, 1989). The traditional religious belief is that the Gods and the spirits of the ancestors live in these groves. The Mawphlang grove close to Shillong town is one of the best preserved, set in a degraded landscape all around. Indeed, the Mawsmai grove in Cherrapunji of about 6 km^2 of protected mixed broad-leaved rainforest, though subject to some disturbance along the peripheral region, is an island in a bleak desertified landscape. Though ceremonies used to be performed regularly in this grove and others in the region, to propetiate the ruling deity, the rituals have been stopped in many of them for the last few years. Removal of plants or plant parts is considered to offend the ruling deity, leading to local calamities. We have recorded in the Cherrapunji region, 21 sacred groves with varied degrees of human disturbance (Khiewtam and Ramakrishnan, 1989).
- Elsewhere in the north-eastern State of Mizoram, there are community woodlots called 'supply forests' from which only regulated harvests permitted; then there are sacred groves which are 'safety forests' from which removal of biomass is strictly prohibited.

(agroecosystems and village ecosystems), leading to the concept of the 'sacred landscape'. Until now, the concept of sacred landscape is only some what vaguely understood. One of the best examples of a sacred landscape that one can visualize is that represented all along the course of the sacred river Ganga, originating from the higher reaches of the Garhwal Himalaya of the northwest, tracing through the plains of Uttar Pradesh, Bihar and West Bengal, before the river drains into the Bay of Bengal. Within this the hill component of the 'sacred landscape' could be a distinct entity. The sacred land, all along the cause of the river, the human habitation and the land-based activities, the temples dating back to antiquity, the sacred cities such as Badrinath, Kedarnath, Rishikesh and Haridwar in the Himalayan and sub-Himalayan tracts, Allahabad and Varanasi in the Gangetic alluvial plains, all together represent a set of inter-connected ecosystem types bound together by the sacred river itself. In this landscape, the humans can be viewed as an integral component of ecosystem/landscape functions (Ramakrishnan, 1995a, this Volume). The concept of the 'sacred landscape' is further concretized with well defined boundaries, for example, in the Demajong' landscape of the Tibetan Buddhists of the Sikkim Himlaya (Box2). The ultimate expression of this is the way in which many societies such as the traditional American Indians have viewed 'nature' itself as sacred in a more holistic sense. The 'Biosphere Reserves' conceptualized under UNESCO's 'Man and Biosphere' (MAB) Programme could be viewed as a modern interpretation of the concept of the 'sacred landscape'. However, human rights related questions are often raised in the Australia-Pacific and North American context about the validity of the biosphere concept impinging upon the local sacred sites and belief. In other words, the major concern is whether international protected areas serve to revalorise and protect local systems of values or rather to dislocate people from their symbolic ties to land? (Hay-Edie and Hadley, this volume). In any case, the 'Biosphere' initiative of UNESCO-MAB and similar initiatives from many other interest groups which emphasize upon traditional ecological knowledge, so called 'Vernacular Conservation' all linked to a 'bottom-up' approach to biodiversity conservation and natural resource management, form the basis for the more recent initiative of UNESCO on 'Cultural Sites, Cultural Integrity and Biological Diversity'.

• Understanding the concept of the 'sacred' from a philosophical perspective (Vannuccii; Saraswati; Hay-Edie and Hadley: this volume) is important to appreciate the socio-cultural value system on which this traditional biodiversity conservation technology is based.

• Burrows's paper (this volume), though it may apparently seem out of tune with the rest, is an illustrative example of some ethical issues involved

Box. 2. 'Demojong', the land of the hidden treasures! (from Ramakrishnan, 1996)

The concept of sacred landscape found holistic expression under the Buddhist philosophy of non-violence kindness to all living beings.

Padmasambhava, who is highly revered and worshipped by the Sikkimese Buddhists is considered to have blessed Yoksum and the surrounding landscape represented by 'Demojong' in West Sikkim District of Sikkim, having placed a large number of hidden treasures ('ter'). Many of these sacred treasures were hidden by Lhabstsun Namkha Jigme in the Yoksum region. It is believed that these treasures are being discovered slowly and will be revealed only to enlightened Lamas, at appropriate time. Conserving these treasures, protecting them from polluting influences is considered important for human welfare.

The area below Mount Khangchendzonga in West Sikkim, referred to as 'Demojong' is the core of the sacred land of Sikkim. Yoksum is considered to be a 'Lhakhang' (altar) and 'Mandala' where the protective deities are made offerings to. No meaningful performance of Buddhist rituals are possible if this land and water is desecrated. Any large-scale human-induced perturbation in the land of the holy Yoksum region would destroy the hidden treasures, the 'Ters', in such a manner that the chances of recovering them sometimes in the future by a visionary will diminish (it is said that the last such discovery was made by Terton Padma Lingpa, when he lived 540 years ago). Any major perturbation to the river system would disturb the ruling deities of the 109 hidden lakes of the river, thus leading to serious calamities (Here they quote the example of the lake 'Khecho-Palri' that is suggested to have moved away from the river during a period of bloodshed, sometimes in the past.

Here, ecological considerations cannot be separated from historical, social, cultural and religious dimensions of the problem. Here is a sacred landscape where the people living in the region are truly integrated within the landscape unit itself, in a socio-economic sense. The variety of natural ecosystem types ranging from the alpine vegetation above the timberline, through the temperate oak and pine forests down below, the sub-tropical moist-deciduous to dry-deciduous forests in the plains and a variety of human-altered ecosystems terraced agriculture and valley land agriculture all along are tightly linked together and controlled by the sacred river and its tributaries in a variety of ways through flooding and silt deposition.

Indeed, the very cultural fabric of the Sikkimese society is obviously dependent upon the conservation of the whole sacred landscape of interacting ecosystems, as was evident during discussions this author had with respected religious leaders and a cross section of Sikkimese society, cutting across religious, cultural and professional backgrounds of the people. The issue here is not merely a question of protecting a few physical structures or ruins. The uniqueness of this heritage site is that the value system here is interpreted in a more holistic sense-the soil, the water, the biota, the visible water bodies, the river and the less obvious notional lakes on the river bed, are all to be taken together with the physical monuments.

in dealing with the concept of the 'sacred'. The commodification of the 'sacred' is indicative of the thin dividing line between what is often distinguished, unfortunately, as the 'profane', constrasted against the 'sacred'.

THE CASE STUDIES

- Equally important are the structural attributes of these systems. A substantial amount of work done deal with 'sacred groves'; the descriptive analysis deal with floristic/faunistic listings, vegetation analysis, social/ anthropological analysis (many papers in Part II). Such descriptive accounts are no doubt very valuable to evaluate the conserved biodiversity in the context of the otherwise heavily exploited larger landscape in the region.

• Socio-cultural insights from a variety of studies suggest;

(i) that a variety of taboos and prohibitions serve to conserve biodiversity within 'sacred groves': (ii) some of the taboos may be internal to the forest such as removal of plants or plant parts from within: (iii) whilst others may be related to sharing the information with the uninitiated outsiders, such as the medicinal value of the biodiversity within: (iv) in many countries in southeast Asia, for example, sacred groves are often linked to the preservation of water springs; (v) amongst many societies all over, sacred landscape is often linked to the mountains–the "Demajong' landscape in West Sikkim (Ramakrishnan, this volume) and the Tongariro landscape of the Maoris in New Zealand (Hay-Edie and Hadley, this Volume) are examples: (vi) among many traditional societies that are animistic, the 'scared groves' are considered to be the abode for their ancestral spirits or their Gods representing natural forces/phenomena; (vii) although 'sacred groves' traditionally are often maintained in a pristine state all over (Hughes and Chandran, this volume), many are disintegrating due to distrubances; (viii) however, it is not uncommon to find a fine distinction being implied between 'small perturbations' versus 'large perturbation', in the case of 'sacred landscapes' where local communities live within and engage themselves in a variety of livelihood related activities, only causing small-scale perturbations-the 'Demojong' landscape in West Sikkim is a typical example of this; however, the society protested against industrial activities that cause large-scale perturbations within (Ramakrishnan, this Volume).

• From an ecological perspective the structural analysis of these case studies lead to the following generalizations: (i) that the 'sacred groves' may range in size from less than a hectare to a few km^2. (ii) the term 'sacred groves' have been used to denote complex relict ecosystems on the one extreme, to a small cluster of planted trees on the other; obviously this is a definitional problem with implications for biodiversity conservation and management; (iii) a major conclusion in almost all case studies is that many of the 'sacred groves' are at varied levels of degradation on the face of onslaught from modernity and neglect; the 'sacred groves' of the ancient Mediterranean area of Europe under Roman and Greek empire and their virtual elimination under christianization of the Roman empire (Hughes, this Volume) is a story which has repeated itself to a great extent in the north-eastern hill region in India (Ramakrishnan, this Volume; (iv) many of the case studies are illustrative of the very limited information available on this very important area of research, across different countries and regions around the world.

Functional Attributes, Conservation, Management and Policy

• While considering natural ecosystems in a purely biological sense, addition or deletion of species, functional or structural groups, and indeed ecosystem components, are sometimes shown to be of no/limited consequence in altering a given ecosystem level process. However what is often missed out is that different functional groups control/govern different process in an ecosystem, with a certain degree of redundancy attached with them, in a manner that others could take over similar functions, though to a limited extent when the system has to cope up with uncertainties in the environment. Biodiversity becomes crucial for the system as a whole and indeed at the landscape level, as a buffer against perturbations, and for maintaining overall ecological integrity in the context of 'global change' which has become a major global concern (Walker and Steffen, 1997). It is in the context of this functional perspective of biodiversity, this Volume on 'Conserving the Sacred: For Biodiversity' (Ramakrishnan *et al.*, 1998) becomes significant.

• Ecologically valuable keystone species performing key functions in the ecosystem and thereby contributing to support/enhance biodiversity, generally are also species that are socially selected and valued by the local communities, for cultural or religious reasons (Ramakrishnan, this Volume). A few selected examples: (a) the Nepalese alder (*Alnus nepalensis*) coming up as an early succesional tree species in the shifting agriculture (jhum) fallow plots and protected by the local communities from slashing and burning during the 'jhum' operations, conserve nitrogen within the system. Being and early successional species, it indeed influences ecosystem processes, in space (where it occurs) and in time (during forest fallow succession); (b) Bamboo species, *Dendrocalamus hamiltonii, Bambusa tulda* and *B. khasiana*, in north-east India, which are important early successional species and which conserve nitrogen, phosphorus or potassium in the system are also culturally valued by all communities, with a variety of uses for village ecosystem functions; (c) the Oaks (*Quercus*), culturally valued in the Kumaon and Garhwal region of the western Himalaya is an important component of the mountain forest ecosystem functioning, for their role in improving soil fertility, promoting efficient nutrient cycling, conserving soil moisture partly through humus build-up and partly through a deeply placed and uniformly distributed root system in the soil profile.

• This linkage that exists between ecologically significant keystone species and socially selected ones has potential for modulating ecosystem level processes, and ensure people's participation in ecosystem management. Many such species are found both within and outside the sacred groves, which could be also used for rehabilitating degraded landscapes (Box 3), and for sustainable development of natural resources including biodiversity. For many traditional societies who are custodians of biodiversity and who use it in a variety of livelihood concerns, managing biodiversity is crucial for their sustainable development (Ramakrishnan, 1996a).

- Understanding the functional attributes of relict ecosystems placed in an otherwise highly degraded landscape, which often is the case, can give clues as to the ecosystem properties relevant to ecosystem rehabilitation. Thus, for example, in the Mawsmai 'sacred grove' forest of Cherrapunji area in north-east India, one could see a dense fine root biomass sitting on top of the nutrient-deficient thin mineral soil, which ensures tight nutrient cycling to support the dense biomass of the highly fragile humid forest ecosystem (Ramakrishnan, this Volume). Once disturbed, this delicately; balanced ecosystem degrades to extensively desertified grassland. From a management point of view, this example emphasizes the role of soil fertility

Box 3. General assumptions related to rural ecosystem rehabilitation (from Ramakrishnan *et al.*, 1994a)

1. Rehabilitation and management would only succeed if short-term economic benefits are assured to local communities, apart from long-term benefits envisaged.
2. If rehabilitation and management strategies are to be effective and successful, womens' participation is necessary.
3. Without a broad understanding of the complexities of the system (through rapid appraisal methodology), rehabilitation strategies may not succeed.
4. Unless ecosystem rehabilitation and management leads to a general improvement and maintenance of soil fertility and water quality it is not sustainable.
5. Ecosystem rehabilitation will be sustainable only if: (a) internal control of processes (e.g. resource recycling) within the ecosystem are strengthened, (b) dependence on external subsidies (e.g., fertilizers) are minimized, and (c) self-regenerating capacities enhanced, to the extent feasible.
6. In order to succeed, ecosystem rehabilitation should have strong community participation in planning, management, implementation, and continuous monitoring of all these parameters.
7. Unless rights and responsibilities of ownership are clearly defined and understood by all the participants, ecosystem rehabilitation is not likely to succeed.
8. If community participation is to be effective, community/user group institutions will have to be built into the rehabilitation strategy.
9. Unless land capability analysis and classification, taking into consideration scientific/traditional knowledge is integrated, rehabilitation work will not be effective and sustainable.
10. Empowerment (training, institutional, access to facilities and resources) of local communities in general and vulnerable sections (landless and women) in particular is crucial for the success of any rehabilitation programme.
11. In order that rehabilitation work is sustainable, surface and ground water resources and its exploitation is monitored and appropriately regulated through institutional mechanism.

and nutrient cycling processes in maintenance of ecosystem equilibrium state and its resilience; this understanding is crucial for designing any rehabilitation strategy for the otherwise balded Cherrapunji landscape.

• Sacred landscapes are often linked to a mountain environment, and invariably linked to water as a resource, such as the Ganga River system in the Garhwal Himalaya, the 'Demajong' landscape in Sikkim with a number of lakes and the Rothang Chu river drainage system and many in south-east Asian region linked to a water source. Even some of the 'sacred' groves in many parts of the world are linked with some kind of spring water source (Tingsabadh, 1995). This has implications for watershed management. Through many of our studies in the Indian region under a monsoonic climate, this author has shown that water could be used as a triggering agent for biodiversity management, rehabilitation of degraded land through accelerated succession and indeed with overall implications for sustainable land use management and rural development (Ramakrishnan, 1992, 1995b, this Volume); Ramakrishnan *et al.*, 1994 a,b,c, 1996a, 1996b). By providing water outside the monsoon period through appropriately designed cheap eco-technology, we have been able to show that a variety of land use systems received a great spurt, right across the Himalayan region, over a range of rainfall situations (less than 200 cm to more than 12 m). Indeed, this traditional knowledge was already encompassed within the concept of the 'sacred'.

• The 'sacred landscape' concept, in a sense, could be visualized as a precursor to the more modern emphasis on watershed as a unit for development and management. Landscape ecology is an emerging area for research in ecology; the inter-connectedness of the landscape data back to antiquity, through the concept of the 'sacred landscape'. Therefore, some of them could be important candidates for priority research.

• With rapid and continued decline occurring in the quality and the number of sacred groves there is an urgent need to document and monitor the existing groves, analyse the scientific basis of these relict ecosystem functional units and evaluate their value for biodiversity conservation and management. The efforts in this direction should be well coordinated between all actors in the play, namely, a variety of relevant governmental agencies, non-governmental voluntary agencies and grass-root level action groups, a variety of socio-cultural and religious bodies and the scientific community at large.

• The research and management should be part of a larger effort, extending beyond the sacred sites themselves, to develop a meaningful action plan for natural resource management at the landscape level, based on traditional knowledge and technology in regions where traditional societies live. In doing this ecological perceptions should be closely knit with socio-economic and cultural concerns of local communities so as to ensure their participation both in conserving and indeed enhancing the quality of these groves. In this context, the locally recognized and accepted land tenure system or institutional arrangements are issues to be taken into account. Joint management committees, with various agencies involved, that may be appropriate for improving the quality of the groves/landscapes, and to use them for local benefit to the extent feasible, should be embedded, to a large extent, within the local traditional values.

• Identifying and conserving selected large groves that are well preserved and which are located in critical ecological zones should be taken up on a priority basis. Indeed, the governments concerned should initiate steps to declare a few of the selected groves representing different ecological zones in the country as 'National Hertiage' sites, for effective conservation and use (e.g., eco-tourism), wherever feasible. The Mawsmai forest in Cherrapunji is an example, because of its geographical significance and ecological fragility. The economically valuable germ plasm, such as medicinal plants, could then be sustainably obtained through cultivation outside the sacred grove, as part of a larger rural development plan.

• All identifiable sacred landscapes, such as the Ganga system, the 'Demojong' of the West Sikkim, Sabarimalai in the Western Ghat region of Kerala that are identified from India, and Adam's Peak in Sri Lanka are some examples, that could be fit candidates for declaration as 'National Heritage' sites, with a view towards the 'World Heritage' status of UNESCO. The conservation of the natural resources contained therein should be linked to a well conceived sustainable development action plan. These cultural landscape with traditional societies living within, and with all natural and managed biodiversity within them forming part of an interconnected and interacting ecological landscape, should be rigorously researched upon with a view to link conservation with sustainable livelihood/development, as part of a landscape management plan (Box 4).

Box 4. Loosely coupled landscape management strategy (Ehrenfeld, 1991)
• Management plan should stimulate loose coupling within the system, rather than maximize tight coupling. Such a strategy would reduce the system's vulnerability to accidents. • Avoid manipulative management plan based on large, prioritized, single-protocol, single theory, generalized scheme. Instead for large protected areas have only a low key management plan, if required. Have individualized management for fragmented landscapes to accommodate singularities and uniqueness of each piece an ensure loose coupling. • In the ultimate analysis, management strategy should be flexible, capable of small-scale operations, information-sensitive, and composed of elements that are integrated and yet independent.

• An issue that also came up again and again during discussions was related to studies centred around management strategies for 'sacred groves/ landscapes. Economic incentives, developing and managing buffer zones around, mobilization of public opinion, strengthening traditional management practices while constantly reviewing and analyzing management plans and incorporating them from time to time, arriving at appropriate legal framework for management networking interested individuals and agencies for exchange of information were some of the identified key issues.

• In the ultimate analysis, enough data exist on conservation biology, but not enough on those aspects linking biological conservation with cultural integrity. Filling this gap is considered crucial as it will provide an additional tool for biodiversity conservation and management.

REFERENCES

Anonymous, 1994. The spirit of sanctuary. Down to Earth, 3:21-36.

Boojh, R. and Ramakrishnan, P.S. 1983. Sacred groves and their role in environmental conservation. In: *Strategies for Environmental Management.* Souvenir Vol. Dept. of Science and Environment of Uttar Pradesh, Lucknow. pp. 6-8

Ehrenfeld, D. 1991. The management of biodiversity: A conservation paradox. In: F.H. Bormann and S.R. Kellert (Eds.) *Ecology, Economics, Ethics: The Broken Circle*, pp. 26-39. Yale Univ. Press, New Haven.

Gadgil, M. and Vartak, V.D. 1976. The sacred groves of the western Ghats in India *Econ. Bot.* 30: 152-160.

Khiewtam, R.S. and Ramakrishnan, P.S. 1989. Socio-cultural studies of the sacred groves at Cherrapunji and adjoining areas in north-eastern India. *Man in India*, 69: 64-71.

Khiewtain, R.S and Ramakrishnan, P.S. 1993. Litter and fine root dynamics of a relict sacred grove forest at Cherrapunji in north-eastern India. *Forest Ecol. Manage.* 60: 327-344.

Ramakrishnan, P.S. 1992. *Shifting Agriculture and Sustainable Development of North-Eastern India.* UNESCO-MAB Series, Paris, Parthenon Publ., Carnforth, Lancs, U.K. 424 pp. (republished by Oxford University Press, New Delhi, 1993).

Ramakrishnan, P.S. 1995a. Biodiversity and ecosystem function: the human dimension. In: F. di Castri and T. Younes. *Biodiversity, Science and Development: Towards a New Partnership.* pp. 114-129. CAB International, London.

Ramkrishnan, P.S. 1995b. Currencies for measuring sustainability: Case studies from Asian highlands. In: M. Munasinghe and W. Shearer (Eds.) *Defining and Measuring Sustainability: The Biogeophysical Foundations.* pp. 193-205, U.N. University, Tokyo & World Bank, Washington, D.C.

Ramakrishnan, P.S. 1996. Conserving the Sacred: From Species to Landscapes. Nature and Resources, UNESCO, 32: 11-19.

Ramakrishnan, P.S. Campbell, J., Demierre, L., Gyi, A., Malhotra, K.C. Mehndiratta, S., Rai, S.N., & Sashidharan, E.M. (Eds.) 1994a. *Ecosystem Rehabilitation of the Rural Landscape in South and Central Asia: An Analysis of Issues.* Special Publication, UNESCO (ROSTCA), New Delhi. 29pp.

Ramakrishnan, P.S. Purohit, A.N., Saxena, K.G. and Rao, K.S. (Eds.) 1994. *Himalayan Environment and Sustainable Development.* Diamond Jubilee Publ., Indian Nat. Sci. Acad., New Delhi. 84pp.

Ramakrishnan, P.S., Saxena, K.G., Swift, M.J. and Seward, P.A. 1994 c. *Tropical Soil Biology and Fertility Research: South Asian Context.* G.B. Pant Institute of Himalayan Environment and Development Kosi & Oriental Enterprises, Dehra Dun.

Ramakrishnan, P.S., Das, A.K. and Saxena, K.G. (Eds.) 1996a. *Conserving Biodiversity for Sustainable Development.* Indian National Science Academy, New Delhi, 246pp.

Ramakrishnan, P.S., Purohit, A.N., Saxena, K.G., Rao, K.S., and Maikhuri, R.K. (Eds.) 1996b. *Conservation and Management of Biological Resources in Himalaya.* UNESCO, G.B. Pant Institute of Himalayan Environment & Development, Kosi-Katarmal, and Oxford & IBH Publ. Co. Pvt. Ltd., New Delhi, 603pp.

Ramakrishnan, P.S. and Saxena, K.G. and Chandrashekara, U.M. 1998. *Conserving the Sacred: For Biodiversity Management,* UNESCO Vol., Oxford & IBH Publ., New Delhi, pp.480.

Rodgers, W.A. 1994. The sacred groves of Meghalaya. Man in India, 74:339-348.

Tingsabadh, C. 1995. Thailand. In: S. Samad, T. Watanabe and S. Kim (Eds.) *People's Initiatives for Sustainable Development: Lessons of Experience.* pp. 215-247. Asia & Pacific Development Centre, Kuala Lumpur.

Walker, B. and Steffen, W. (Eds). 1997. The Terrestrial Biosphere and Global Change: Implications for Natural and Managed Ecosystems—A Synthesis of GCTE and Related Research. IGBP, Sweden. 32 pp.

List of Contributors

Mr. A. Pandey, Indian Institute of Forest Management, Bhopal, India

Mr. Ahmed Emdadul Hoque, Ecology Laboratory, Department of Botany University of Dhaka, Bangladesh

Prof. Anil K. Gupta, Indian Institute of Management, Vastrapur, Ahmeddabad - 380 015, India

Ms. Aparna Watve, Applied Environmental Research Foundation, 917/7 Ganeshwadi, Near British Council Library, Pune - 499 004, India

Dr. Archana Godbole, Applied Environmental Research Foundation, 917/7 Ganeshwadi, Near British Council Library, Pune - 499 004, India

Dr. Arvind G. Raddi, PCCF (Rtd.), 809, Bhandarkar Institute Road, Pune - 411 004, India

Dr. Asghar Askari Khaneghah, Anthropologist, Associate Professor, Facutly of Social Sciences, University of Tehran, Tehran

Dr. S.K. Barik, Centre for Environmental Studies, North Eastern Hill University Shillong - 793 014, India

Mr. Bhaskar Sinha, School of Environmental Sciences, Jawaharlal Nehru University, New Delhi-110 067, India

Dr. Beth Burrows, President/Director, The Edmond Institute, 20319, 92nd Avenue West, Edmonds, Washington 98020, USA

Dr. S. Chand Basha, Principal Chief Conservator of Forests (Retd.), B-503, Kanchanjunga Apartments, Palarivattom, Kochi-25, Kerala, India

Mr. S. Chandrasekaran, Department of Plant Sciences, Madurai Kamaraj University, Madurai - 625 021, Tamil Nadu, India

Dr. U.M. Chandrashekara, Scientist, Kerala Forest Research Institute, Peechi, 680 653, Trichur, Kerala, India

Dr. J. Donald Hughes, John Evans Professor of History, University of Denver, Denver, Colorado - 80208, U.S.A.

Dr. S. Dury, Delegation CIRAD Cameroon, BP 2572, Yaounde, Cameroon

Prof. ABM. Enayat Hossain, Department of Botany, Jahangir University, Savar Dhaka, Bangladesh

Dr. M.G. Gogate, Chief Conservator of Forests (Wildlife), Maharashtra State Forest Department, Jaica Building, Civil Lanes, Nagpur-440 001, India

Dr. Gopal Shankar Singh, House No. 289, Sector-8, R.K. Puram, New Delhi, India

Dr. Harsh Vardhan, World Wide Fund for Nature-India, Pirojeshah Godrej National, Conservation Centre, 172-B, Lodi Estate, New Delhi-110 003, India

Dr. Hemantha Withanage, Senior Environmental Scientist, Environmental Foundation Ltd., No. 3 Campbell Terrace, Colombo-10, Sri Lanka

Mr. A.H. Hussain, Ministry of Planning, Human Resources and Environment Male (Republic of Maldives)

Mr. Jayant Sarnaik, Applied Environmental Research Foundation, 917/7 Ganeshwadi, Near British Council Library, Pune - 499 004, India

Prof. Kailash C. Malhotra, Professor of Human Ecology, Indian Statistical Institute, 203 BT Road, Calcutta - 700 035, India

Mr. Krishna Kumar, World Wide Fund for Nature-India, Pirojeshah Godrej National, Conservation Centre, 172-B, Lodi Estate, New Delhi - 110 003, India

Dr. P.N. Krishnan, Tropical Botanical Garden and Research Institute, Thiruvanathapuram - 695 562, Kerala, India

Prof. Madhav Gadgil, Centre for Ecological Studies, Indian Institute of Science Bangalore - 560 012, India

Dr. R.K. Maikhuri, G.B. Pant Institute of Himalayan Environment and Development, Garhwal Unit, Srinagar (Garhwal), India

Dr. Malcolm Hadley, Division of Ecological Sciences, UNESCO, 7, Place de Fontenoy, F-75700 Paris, France

Ms. Manju Jha, World Wide Fund for Nature-India, Pirojeshah Godrej National Conservation Centre, 172-B, Lodi Estate, New Delhi - 110 003, India

Dr. Marta Vannucci, C/O UNDP, 55, Lodi Estate, New Delhi - 110003, India

Dr. G. Michaloud, ICEM (UMR 5554), CNRS et Universite de Montpellier II, Laboratoire de Botanique, 163 rue Auguste Broussonet, 34090 Montpellier, France

Mr. Mohamed Anwarul Islam, Department of Zoology, Jahangir Nagar University, Savar, Bangladesh

Mr. Mohamed Zaman, Member, Academy of Scinces of Afghanistan, Chemistry and Biology Department, Kabul (Afghanistan)

Dr. A.K.M. Nazrul Islam, Ecology Laboratory, Department of Botany, University of Dhaka

Dr. P. Pushpangadan, Director, Tropical Botanical Garden and Research Institute, Thiruvanathapuram - 695 562, Kerala, India

Mr. Raghunandan Velankar, Ranwa, 16, Swastishree Society, Ganesh Nagar, Pune - 411 052, India

Dr. M. Rajendraprasad, Tropical Botanical Garden and Research Institute Thiruvanathapuram - 695 562, Kerala, India

Prof. P.S. Ramakrishnan, School of Environmental Sciences, Jawaharlal Nehru University, New Delhi - 110 067

Dr. K.S. Rao, G.B. Pant Institute of Himalayan Environment and Development Kosi-Katarmal, Almora - 263643, Uttar Pradesh, India.

Dr. Sanjay Deshmukh, Principal Scientist, Bombay Natural History Society Hornbill House, Shaheed Bharat Singh Marg, Mumbai - 400 023, India.

Dr. S. Sankar, Scientist-in-Charge, Division of Agroforestry-cum-Publicity, Kerala Forest Research Institute, Peechi, 680 653, Thrissur, Kerala India

Prof. B.N. Saraswati, UNESCO Professor, Indira Gandhi National Centre, for the Arts, Janpath, New Delhi - 110 001, India

Dr. A.R.K. Sastry, World Wide Fund for Nature-India, Pirojeshah Godrej National, Conservation Centre, 172-B, Lodi Estate, New Delhi - 110 003, India

Dr. K.G. Saxena, School of Environmental Sciences, Jawaharlal Nehru University, New Delhi - 110 067, India

Dr. Sayeeda Wafar, Biological Oceanography Division, National Institute of Oceanography, Dona Paula - 403004, Goa, India

Dr. M.D. Subash Chandran, Professor, Dr. Baliga College, Kumta, North Canara, Karnataka, India

Dr. P.S. Sudhakar Swamy, Department of Plant Sciences, Madurai Kamaraj University, Madurai - 625 021, Tamil Nadu, India

Mr. Sudipto Chatterjee, World Wide Fund for Nature-India, Pirojeshah Godrej National, Conservation Centre, 172-B, Lodi Estate, New Delhi - 110 003, India

Mr. S.M. Sundarpandian, Department of Plant Sciences, Madurai Kamaraj University, Madurai - 625 021, Tamil Nadu, India

Dr. Suprava Patnaik, Indian Institute of Forest Management, Nehru Nagar, PO. Box - 357, Bhopal - 462 003, India

Mrs. Swapna Prabhu, Applied Environmental Research Foundation, 91717, Ganga-Tara Apts., Ganeshwadi, Near British Council Library, Pune-411 004, India

Mr. Terence Hay-Edie, Division of Ecological Sciences, UNESCO, 7, Place de Fontenoy, F-75700 Paris, France

Dr. B.K. Tiwari, Centre for Environmental Studies, North Eastern Hill University, Shillong - 793 014, India

Dr. Thomas Schaaf, Programme Specialist, Division of ecological Sciences UNESCO, 7, Place de Fontenoy, F-75700 Paris, France

Dr. R.S. Tripathi, Department of Botany, North Eastern Hill University, Shillong - 793 014, India.

Dr. A.G. Untawale, Biological Oceanography Division, National Institute of Oceanography, Dona Paula - 403004, Goa, India

Mrs. Uranbaigal Gongorin, Specialist, Land Management Department, Ulaanbaatar City, Ulaanbaatar-46, Jigjidjab street-4, Mongolia

Dr. M. Wafar, Biological Oceanography Division, National Institute of Oceanography, Dona Paula - 403004, Goa, India

Mr. Yogesh Gokhale, Ranwa, 16, Swastishree Society, Ganesh Nagar, Pune - 411 052, India

Part 1

CONCEPTUAL, HISTORICAL AND SOCIO-CULTURAL

1

Conserving the Sacred for Biodiversity: The Conceptual Framework

P.S. Ramakrishnan

School of Environmental Sciences, Jawaharlal Nehru University
New Delhi 110067, India

INTRODUCTION

Many traditional societies all over the world value a large number of plant species from the wild for a variety of reasons, for food, fibre, shelter or medicine. Partly, perhaps, arising out of this close human-forest linkage and partly because of anemistic belief system of the forest dwelling traditional societies, the protected refugia of the natural ecosystem in a given region has existed as 'sacred groves' in many societies all over the world (Hughes and Chandran, this Volume). These groves are small patches of native vegetation types traditionally protected and managed by the local communities, though a wide range of management practices are possible. They may range in size from less than a hectare to up to a few km^2. Much of these groves have disappeared more or less completely as in most of Europe except for a few exceptional ones reported in the Kanin tundra (Hughes and Chandran, this Volume). Elsewhere, they are at varied levels of protection. On the other extreme, planted trees being considered as 'sacred groves' as in many religious temple lands of India and elsewhere, obviously is a more recent development. This, in a sense, is an elaboration of the concept of 'sacred species' discussed here.

The traditional pre-Colombian native American belief that the Earth with all the living creatures contained there in are all sacred (Hughes and Chandran, this volume) is a more holistic view point of Ecology. However, the traditional north American view point where the natives believed that nature itself is sacred has seen a complete reversal in the value system

(Burrows, this Volume). In any case a more holistic view point of nature as perceived by the American Indians is also prevalent in many parts of Asia and Africa through the concept of the 'sacred landscape', though in a more concertised way with boundaries clearly demarcated. The purpose here is to set the tone for a whole variety of issues discussed in the subsequent Chapters through an ecological analysis of the 'concept of the sacred' from a biodiversity perspective, in the Indian context, largely drawing upon my experiences from the better understood north-east Indian situation.

CONCEPT OF THE SACRED IN AN ECOLOGICAL CONTEXT

Stretching into the pre-historic times, the concept of sacred grove in India has its roots in antiquity, even before the Vedic age, the Vedas representing the only recorded remains of the thoughts of the ancient Aryans who migrated into this sub-continent. In their migration from the steppes of the Central Asia through Balkh in Khorassan (Dandekar, 1979) to the Indian sub-continent, the ancient Vedic people of pre-historic times assimilated new environmental values; they also incorporated into their value system the concept of the 'sacred grove' from the original inhabitants of the Indian sub-continent (Vannucci, 1994). Though sacredness attached to species is perhaps more recent, being part of the post-vedic Hindu ritualism. Thus the already existing ecosystem level concept of the 'sacred grove' of the original pre-Vedic inhabitants of India was extended by the Vedic migrants down to the 'species' level on one extreme of the scale and to the level of the 'landscape' on the other extreme (Ramakrishnan, 1996).

Buddhism and Jainism initially branching out as revivalistic religious off-shoots of Hinduism lead to revivalism of conservation practices too. On one extreme it lead to a sect of Jains, the 'Digambara Jains', set dead against killing of all living organisms; at the other end of the spectrum is the sacred landscape of the Sikkimese Buddhists based on a holistic ecological philosophy.

Here we look at the conceptual framework determining the evolution of the sacred in the Indian context, using selected examples to illustrate this.

The Sacred Groves

Distributed throughout India, the current level of protection and conservation of sacred groves in India is hampered by erosion in traditional value systems. It is gratifying to note, however, that these traditional methods of social fencing of ecosystem types as conservation patches is being rediscovered. Whilst there are many reports on general aspects of the sacred groves (Box 1; refer to the papers in this volume), the ecological value of the species within them and the ecosystem function of most of these are yet to be fully evaluated. In many landscapes, the sacred grove is a representative of the relict climax vegetation which existed once upon a time and is now destroyed through a variety of perturbations caused through human activities. In that sense, these islands of biodiversity have the potential for biodiversity conservation in a variety of ways. The following account of a typical grove from Meghalaya in north-east India is indicative of the conceptual framework within which one could ask questions while dealing with them for biodiversity management.

Sacred Groves in North-east India

North-eastern hill regions of India constitute an important component of the Eastern Himalayan complex, a 'hot spot' of biodiversity (Ramakrishnan, 1992). This region has a rich ecological diversity, ranging from the tropical humid forests of lower elevations, through the sub-temperate forests of mid-elevational ranges, to the alpine vegetation of high elevations. This ecological complexity is compounded through a rich human cultural diversity with over 100 different tribes originating from different parts of south and south-east Asia and bringing with them their own distinct cultures and having evolved and adapted to local situations in isolation from one another due to linguistic barriers. It is in this context of bio-cultural diversity that the concept of the 'sacred groves' is embedded in the agro-forest landscape. Whilst we know so little about this eco-cultural interface, the Meghalayan situation is indicative of the yet to be documented large body of information embedded in the tribal culture.

Sacred groves have always been part of the cultural life of the people of Meghalaya (Khiewtam and Ramakrishnan, 1989; Tiwari, et al., this Volume). Whilst, sometimes before the advent of Christianity in Meghalaya, each village is suggested to have had a sacred grove, at present there are only about 79 groves reported in varied levels of protection (Tiwari, et al., this Volume).

A large number of them are located in the Cherrapunji region of the East Khasi Hills District. Here, there are two types of groves: (a) those that are considered traditionally sacred and therefore ideally not even a fallen twig could be removed from within and where a whole variety of religious beliefs and practices exist and (b) those that are protected reserve forest groves, declared to be so in recent times by the village headman with the concurrence of the village council, where limited extraction of timber is permitted (Kiewtam and Ramakrishnan, 1989).

Mawsmai Sacred Grove: A Typical Example

The climax forest in Cherrapunji in Meghalaya in north-east India, which is now represented only by a relic sacred grove forest, protected by the local Khasi tribe for religious/cultural reasons, has developed in one of the wettest spots in the world with an annual average rainfall of about 10370 mm which in an exceptional year may increase to 24550 mm, as in 1974. Obviously, the humid sub-tropical forests which were extensive in the past, are developed on highly leached nutrient deficient soils and are therefore very fragile (Kiewtam and Ramakrishnan, 1993). The fragility of these humid forests are compounded by the limestone formations beneath them, that have extensive underground tunnels of stalactites and stalagmites (Kratz topography); in spite of the limestone formation the soil is highly acidic (pH 3.9–5.2) and with poor nutrient supply (Ramakrishnan, 1992).

Indeed, perturbation of these forests results in degraded seral grasslands representing arrested successional stages (Ram and Ramakrishnan, 1992), and in extreme situations ending up in a balded landscape. It is not surprising then that the landscape of Cherrapunji, except for the protected forest patches or the only relict sacred grove at Mawsmai, is largely balded or transformed to arrested seral grassland types (Ramakrishnan and Ram, 1988). The Cherrapunji ecosystem that now stands desertified due to deforestation inflicted sometime in the distant past now refuses to recover to its original state. The fact that shifting agriculture (locally called 'jhum'), which is the predominant landuse in the region, around Cherrapunji remains banned by the village council for quite sometime now is suggestive of the part played by this landuse in creating the present landscape. It is in this context of extreme case of ecosystem fragility that the Mawsmai sacred grove gains added significance.

Linked with the drastic loss in biodiversity is the human suffering which now is immense (Ramakrishnan, 1992). Water is a scarce commodity during the dry months in this high rainfall spot of the world, a contender for being the wettest spot on Earth along with the nearby Mawsengram in the Khasi hills of Meghalaya and the far-away spot in the Hawaiian islands of the Pacific. All the water flows away into the plains down below, as here there is no vegetation to hold the water and the soil, and there is no soil to recover the forest. It is a vicious circle!

For fuelwood, the tribal villager has to trek long distances. The ruins of villages all around remind one of the yester years and migrations of the past. The tribal, who is traditionally bound to the land and forests has been forced to seek other avenues for survival.

Though the sacred grove forest vegetation itself is somewhat stunted because of unbalanced soil characteristics and extreme climatic conditions, the species diversity is very high with a variety of tree, shrub, herb, climber and epiphyte species, forming a tightly packed system. Among the trees in the grove, *Engelhardtia spicata* is the dominant species followed by tree species such as *Syzygium communii, Echinocarpus dasycarpus, Drymycarpus racemosus* and *Elaeocarpus lanceaefolius*. Important shrubs are *Clerodendron nutans, Phlogocanthus* spp., *Randia* spp., *Thelypteris* spp., *Strobilanthes coloratus* and *Boehemeria platyphylla*.

With a very high annual litterfall and with rapid litter decomposition rate comparable to lower montane rainforests elsewhere, nutrient release rate in the grove is very high (Khiewtam and Ramakrishnan, 1993). The soil itself has little nutrients to offer to support a large biomass of the sacred grove. The fine root mat (Fig. 1) developed on the surface layers of the soil and that particularly located over the mineral soil are important for supporting the large aboveground biomass and for tight cycling of nutrients. The nutrients released from the decomposing litter is short-circuited back into the living biomass through the surface root mat, even before the nutrients could leach out into the mineral soil. Operating under highly stressed environmental conditions, this delicately balanced sacred grove forest ecosystem is highly fragile. The sharp boundary between these sacred grove and the balded landscape indicates that the system will never recover through natural processes of revegetation. Artificially, the cost for ecosystem rehabilitation could be enormous. What one sees now around the sacred grove is stunted grassy growth protected within prock crevices like pock-marks on a rocky landscape.

Fig. 1. A Sacred Grove in the north-east.

The Sacred Landscape

Elaboration of the concept of the sacred grove representing a given ecosystem type leads into a set of interacting ecosystem types, both natural (terrestrial, wetland and/or freshwater systems) and human-made (agroecosystems and indeed village ecosystems), in which humans can be viewed as an integral component of ecosystem/landscape functions. The ultimate expression of this is the way in which many societies such as the traditional American Indians have viewed 'nature' itself as sacred in a more holistic sense.

Until now, the concept of sacred landscape is only some what vaguely understood. One of the best examples of a sacred landscape that I can visualise is that represented all along the course of the sacred river Ganga, originating from the higher reaches of the Garhwal Himalaya of the northwest, tracing through the plains of Uttar Pradesh, Bihar and West Bengal, before the river drains into the Bay of Bengal. The sacred land, all along the course of the river, the human habitation and the land based activities, the temples dating back to antiquity, the sacred cities such as Badrinath, Kedarnath, Rishikesh and Haridwar in the Himalayan and sub-Himalayan tracts, Allahabad and Varanasi in the Gangetic alluvial plains, all together represent a set of inter-connected ecosystem types bound together by the sacred river itself. The variety of natural ecosystem types ranging from the alpine vegetation above the timberline, through the temperate oak and pine forests down below, the sub-tropical moist-deciduous to dry-deciduous forests in the plains and a variety of human-altered ecosystems terraced agriculture and valley land agriculture all along are tightly linked together and controlled by the sacred river and its tributaries in a variety of ways through flooding and silt deposition. The sacred groves of the Bishnois in Rajasthan (Box 1) viewed as units around each Bishnoi village could indeed be enlarged to encompass a cluster of villages forming a landscape unit to include interacting ecosystem types such as agriculture, animal husbandry and domestic units of the village ecosystem, natural water bodies, and the protected natural ecosystem ! In a conceptual sense, this is an area which needs further exploration for its value as a conservation tool.

The concept of sacred landscape found further holistic expression under the Buddhist philosophy of non-violence and kindness to all living beings. Whilst the concept of sacred grove is now well recognised through many studies from Asia and Africa, I am not aware of any documentation of the 'sacred landscape' concept. It is in this context, the example detailed below assumes significance.

Box 1. Distribution of Sacred Groves in India

In India sacred groves occur under a variety of ecological situations. They have evolved under resource-rich situations such as in Meghalaya in north-eastern India, western Ghat region in southern India or in the Bastar region in the State of Madhya Pradesh in central India. In that sense, it contradicts the view point that forest conservation measures always follow from some perception of resource scarcity, because all the above examples are from resource-rich situations, unlike the situation elsewhere, as in arid regions of Rajasthan or elsewhere in East Africa (Rodgers, 1994). Religion and culture are the over riding considerations in the Indian context.

- According to a report by the Centre for Earth System Studies in Kerala, at present there are 240 groves in the Western region of Kerala, in southern India. The largest of these (20 ha) is the famous Iringole Kavu. Many of them are linked to temples premises. These 'kavus' contain all the rare and endangered species of the western Ghats and rare medicinal plants of the region.
- Elsewhere in the Maharashtra region of the Western Ghats further north, Gadgil and Vartak (1976) report an equal number of groves locally called 'deorais', many of them being degraded by humans due to a general decline in the value system, as elsewhere in the country.
- Known as 'Sarnas', the groves of Chhotanagpur region of Bihar in northern India was established around 600 BC by the 'Munda' tribe, as an abode of their Godly spirits. Because removal of trees could be done against sacrificial offering of animals and with permission of the village priest, many of the groves are degraded.
- Variously called 'Vanis', 'Kenkris', 'Orans' or 'Shamlet Dehs', the groves of the arid region of Rajasthan in northwestern India act as refugia of biodiversity for the people of the desert.
- Then there is the story of a sect, 'Bishnois', founded about 500 years ago in the Rajasthan desert that led to absolute protection not only to the 'khejadi tree (*Prosopis cinerarea*), a multipurpose legume tree valued by the villagers, but also promote plant and animal biodiversity within their village ecosystem boundary. These trees are valued by the local people for pods for food, leaves for fodder and manure and branches as construction material. It is said that some 350 years ago, many Bishnois even laid down their lives when the Prince of Jodhpur tried to fell Khejadi trees for his lime kilns.
- In Meghalaya in the north-eastern hill region, many sacred groves are still well protected, inspite of a rapid decline in the traditional value system, with the advent of Christianity (Boojh and Ramakrishnan, 1983; Khiewtam and Ramakrishnan, 1989). The traditional religious belief is that the Gods and the spirits of the ancestors live in these groves. The Mawphlang grove close to Shillong town is one of the best preserved, set in a degraded landscape all around. Indeed, the Mawsmai grove in Cherrapunji of about 6 km^2 of protected mixed broad-leaved rainforest, though subject to some disturbance along the peripheral region, is an island in a bleak desertified landscape. Though ceremonies used to be performed regularly in this grove and others in the region, to propitiate the ruling deity, the rituals have been stopped in many of them for the last few years. Removal of plants or plant parts is considered to offend the ruling deity, leading to local calamities. We have recorded in the Cherrapunji region, 21 sacred groves with varied degrees of human disturbance (Khiewtam and Ramakrishnan, 1989).
- Elsewhere in the north-eastern State of Mizoram, there are community woodlots called 'supply forests' from which only regulated harvests permitted; then there are sacred groves which are 'safety forests' from which removal of biomass is strictly prohibited.

Demojong in West Sikkim District—An Unique Example of a Sacred Landscape

Sikkim has a long tradition of Buddhist religion. Buddhism is practised by about 25% of the local population and the majority religion is Hinduism (70%). Because of the rule by the Chogyal dynasty, since the time the first Chogyal (king) of Sikkim was crowned in 1642 at Norbugang in Yuksom. Buddhist traditions are deeply ingrained into the psyche of the Sikkimese people. This is evident in all walks of life—a rich tapestry woven with Buddhist symbolisms, legends, myths, rituals and festivals, the typical Sikkimese building architecture, and the large number of monasteries and 'Stupas' doting the landscape through out the State. It is important to note that these traditions are shared by all the three communities; the unique culture so developed is a blend of the Buddhism of the 'Lepchas' and the 'Bhutias' and the Hinduism of the majority 'Nepalis'.

Of the four Buddhist sects, the Nyngmepa, Kagupa, Gelugpa and Sakyapa, represented in the State, the Nyngmepa sect initiated by the Buddha incarnate, Maha Guru Padmasambhava is most significant. Whilst Sikkim as a whole is considered to be sacred by the Sikkimese Buddhists, according to the sacred text 'Nay Sol', the area below Mount Khangchendzonga in west Sikkim, referred to as 'Demojong' is the most sacred of all, being the abode of Sikkim's deities. Interestingly the air, soil, water and the biota are all sacred to the people, because of the interconnections that they perceive to exist. Any human-induced perturbation is considered by the Sikkimese Buddhist's to spell disaster to Sikkim as a whole, because of the disturbance caused to the ruling deities and the treasures ('ters') placed in the landscape (Box 2). Interestingly, it is believed that there is no way of knowing where the 'ters' are hidden, as they will be revealed only to the right person when the right time comes.

This region has a number of glacial lakes in the higher reaches. These are sacred lakes. The Rathong Chu, itself a sacred river, is said to have its source in nine holy lakes of the higher elevation, closer to the mountain peaks. Besides, the river in the Yoksum region itself is considered to have 109 hidden lakes (Fig. 2). These visible and the less obvious notional lakes identified by the religious visionaries are said to have presiding deities, representing both the good and the evil. Propitiating these deities through various religious ceremonies is considered important for the welfare of the Sikkimese people. Indeed, conserving this rich tradition is considered to be significant for peace, harmony and welfare of not only this area but also for the region as a whole.

It is no wonder that Rathong Chu is the focus of religious rituals. During the 'Bum Chu' ritual, considered holiest of all festivals, held annually

Box 2. 'Demojong', the land of the hidden treasures!

Padmasambhava, who is highly revered and worshipped by the Sikkimese Buddhists is considered to have blessed Yoksum and the surrounding landscape represented by 'Demojong' in West Sikkim District of Sikkim, having placed a large number of hidden treasures ('ter'). Many of these sacred treasures were hidden by Lhabstsun Namkha Jigme in the Yoksum region. It is believed that these treasures are being discovered slowly and will be revealed only to enlightened Lamas, at appropriate times. Conserving these treasures, protecting them from polluting influences is considered important for human welfare.

The area below Mount Khangchendzonga in West Sikkim, referred to as 'Demojong' is the core of the sacred land of Sikkim. Yoksum is considered to be a 'Lhakhang' (altar) and 'Mandala' where the protective deities are made offerings to. No meaningful performance of Buddhist rituals are possible if this land and water is desecrated. Any large-scale human-induced perturbation in the land of the holy Yoksum region would destroy the hidden treasures, the 'Ters', in such a manner that the chances of recovering them sometimes in the future by a visionary will diminish (it is said that the last such discovery was made by Terton Padma Lingpa, when he lived 540 years ago). Any major perturbation to the river system would disturb the ruling deities of the 109 hidden lakes of the river, thus leading to serious calamities (Here they quote the example of the lake 'Khecho-Palri' that is suggested to have moved away from the river during a period of bloodshed, sometimes in the past.

Indeed, the very cultural fabric of the Sikkimese society is obviously dependent upon the conservation of the whole sacred landscape of interacting ecosystems, as was evident during discussions this author had with respected religious leaders and a cross section of the Sikkimese society, cutting across religious, cultural and professional backgrounds of the people. The issue here is not merely a question of protecting a few physical structures or ruins. The uniqueness of this heritage site is that the value system here is interpreted in a more holistic sense the soil, the water, the biota, the visible water bodies, the river and the less obvious notional lakes on the river bed, are all to be taken together with the physical monuments.

at Tashiding, Rathong Chu is said to turn white and start singing, and this is the water to be collected at the point where Rathong Chu meets Ringnya Chu. Attracting thousands of devotees from the State and the neighbouring region, 'Bom Chu' ritual is predictive in nature, in that it is suggested to be indicative of coming events-possible calamities and prosperity for the people of Sikkim. The water kept in vases, if it overflows is indicative of prosperity. Decline in water level is indicative of bad events such as drought, diseases, etc. Turbid water is indicative of unrest and conflicts.

More generalised rituals, such as the one done throughout Sikkim by the Buddhists during 'Pang-Lhabsol' to propitiate the various ruling deities of the mountain peak of the Khangchendzonga, the midlands represented by the Yoksum region, and the lowlands down below, is indicative of the widespread respect with which this sacred landscape region is worshipped by the people.

Of the total catchment area of 328000 ha of Rathang Khola, 28510 ha is under snow cover. The vegetation is varied ranging from alpine scrub vegetation at higher reaches to sub-tropical moist evergreen forests down

Fig. 2. Damojong Land of Hidden Treasures: Pictorial depiction of holy sites in West Sikkim, according to the 'Neysol' text of Tibetan Buddhism. Names with 'Tso' attached signify a lake, while 'Chu' in the term used for local rivers. The sacred landscape of Damojong stretches from Khangchendzonga peak down to sub-tropical forest below, and includes forests, lakes and rivers, as well as historical monuments and systems of terraced agriculture. (Drawing: courtesy of Concerned Citizens of Sikkim, Gangtok.)

below. Afforestation is essential over an area of 4290 ha catchment area of the Rothang Chu. This task is crucial for conservation of the sacred landscape as a whole and for its ecological integrity. Implementing the above mentioned activity is crucial for controlling erosion and flash floods. According to the data collated by the Himalayan Nature & Adventure Foundation, Siliguri, the Ouglthang and Rathong glaciers are retreating rapidly, with size reduction. Retreating glaciers create several moraine dams, containing sizable quantities of water. Increased snow melt and exceptionally high rainfall in a given year could result in dam bursts and flash floods in the lower regions. The 1988 dam burst is an example that could recur, if adequate catchment treatment measures are not initiated immediately.

Sikkim Himalaya is richly endowed with biological resources spread over a variety of ecosystem types over a range of altitude, from the alpine *Rhododendron* dominated scrub forest through conifer forests with *Abies densa* and *Tsuga demosa* getting down to mixed evergreen forests dominated by species such as *Castanopsis* spp., *Quercus lamellosa, Lithocarpus spicatus, Elaeocarpus lanceaefolius, Machilus edulis, Michelia* spp., etc. The region under consideration here has all these types over a very short transect running down from the alpine to the sub-tropical zone. Orchids are abundant. With rich wildlife represented by Himalayan Black Bear, Musk Deer, Fishing Cat, Leopard Cat, Black Capped Langur and a rich bird life, this unique landscape unit should be protected.

The region is rich in medicinal plants of value to traditional Tibetan pharmacopoeia, nurtured in Sikkim by the Buddhist monasteries (Tsarong, 1995). Conserving these plants and their cultivation would ensure the survival of one of the oldest system of medicine stretching back to more than 2500 years. With recent attempts to revive traditional medicine, possibilities of providing sustainable livelihood to local communities through cultivation of medicinal plants are immense.

Yoksum is an area which what Sikkimese perceive as the very basis for the present Sikkimese culture. The entire region right from the Kangchendzonga to the Yoksum lowlands (the sacred region of Demojong) is most appropriate to be declared a 'National Heritage Site', with all its people ecological and cultural heritage—the land and the landuse systems (the traditional terraced agricultural system included), all the water bodies (the obvious and notional lakes), the Yoksum Chu, the monasteries, the historical sites, and the rich biodiversity for conservation in a truly holistic spirit.

The Sacred Species

A reductionsitic view of the sacred grove concept will lead to the concept of the 'sacred species'. The concept of the sacred species is well recognised by many traditional societies all over the world. Thus, it is reasonable to assume that the traditional Hindu society recognised individual species as objects of worship, based on accumulated empirical knowledge and their identified value for one reason or the other. Thus, *Ficus religiosa* (the Peepal tree) and other species of the same genus form components of a variety of ecosystem types and support a variety of plant and animal biodiversity. Species belonging to the Fig family are culturally valued by many societies in the Asian and the African continent (Micheloud, this Volume). The sacred Basil called 'Tulsi' *(Ocimum sanctum)* is worshipped in all traditional homes as a Goddess, and indeed is a multipurpose medicinal plant according to the traditional Indian pharmacopoeia.

Others may not be worshipped in a religious sense, but form part of the socio-cultural traditions. The socially valued multi-purpose *Quercus* species of Kumaon and Garhwal region of the western Himalayas are important fodder and fuelwood species. They are also considered to serve a variety of functions as an important component of the mountain forest ecosystem function. They help in improving soil fertility through efficient nutrient cycling, conserving soil moisture partly through humus build-up in the soil and partly through a deeply placed root system which has root biomass uniformly distributed throughout the soil profile. Consequently, they support a rich biodiversity in the ecosystem. It may be noted here that Oaks are the climax species in the mid-elevation regions in the western

Box 3. The 'Chipko' Movement in the Himalayas

The now well known 'Chipko movement' (the movement of hugging the trees) is a unique grassroot level conservation movement against wanton destruction of forests, originating in the western Himalayan Garhwal region. Started off as a small protest movement in the sleepy small town of Gopeshwar against cutting a few ash trees by a sports factory, the women of the region got involved in a big way, the movement reaching its climax in the mid seventies. Hugging the trees against cutting it is a unique protest movement where women, who are worst affected because of scarcity of natural resources caused by deforestation (all strenuous household chores such as collection of water, fodder and fuelwood), eventually took the lead role, though the movement itself was initiated by a grass-root level social worker in Mr. Chandi Prasad Bhatt and propagated by the environmentalist, Mr. Sunderlal Bahuguna.

The genesis of this movement could be traced to both ecological and economic considerations. Conversion of oak forests of socio-economic value to pine plantations of industrial value or to barren lands were important considerations for the local communities. Rapid decline in soil fertility, increased soil erosion often caused by flash floods, and the local perception associating these with large-scale forest conversions had already reached alarming proportions. Though 'Chipko' movement had the value of raising environmental awareness amongst the local communities to its highest pitch, unfortunately, after more than 20 years of raised expectations through the movement, positive action towards sustainable development of the Himalayan region needs much to be desired (Ramakrishnan, 1992).

Himalayas. Being valuable timber species, there has been large scale extraction of timber during the past few decades, and the subsequent large-scale conversion of the climax oak forests to early successional pine forests; pines are favoured by the foresters being fast growing and in demand for resins. This has done considerable ecological damage in the region, making the soil more acidic, adversely affecting nutrient cycling and soil fertility, almost forming a monoculture with little undergrowth and low biodiversity. No wonder that oaks are socially valued and is the focal point for much of the folk music and dance of the Kumaonis and the Garhwalis. The genesis behind the now well known 'Chipko' movement (Box 3) could be traced to the felt hardships of the local people, as a consequence of large-scale conversion of oak to pine forests.

REFERENCES

Boojh, R. and Ramakrishnan, P.S. 1983. Sacred groves and their role in environmental conservation. In: *Strategies for Environmental Management*. Souvenir Vol. Dept. of Science and Environment of Uttar Pradesh, Lucknow. pp. 6-8.

Burrows, B. Where nothing is sacred. In this volume.

Dandekar, R.N. 1979. *Vedic Mythological Tracts. Select Writings, I & II.* Ajanta Publ., Delhi.

Gadgil, M. and Vartak, V.D. 1976. The sacred groves of the western Ghats in India. *Economic Botany, 30:* 152-160.

Hughes, J.D. and Chandran, M.D.S. Global spread of sacred groves. In this volume

Khiwtam, R.S. and Ramakrishnan, P.S. 1989. Socio-cultural studies of the sacred groves at Cherrapunji and adjoining areas in north-eastern India. *Man in India, 69*: 64-71.

Khiewtam, R.S. and Ramakrishnan, P.S. 1993. Litter and fine root dynamics of a relict sacred grove forest at Cherrapunji in north-eastern India. Forest Ecology and Management,60: 327-344.

Michaloud. Sacred groves in Africa In this volume.

Ram, S.C. and Ramakrishnan, P.S. 1992. Fire and nutrient cycling in seral grasslnads of Cherrapunji, *International Journal of Wildland Fire, 2*: 131:138.

Ramakrishnan, P.S. 1992. *Shifting Agriculture and Sustainable Development of North-Eastern India*. UNESCO-MAB Series, Paris, Parthenon Publ., Carnforth, Lancs. U.K. 424 pp. (republished by Oxford University Press, New Delhi, 1993).

Ramakrishnan, P.S. 1995. Biodiversity and ecosystem function: the human dimension. In: F. di Castri and T. Younes (eds.). *Biodiversity, Science and Development: Towards a New Partnership*. CAB International, London (in press)

Ramakrishnan, P.S. 1996. Conserving the sacred: from species to landscapes. *Nature and Resource. 32:*11-19.

Ramakrishnan, P.S. and Ram, S.C. 1988. Vegetation, biomass and productivity of seral grasslands of Cherrapunji in north-east India. *Vegetatio, 74*: 47-53.

Rodgers, W.A. 1994. The sacred groves of Meghalaya. *Man in India, 74*: 339-348.

Tiwari, B.K., Barik, S.K. and Tripathi, R.S. Sacred gorevs of Meghalaya. In this volume.

Tsarong, T.J. 1995. Tibetan medicinal plats: An agenda for cultivation. In: R.C. Sundriyal and E. Sharma (eds.) *Cultivation of Medicinal Plants and Orchids in Sikkim Himalaya*. G.B. Pant Inst. of Himalayan Environment & Development, Kosi, Almora. pp. 75-79.

Vannucci, M. 1994. *Ecological Readings in the Veda*. D.K. Printword (P) Ltd., New Delhi.

2

Sacredness and Sacred Forests

M. Vannucci

International Society for Mangrove Ecosystems (ISME)
College of Agriculture, Univeristy of the Rhukyus, Okinawa, Japan

THE CONCEPT OF SACRED

The word sacred in English, from the Latin Sacrum, comes from the same root as *sachch*: Truth in the Indian tradition. Truth is the self-existent, the Absolute, the all in all; Truth is what exists in the Cosmos, it is the Cosmos itself, both material and immaterial. The knowledge of Truth is not only God-given, it is also first and foremost the result of diligent study, observation, logical deductions as well as meditation and introspection. Knowledge is necessary for understanding whatever we wish to understand. The perception and knowledge of Truth, or in Vedic terms: "the unveiling of Truth", is the path that leads to Wisdom. Without wisdom there can be no correct behaviour; therefore, to put it in contemporary language, there cannot be rational use and management of whatever on Earth, living and non living.

Because of this thinking, Truth is sacred by definition. The concept of sacredness therefore also implies the perception of the existence of something, or much, not yet fully understood. Man knows that he is part of the whole of nature, but the whole is awe inspiring, imperfectly understood and must therefore be treated with care and respect. It would be out of place to discuss here the difference between sacred and divine, but it should be noted that, for instance, it would be incorrect to speak of "divine groves" or "divine forests". Sacrality is different from divinity.

There are many aspects of Truth, some immaterial such as spiritual Truths, or the concept of Life and Living, some material such as the act of living or the existence of living and non living bodies and matter. As a logical consequences, all living bodies and things or associations of living beings are to be respected as sacred manifestations, not just symbols, and also as embodiments of the Eternal mysterious. Being, the *Esse* or Ultimate Reality, the one that *IS*, without a second. The Absolute Truth is manifested and regulated by *Rita*, the Cosmic Law and Order to which we are all bound by invisible fetters: the fetters held by the very ancient, God, Asura Varuna who guides and regulates, rewards and punishes for good or for bad, everything in the Cosmos, living and non-living. In present day language we would say that the laws of Nature, the laws of the physical and biological worlds regulate all and every phenomenon; the laws of physics, of matter and energy are inescapable.

Modern as well as ancient science is not capable of stating clearly what makes the difference of a living body before and after death. In other words, the essence of the abstract concept expressed by the words Life and Living are still now undefinable with exactitude. This is obvious, if nothing else, in the present day discussions on the determination of the onset of death of the human body for the purpose of organ transplants. Experimental sciences searches for tangible, measurable proofs; ancient man revered the unexplainable phenomena of nature with awe and fear of punishment if he dared trespass the laws of *Rita*, which he perceived and knew only empirically. Whenever he could understand the relations of cause and effect he performed meaningful rites that were propitiatory and at the same time enabled him to participate in the dynamics of the cosmos (Vannucci, 1995b).

For the ancients, everything that was not fully understood was sacred, vastly because it was mysterious; since the concepts of Life and Living are vastly unexplained, as a corollary, all living things had to be sacred. Sacredness begins with Fire, Lord *Agni*, the primeval God (Vannucci, 1995b). *Asura Agni* is very much alive, since he is energy itself, manifested under many different forms including life energy which is a special manifestation of Fire and as such has a special name: *Matarisvan* while the personification of the concept or abstract idea of life, is *Savitr*, the Great Impeller, *Savitr* is sometimes identified with the Sun, *Surya*, as for instance in the current understanding of the *Gayatri mantra*, because the Sun is recognized as the giver of life on Earth.

ECOSYSTEMS OF SPECIAL SIGNIFICANCE

Some natural elements, ecosystems and renewable resources are recognized as of special importance for the maintenace of man and other creatures. Such elements should be specially studied "for the purpose of understanding" as *rishi* Vasistha says (RigVeda, mandala VII) and only then can they be used and managed appropriately. On land, trees and forests are as important as earth and water. In the littoral zone of north Australia there are sacred places in the sea (*dhuya wanga*) and in the Tokelau atolls there are sacred fishes and turtles called *ika ha* that conceptually correspond respectively to the sacred forests and sacred trees of people inhabiting large islands or inland areas. The sea is of primeval importance for islanders, specially those of small islands.

Since man is part of the nature in which he lives, man realized very early that he belongs to a complex network of exchange of matter and energy within the system and of the system with other ecosystems. If any element of the system fails to function correctly, according to the laws of *Rita*, the whole system may collapse. For instance, if rains fail, if rivers and ponds dry out, one by one the other elements of the system either die out or are forced to emigrate out of their system into a new one.

As mentioned, some ecosystems are particularly meaningful: such are forest and the marine environment for island and coastal people. A grove of *Prosopis* spp. with a few other associated species of plants and animals, are as important to the Vishnoi of the Rajasthan desert as the rich mangrove forests are to the Bengalis and Santals of the Sunderbans. Both are sacred to the human dwellers, though each has different presiding deities, each is prayed to in a different manner and inevitably each is used and managed in a particular manner that can range from total protection and untouchability throughout the year or only at definite seasons, to actual intensive use, according to customary habits. An oasis in the Sahara is a much more vulnerable and fragile ecosystem than a rich evergreen forest south of the Sahel; though not specifically designated as "sacred forests" or" sacred landscapes" oases are protected by traditional law and customs which no traveller would dare disregard. Customary law acquires a sacredeness of its own and may be even more stringent than religious law, as implied by the term "sacred". Thus violation of a *dhuyu wanga* (sacred place) in the littoral zone inhabited by the yolngu of North Australia calls for action by the Aboriginal Custodian, with immediate consequences for the trespasser that range from mild sickness to death, as recorded of many incidents (Davis, 1985). Punishment for trespassing the laws of *Rita*, a concept nowadays partially covered by the word Nature, are the more stringent, the ecosystem concerned is more marginal and fragile and they are respected even in the time of war. What more efficient than religious injuction, traditional law and social ostracism to ensure obedience to the laws of nature where and when they have been well defined by the clarification of the relationships between cause and effect. If our forefathers had not been knowledgeable and wise, we would not be here today.

Hymn RigVeda X: 146 is addressed to the Lady of the Forest, *Aranyani*, "she who tills not yet has stores of food"; a sentence that clearly expresses the concept of sacred forest; she becomes fearful and destructive towards those who trespass her law. *Aranyani* is the invisible personification of the forest she "seems to vanish from sight" as one advances in the forest; she is the spirit, the concept, the form or essence, the energy, the virtual reality and intellectual formulation of the forest. She is feminine, because as a mother she gives her all asking only for devotion, care and love in return. She is unforgiving towards those who harm her and through her, harm her obedient children. She is the idea of the unending exchange of matter and energy that goes on in the forest and in the clearings where the cattle graze, she rejoices with the sounds of life in herself, she has stores of food which she gives off freely to the deserving ones, according to the law. There can hardly be a better rendering of the concept of the forest ecosystem in poetical yet correct terms.

The tree is the most prominent element of the forest, hence it becomes the materialized symbol of the bounties offered by the forest. In India, already before the invasion by the Indo-Europeans, the local people worshipped the trees, some of them in a particular manner, the pipul (*Ficus religiosa*) being the most important among them, as a symbol of life and regeneration. In all cultures, certain aspects of knowledge become stereotyped and the original meaning or causes of the belief are forgotten with time. Many of these are labelled as superstitions, or "habits' devoid of meaning. Among the local people of India, there is a legend that refers to the '*kalpavriksha*' or tree of plenty. This is particularly alive among a tribal community that performs special rites at appointed seasons in honour of the Tree who grants all needs and desires to its honest and respectful worshippers. The symbolism of the tree plays an important role also in Buddhism, not only because Siddhartha of the Lichchhavi clan of Aryans received enlightenment under the bodhi tree (the tree of wisdom) and became Gautam Buddha, but mostly because the *kalpavriksha* fulfills all desires, both material and spiritual, to the sincere seeker. But there is more to the forest than just trees; vines, creepers and herbs are equally given due recognition; creepers and trees in Buddhist and Hindu iconography are present even on coins and at Shravanabelgola where the naked Jain saint Bahubali is shown entwined by creepers grown during his long standing penance.

Forest people have the tree as a symbol of plenty and of life, the Aryans who came to the Indian sub-continent in successive waves and who were originally shepherds and agricultural migrants, had the '*Kamadhenu*', or the cow of plenty who would grant all the food and all the desires to its owner and, even more important, she would grant or deny at will also intellectual capability according to her judgement.

Both Kalpavriksha and Kamadhenu are symbols of nature's generosity, if, and only if nature is revered, respected, obeyed and used according to its own rules and laws. We should note here that there is no "Sacred agriculture", because the rules, regulations and laws of agriculture are manmade. However, the so much discredited jhum or shifting agriculture system has been shown not to be damaging when practiced in the age-old traditional manner, according to traditional wisdom and empirical knowledge handed down generation to generation and only in specific areas that could take such a system of cultivation. For instance in the evergreen humid forests of the north-east states of India, lasting damage and degradation occur when the rotation cycle is shortened from 30 or more years to fifteen, seven or even shorter periods (Ramakrishnan, 1992). There is not instance of "sacred agriculture" as such that I know of, though important plants may be cultivated with special respect and even devotion, such as barley used for offerings to the gods in preference to wheat, medicinal and aromatic plants as for instance the basil (*Ocimum sanctum*). One reason why barley has kept its sanctity in preference to wheat, may be that it was domesticated by man much earlier. An exception to the above and an example of "sacred agriculture" could be the cultivation of hexaploid wheat in the plains of Attika (Greece) where, at Eleusys, goodess Demetria taught the mysteries, the celebration of which lasted for over 2000 years. Demetria instituted her own cult and introduced methodical agriculture in Greece and, as I have shown, she probably introduced hexaploid wheat in that part of the world (Vannucci and Bieloric, in press). The river Ilyssos that waters the land was considered to be sacred and I wonder whether this was not the beginning of water management practices in the lands of the northern rim of the Mediterranean Sea.

In conclusion, we may say that biological species and ecosystems of particular significance acquire a sanctity of their own above all others.

ALL NATURE BEING SACRED, WHY "SACRED FORESTS"?

Sacredness implies a notion of distance from all bodily sense organs, except the mind. However, Hindus are spiritually intimate with their *Ishtadevata* (personal deity), Christians are intimate with Christ, the Vergin Mary and their chosen saint. Islam precludes all intimacy with Allah, but frequently one of the thousand names of Allah is in people's speech and thought, as a spiritual testimonial.

In Hinduism there is a deeply ingrained feeling of the belonging of man in the nature of which he is a part. In other faiths, the people see themselves as part of their environment, though the philosophy of their official faith teaches that the use of nature's bounties is a matter of fact to the extent that man was made the centre of the universe with unconditional rights to exploit all and everything in nature that could satisfy his needs and greeds. We now witness the results of this foolish attitude. In practice, to this day, village and mountain people and those who live in close intimacy with nature know better and respect the laws of nature, alternatively called *Rita*, Traditional Law, Habit, *dhuyu wanga* or whatever. A practice that regrettably is quickly disappearing under the pressure of modernisation and development.

The more sacred the concept and more distant and impersonal the idea, greater the human need for a tangible object or just a thought that can yoke him to the sacred, the unthinkable, the unexplainable. The instinctive fear of "night, darkness and cold" (slumber and insecurity, ignorance and death) that froze *Indra* when he pursued the demons that had taken refuge in night be the winner in the "struggle for like" (Avesta); but man also needs the joys that only light, knowledge and life can give. Inevitably man, who is not particularly gifted physically, need to explore nature's gifts of food and shelter for rearing healthy offsprings he also must bow to natural selection to participate in the requirements of the cycling of matter and energy, hence the sacrifice, *yajna*, which is essentially a participatory rite that like all rites and celebrations brings the joys that life can give. The Holy Communion of the Christians is one such participatory rite celebrating the oneness of man and nature. Wheat, a substitute of barley and wine made of grapes are the sacred elements of nature of which all Indo-European cultures partake. I have good reasons to believe that the Vedic *soma* is grapes' wine, but this will be the object of another paper.

Although human foolishness (called Human Stupidity by H.G. Wells) is immense, streaks of wisdom have remained alive everywhere in all cultures. I would like to recall only a few of the innumerable examples of habits that are reminiscent of ancestral wisdom. As a child in Tuscany, Italy, we spent our vacations on the family farms, near Florence. I remember asking my great-grandmother Palmira, why we kept horse-chestnuts in addition to good chestnut trees and why Jesus Christ was welcomed on his arrival at Jerusalem with palm tree fronds on Ester Eve, while we collected olive tree fronds instead. The answers were of this type: "the world would not be so beautiful if there were only one kind of trees" -or : "in Palestine palm trees grow in the desert, here the olive tree grows also on harsh rocky soil; both show that the Divine Providence can produce good fruit even from poor places". Traditional wisdom respected and celebrated biodiversity all over the world.

Ancient India was mostly covered with forests until when the Mogul emperors considered it meritorious to hunt wild animals and clear the forest for the "protection" of the people. Intensive deforestation continued under British colonial rule which legalised the practice by the Forest Act of 1878; this in practice transferred the ownership of the forests from the people to the State. Forests were sacked, plundered and finished off first for making sleepers and for firing the boilers of the locomotives when the railways were introduced in India, from 1853 onwards, and then for building the large wooden ship fleet that reached its peak during the Second World War. During the war period alone, over 6,000 square miles of forest were devastated; were later turned into agricultural lands, others were abandoned to their own sort; many are now listed as "wastelands" though it is not mentioned that the wastelands are man made more often than not. India was not the only sufferer, Nepal and Burma also contributed actively to such "developmental" and "war" efforts. The distant areas of Kafiristhan, later called Nuristhan, Gilgit, Hunza, were until very recently still heavily forested with oak trees and the *deodar* (*Cedrus deodara*), or "god's tree"; they were preserved and sparingly used according to traditional law and custom and the forest was sacred, managed since centuries and millennia by the local community. Unfortunately, those areas have succumbed to the greed of commercial powers: the rivers are no longer rivers of water but rivers of impressive logs that took centuries to be formed, pushed downstream by the same people who earlier lived by the forest and who are now exploited helpless destroyers of what was their own wealth.

Perhaps even worse than that ruthless destruction, is the loss of respect for the forest and the loss of knowledge and understanding of the structure and dynamics of the ecosystem. Sad consequence of the growing greed of "developers" and commercial enterprises. People's wisdom, where some may still be found, goes unheeded and purposely brushed aside. Now there is a sense of awakening, and villagers beg for help to reforest their hills, because the "Himalayas are coming down".

Tradition recognises forests as one of the main life supporting system and as a climate regulator; it recognises the direct and indirect benefits they give to mankind, such as food and materials, but mainly for the role they play for soil conservation and in the water cycle. Forests continue to be the object of reasoned and varied forms of protective worship. Even in the outskirts of Islamabad in Pakistan, I have seen garlands of flowers resting on *pipul* branches or other trees. Early morning tree worship is currently done everywhere in India even in the heart of the developed urban environment.

Why then sacred, or specially sacred forests? Simply because to people living on land the ecological role of vegetation cover is well known and forests are the most impressive and the primeval vegetation cover.

SPECIALLY SACRED FORESTS

These are the forests that have remained relatively or totally unexploited and are still in a pristine condition, specially rich in biodiversity.

If all nature is sacred, how can there be some specially sacred forests? Sacredness implies mystery, secrecy and untouchability. The Cosmos is full of mysteries, meaning acts and phenomena that are beyond the understanding of Man the Observer. Gradually Man the Thinker unveils the secrets of nature. He becomes Man, the Doer when he uses his knowledge of nature; thus when man understands the *how* of many natural phenomena even if he does not understand *the why*, he learns to act accordingly: he uses the environment and its elements to the best of his ability, including the notion of preservation for future needs and future generations. Man, aided by *Vak*, (the word or speech same root as "vocal" in English, from the Latin: *Vox, vocis*: voice) teaches the new generation not only by example but also by the spoken, sung and written word. Much of the success of the *Chipko* movement in India was due to the songs that accompany the teachings: no better way to spread the message. As Truth is gradually incorporated into everyday's knowledge, it still retains its sacredness, because there always is a secret un-explained element in every bit of knowledge. Man has never been able so far to explain or relate anything up to its ultimate cause.

Logically those forests that have remained relatively untouched, are most sacred, one reason being that the original variety of plants and animals remains intact. Here I would recall Andrewartha's definition of Ecology: "the study of distribution and abundance of species". In practical terms, the "degree" of sacredeness, if I may say so, in many places is expressed by the permanence of the original species diversity, of both plants and animals, which are naturally interrelated. This is obvious whenever listening to local people. Each species of plant has its own usefulness, whether for practical purposes, such as making windows, rafters, hoes and ploughs, for medical purposes, food or just for beauty. *Aranyani* recalls the "cici-cici" of the cicadas as an element of the joy of living; in fact each species of plant or animals plays its own role and, as people know, the more ecological niches in an ecosystem, greater the number of different species and greater the "richness" and stability of the system. The concept of "weed" is uncommon and I guess that it was developed with agriculture. Even nowadays wild sugarcane, a "weed" hybridizes freely with the cultivated species. After all, this is how the most valuable contemporary species of wheat, *Triticum aestivum* which is hexaploid (AABBDD) most probably originated (Feldman, et al., 1986) in some fields of the eastern Fertile Crescent, which is now Iran. All hexaploid wheats are cultivated, and probably never existed in the wild (Porceddu et al., 1986). Haxaploid wheat originated in Asia by spontaneous hybribization with wild varieties; this could have taken place in the hills above the river Tigris or in any other place between there and the northern Indian subcontinent, perhaps even in more than one place (Harlan, 1971; Hancock, 1992). Hexaploid wheat is so important that it most probably was the "great gift" that Demetria took to Greece in the middle of the second millennium B.C., during the reign of the king of Athens called Pandian. Spontaneous hybridizations allopolyploidy and gene as well as somatic mutations probably continue to take place in the forest and contribute to the changes over time of the species diversity of the forest. Untouchability of the forest in obviously a very wise practice.

Different environments bear different plant composition and are used differently by man. Thus the mangrove forests of the Solomon islands are used for the disposal of dead human bodies, while the mangrove forest that still exists between the temple complex at Chidambaram (Tamil Nadu, India) and the sea, is a sacred site. Remains of the shrine from the forgotten past can still be found and the forest may have been considered sacred because of the large number of trees of the species *Excoecaria agallocha* of the family. Apocynaceae whose latex contains alkaloids and that was and still is used to alleviate the pains and reputedly to cure leprosy. We may recall that the magnificent Sun temple at Konarak was built by a king affected by leprosy and was the refuge of lepers. It also stands in an area where there are now degraded mangroves that must have been well developed mangrove forests in the past. Mangrove forests of the rivers of Casamance, south Senegal are sacred because lengthy and elaborate rites of initiation of boys are practiced there, they include teaching the traditional lore and knowledge, not only about the forest but of the people's culture in general. In Guinea, the same is true and shamans are usually trained in the forests that, logically, could be nothing else than sacred.

It is particularly important to preserve mountain forests for obvious ecological reasons well known to the people. Logically, therefore, the whole area and in particular certain mountains with their intact forest cover, are sacred. In the island of Okinawa, in the Ryukyus archipelago of southern Japan, there is a sacred mount Onna that is the stage of several interesting folk tales; at present, one of the many serious objections of the people of Okinawa to the continued presence of the American troops on their island, is that the sacred Mount Onna is defaced and desecrated by the live-shell artillery exercises of the American military still occupying parts of the island. Further south, in the Yaeyama group of small islands of the same archipelago the Ryukyus, it is the sea that acquires connotations of special care and respect.

In the context of Sacred Forests, the island of Bali, Indonesia, where an ancient form of Hinduism is practiced, offers several instances worth recalling here, though dealt in more details elsewhere (Vannucci, 1995a). All temple compounds, large and small, are surrounded partly or completely by a sacred forest that has a variety of trees, shrubs and herbs. Mostly these are reduced to a minimum from what clearly were much larger sites in the past. Encroachment by agriculture and habitations is visible almost everywhere. Unfailingly in every temple enclosure that I visited in Bali, there are some Palmyra palms, the leaflets of which are made into writing material and books, some rudraksha trees (*Eleocarpus ganeti*) that symbolizes worship and divinity since it has the aspect of *Shiva's* third eye, and mangostin and coconut trees that provide food for the body. In other words (Vannucci and Bielovie, in press) these are trees that symbolise the essentials for human livelihood, food for the body, for spirit and for the mind and transmission of knowledge. The most spectacular Sacred Forest of Bali is Sangeh Holy Forest that has, in addition to the species mentioned, a dense forest of *Dipterocarpus* trees that grow up almost beyond sight. In the navel of the forest, where two paths cross each other at right angles, is situated a small shrine that marks the four directions of space. It is a very sacred, though unusual type of sacred forest, from which nothing can be removed. I was not able to find out the origin or story of the Sangeh Forest, whose origin is clouded in mystery, though the temple complex annex to it is relatively recent. Sangeh Forest is the realm of the most cunning monkeys I have ever seen, they jump on people's shoulders, speak to them and ask for food and if their demand is not fulfilled, they steal people's spectacles only to trade them back for food and sweets; and many other intelligent mischievous actions.

Countless pages have been written on the ecological and other roles of forests, specially their uses and products for everyday life in India. We may recall here that traditionally, and to a certain extent even at present, forest called *shrivan* or: "fortunate, happy forests" that are associated to villages and are known to generate good water and maintain the soil. They are usually to be found on land belonging to the village and they are common property, the management of which is the responsibility of the *panchayat* (five village elders). Thus, for instance a tree or a species used to make house rafters, will only be felled when mature and when there is enough demand for all the timber for general repairs of all the village houses that need it. I remember following the many month long discussions, which actually lasted almost two years, about felling a mature tree that grew at the border and marked the boundary of someone's field in the westerns Himalayas, in Pithoragarh district. The tree belonged to the owner of the field where it grew, but the owner of the adjoining field objected to its felling because of the benefits the tree provided to his own field. Finally the *panchayat* decided that the tree could be out, that a defined part of it should be given to the neighbour who would help in the work and that so many saplings (I forget how many, five or seven) should be planted by the owners of the tree to replace the old one. This is traditional law and Wisdom in act.

CONCLUSION

In conclusion, the concept and practice of maintenance of sacred forests is scientifically valid. Man, from the Arctic to the Antarctic and round the world, considered sacred, meaning: not to be tampered with, those ecosystems which he realized were important for his survival and well-being. It could be a breathing hole in the ice for a seal-hunting eskimo, or a sacred fishing ground for a Polynesian, a tree or a forest for land people.

All nature is sacred in the sense that its laws are to be recognized and respected, failing which disaster occurs. But as knowledge grows and the understanding of the ways by which nature operates also grows, norms to be followed for the use and management of the forest are established. The forest is then framed in its sacredness. Inevitably, the norms are site-specific. The enforcement of proper use is determined by the sacredness of the place and by the traditional law and custom. Sacred forests do not loose their sacredness if they are used according to the knowledge and understanding acquired by man over time, generation after generation; the use is dictated by the experience and wisdom of the elders accumulated generation after generation. The most sacred Forest must remain unused for ever, or for long periods of many years, or for just some seasons.

Each forest has its own peculiar characteristics, it is impossible to speak of generally valid laws for the correct management of forests; the commonest general rule is that sacred forests should remain unused, untouched forever, or for cycles of prescribed duration. The main purpose of the rule is to preserve the maximum biodiversity possible under the local ecological conditions. The snag to this rule is that, since everything always evolves and changes over time, even the species composition of an undisturbed forest will change over time, specially so if the area covered by the forest is not very large. Even in large forested areas, disastrous episodic events may cause unpredictable changes in the structure and dynamics of the ecosystem.

In usual practice, at present, people can extract or add to sacred forest some items at prescribed times, after performing propitiatory rites and fulfill the sacred promise to abide by *Rita*, the Traditional or Customary Law. Traditional Law and Wisdom is usually scientifically valid, even if not formulated or explained in contemporary scientific jargon.

It would be useful to teach people's wisdom as a subject of training in regional Forest Research Institutes. This would minimize such blunders as I have seen: once a well meaning Scandinavian expert forester planted pine trees in a mangrove area of the western Sunderbans. The poor man can be excused for his foolishness, he had never even seen a mangrove forest before being deputed to Bengal. He was, nevertheless, the laughing stock of all the poker face forest guards and villagers who did not know that I could follow what they were saying in Bengali ! Needless to say the whole area, fortunately not too large, was ruined and some ten years later, after the "experiment", it still exhibited suffering pine trees struggling hopelessly for their survival while useless vegetation was encroaching on the plots, preventing the natural restoration of the earlier mangrove forest.

Everywhere in the world, what is needed is appropriate ecological management. Traditionally man knows that he is part of the ecosystem in which he lives and that he has learned to manage as appropriately as he manages his own species, *Homo spiens* for the imperative need of survival. This is the basic knowledge for understanding and adjusting to *human ecology* realities, whether in mountains, plains or borders of the sea. What has to be done is to apply contemporary scientific knowledge to study dispassionately people's traditional, habits and practices to support and improve on the usefulness of such practices. One good example of this procedure is the cultivation of the Nepalese alder (Ramakrishnan, 1993).

To close, I would recall a Seminole Indian saying: Nature is the father, Indian is the son. Father takes care of son, son must respect father.

REFERENCES

Davis, S. 1985. Traditional Management of the Littoral Zone among the Yolngu of North Australia. In: *The Traditional Knowledge and Management of Coastal Systems in Asia and the Pacific.* UNESCO/Jakarta, pp. 101-124.

Feldman, M., Galili G. and Levy, A.A. 1986. Genetic and Evolutionary aspects of Allopolyploidy in Wheat. In: C., Barigozzi, (ed.). *The Origin and Domestication of Cultivated Plants.*

Hancock, J.F. 1992. *Plant Evolution and the Origin of Plant Species.* Prentice Hall, New Jersey.

Proceddu, E. and Lafiandra, D. 1986. Origin and Evolution of Wheats. In: *The Origin and Domestication of Cultivated Plants,* C. Barigozzi, (ed.), pp. 143-178.

Ramakrishnan, P.S. 1992. *Shifting Agriculture and Sustainable Development of North-Eastern India*. UNESCO-MAB Series, Paris, Parthenon Publ., Carnforth, Lancs, U.K.

Vannucci, N. and Bielovic Z. 1995 (In press). *The Origin of the Cult of Demetria*. World Heritage Press, Quebec, Canada.

Vannucci, N. 1995a. Sacred Groves or Holy Forests, *In: Concepts of Space. Ancient and Modern*. Indira Gandhi National Centre for the Arts, New Delhi. pp. 323-332

Vannucci 1995b. *Ecological Readings in the Veda* Print world, New Delhi.

3

The Logos and the Mythos of the Sacred Grove

B. Saraswati

Indira Gandhi National Centre for the Arts, New Delhi, India

INTRODUCTION

It has become increasingly obvious that mankind is rushing towards an ecological disaster. That fear(!), or rational response, has lead to believing that the hope of human survival lies in the "sacred grove". Ecologists do not believe the myth, but many of them have given a straightforward account of the *legin* of the sacred grove, reducing it to its literary account. Clearly, there is a whole *methodic* latent here. Unless the *logos* of the myth is analysed the subtleties of traditional attitude to natural environment cannot be comprehended.

This essay attempts to show that ecology was a "sacred science" for the ancients who lived in a world of rich and vivid experience. Ecology needs a language which the philosophers, poets and prophets know. Language is a body, a part of oneself. There cannot be a real representation of the ancient thoughts of the sages in another body. What is provided in the following pages is an imitation of imitation.

THE TREE OF THE COSMOS

The *Axis Mundi* owes to the Old World. The fundamental symbol in this conception is the Centre of the Universe, the Cosmic Tree. The Upanishadic seers spoke of the Cosmic Tree rooted in Brahman, the Ultimate:

> Its Root is above, its branches below-
> This Eternal Fig Tree!
> That (Root) indeed is the Pure. That is Brahman.
> That indeed is called the Immortal.
> On it all the worlds do rest,
> And no one soever goes beyond it.
> This verily is That (in Katha Upanishad)

For the Vedic man none was more full of symbolic power than the Cosmic Tree (also called the Tree of Life, the Poles of the Sacrifice). There is a connection between the tree-image of the Cosmos and the Sun which also embrances the entire Universe. The Sun directs his beams down towards the Earth and keeps his Roots in Heaven. The Cosmic Tree is the three-quartered Brahman whose branches are space, wind, fire, water, earth, and the like (In Maitri Upanishad). The tree-image symbolizes Life:

> Of this Great Tree, my dear, if some one
> should strike at the root, it would bleed,
> but still live. If some one should strike at
> its middle, it would bleed, but still live.
> Being pervaded by *Atman* (soul), it
> continues to stand, eagerly drinking in
> moisture and rejoicing (In Chandogya Upanishad).

The Old World has used the symbol of the Cosmic Tree to describe the world-ground, the world-producer: "The Egyptians worshipped the sacred sycamore fig tree, the Aztecs of Mexico had their sacred agave plant and the ancient Sumerians of Eridu tell of a wondrous tree with its roots of white crystal stretched towards the deep, its seat the central place of the earth, its foliage the couch of the primeval Mother. In its midst was *Tammuz*" (Prem, 1982). In the Maya World, the silkcotton tree is presented as *Axis Mundi,* as the Tree of Life, with Roots down into the lower world and Trunk and Branches passed through different layers of the heavens.

On the symbolic representation of the Cosmic Tree, it is said: "This Tree is the Great World Mother, the Goddess of Nature who nourishes all life with the milk of her breasts. Hence the choice by the Egyptians of the sycamore fig with its milky juice and hence the fact that these most sacred trees of the ancient Indo-Aryans were the *ashwatha*, the *bat* or *banyan* and the *udumbara*, all of them being species of fig tree. There is, however, one peculiarity of the Indian tradition, namely, that the Root of the Tree is said to be above and the Branches beneath, whereas all the other World Trees have their Roots in the underworld and their Branches in the Sky" (Prem, 1982).

The Mother-symbol is most appropriate, even to reach the metaphysical level: "The tree is, in fact, rooted in the unmanifest Darkness of the Parabrahman, which is usually symbolised as "above", though in fact it is no less truly "beneath" as well. From that transcending rooting-place it sends down its trunk of manifestation through the worlds, and on that trunk are seven main branches each of which splits into countless branches and twigs. Nevertheless the Trunk itself is one, for it is the Tree of the Mother, the great *Mula-prakriti*, the substance of the universe. The sap that runs in its veins is the very Life of all beings and on its Branches hang the stars themselves" (Prem, 1982).

THE TREE OF THE HUMAN PSYCHE

Here we have a fine example of the archetypal character of the Fig Tree: "This Tree exists and therefore in man, the microcosm, we find in the structure of the nervous system which resembles a tree, rooted in the brain and ramifying all over the body. This, however, is on the purely physiological level and concerns man"s physical body alone. Of far more significance is the Tree of the human psyche, the Tree which has its Roots "above" in the pure consciousness of the *Atman* and its Branches in the thoughts, feelings and sensations of normal psychic life. Here again the Trunk is the Middle Pillar of *Manas*, that which stands connecting the ramifying network of Roots above with the similarly ramifying Branches beneath. To reach the Roots of our being, therefore, we have to leave the Branches that wave in the breeze of the outer world, those Branches laden with the sticky sprouts of sensation, climb the unitary central Trunk and thus reach the Roots that dwell in the Ocean of Life" (Prem, 1982).

There is a mystical tie between trees and men. Man resembles a tree growing from Brahman, one of the Upanishads expresses this idea vividly:

As a tree of the forest
Just so, surely, is man.
His hairs are leaves,
His skin the outer bark.

From his skin blood,
Sap from the bark flows forth.
From him when pierced there comes forth
A stream, as from the tree when struck.

His pieces of flesh are under-layers of wood.
The fibre is muscle-like, strong.
The bones are the wood within.
The marrow is made resembling pith.

A tree, when it is felled, grows up
From the root, more new again;
A mortal, when cut down by death-
From what root does he grow up?

Say not "from semen",
For that is produced from the living,
As the tree, forsooth, springing from seed,
Clearly arises without having died.

If with its roots they should pull up
The tree, it would not come into being again.
A mortal, when cut down by death-
From what root does he grow up?

When born, indeed, he is not born (again).
Who would again beget him?

Brahman is knowledge, is bliss,
The final goal of the giver of offerings,
Of him, too, who stands still and knows it (in Brihadaranyaka Upanishad).

A fundamental human experience of the Tree is that of patterned vibrations. An Australian Aborigine also speaks of the unifying resonance between man and Tree (Lawlor,1991):

Tree...
he watching you.
You look at tree,
he listen to you.
He got no finger,
he can't speak.
But that leaf....
he pumping, growing,
growing in the night.
While you sleeping
you dream something.
Tree and grass same thing.
They grow with your body,
with your feeling.

The tree-symbol affords communication between human and divine. One cannot understand the value of the Tree on the merely mental plane. A symbol symbolizes something, stands for something. The level at which the tree-symbol can be assimilated with the cosmovision is the *logos*.

THE TREE OF THE GODS

As the sages have said, the Cosmic Tree shelters the Gods in its branches, spreads on the surface of the earth, and provides lodging for the whole of reality, including Nonbeing. It mediates between man and God. It encompasses all and, hence, all trees are sacred-effective and all trees have curative power. Traditions may, however, differ in respect of attributing specific significance to a particular tree or a particular configuration of trees. *Ficus benghalensis* (Banyan tree; Vata), *Ficus religiosa* (The peepal tree), *Saraca asoka* (The Ashoka tree), *Aegle marmelos* (The Bel tree) are some of the sacred trees in Indain tradition.

In Sanskrit *vata* is a common name for tree. Hence these five great trees are called *pancavata* (*panca* five, *vata* tree). There are other combinations of the *pancavata*. Tulsidasa"s *Ramachritamanas* describes a grove of trees. In the middle of the grove is a *banyan* tree under which Lord Rama is seated and it is surrounded by *peepal*, mango, blackberry and *pakhara*, *Ficus infectoria*. Valmiki's *Ramayana* includes in this grove such trees as *sal*, *tal*, palm, jack fruits, *champa,* sandal, mango, *ashoka*, *kadamba*, and *patal*.

In the Indian worldview, five is a cardinal number; it stands for the entirety. The universe consists of "five" primordial elements (*panca tattva*). Man, the microcosm, is also made of "five" elements and enveloped by "five" sheaths (*panca kosa*). *Pancavati* means a sacred forest in which there are "five" (kinds of) Trees.

There are regional variations in the consideration of the "five" Trees. For example, in coastal area the coconut tree would be counted as one among the five main trees. In many regions the banana plant is included as one of the main trees among the "five" trees. At other places the lemon tree and the *amla* tree, *Phyllanthus emblica*, are included in this category. It is sheer custom which ensures that the main trees are planted, grown and looked after in the society.

Other configurations of trees and plants are linked to the divinities. The *rudraksa* is sacred to Shiva. The *kadamba, Galicon tricorne*, is associated with Lord Krishna's *rasa-nrtya*. The *tulasi*, basil, cures all ailments in the world, it is worshipped daily and is sacred to Vishnu. The *aravinda*, lotus, specially *nilotpal*, the blue lotus, is a symbolic arrow of Kamadeva, the God of Love. It is believed to have originated from the navel of Lord Vishnu. The *neem, Melia azadirachta*, is associated with Goddess Shitala. "In Mithila, the *amla* tree is worshipped in the month of *Kartika* (October/November) and after that people pray and feast under that tree. Also, the *mahua* and the mango trees are worshipped at the time, of wedding. The bride performs a symbolic marriage of (with!) these trees before she is given in *kanyadana*, gifting away of the daughter, to the bridegroom" (Saraswati, 1985).

The message of the tree-Gods is loud and clear. The instructions (*logos*) of the myth is to be taken literally, that is, as meaning what it says: Gods and trees are not separate from each other. The *mangala*, spiritual and material wellbeing, of mankind lies in preserving his primordial link with the world of trees and plants. All trees and plants are sacred, all endowed with curative powers, and, hence, are worthy of worship.

THE WORLD OF THE SACRED GROVE

The world of man is *in* and *with* the sacred grove. Man cannot live in disjunction from trees and plants, his natural environment. The ancients knew that man's relationship with nature is not a technical one. Nor a relationship of dominion or exploitation. The sages praised and glorified the forest as ecological redresser:

> Sprite of the Forest, Sprite of the Forest,
> slipping so quietly away,
> how is it that you avoid peoples" dwellings?
> Have you no fear all alone?

When the cicada emits his shrill notes
    and the grasshopper is his accompanist,
it"s the Sprite of the Forest they hail with their praises,
    as with cymbals clashing in procession.

Cows seem to loom up yonder at pasture,
    what looks like a dwelling appears.
Is that a cart with creaking wheels?
    The Sprite of the Forest passes!

Hark! there a man is calling a cow,
    another is felling a tree.
At evening the guest of the Sprite of the Woods
    fancies he hears someone scream!

The Sprite of the Forest never slays,
    unless one approaches in fury.
One may eat at will of her luscious fruits
    and rest in her shade at one"s pleasure.

Adorned with fragrant perfumes and balms,
    she needs not to toil for her food.

Mother of untamed forest beasts,
    Sprite of the wood, I salute you! (in Rigveda)

This hymn dedicated to Aranyani, the Sprite of the Forest, portrays the nightlife of the forest and depicts the hospitable forest which is always scented and generous in its provision of food and restful hiding places.

The Aranya, a Place of No War

For the sages *aranya*, forest, was a world of wisdom and peace. The term *aranya* is made up of two parts: a and *ranya*, a means 'no' and *ranya* means 'war'. Thus derivatively, *aranya* means a place of no war, nonviolence, that is, a place where violence is forbidden. The sages lived in the *aranya* a life of peace and contemplation. One of the four parts of the Vedic literature is the "Aranyakas", forest treatises, which deal with the speculations and spirituality of the forset dwellers (*vanaprasthi*), renouncers of the world. They made to rest a place for interiorization.

The Tapovana, a Place of Penance

Like *aranya*, the *tapovana* is a forest where the hermits, the recluses, and the monks meditated and performed *tapas*, spiritual fervor or ardour. One of the Upanishads (Mundaka Upanishad) says:

Those who in penance and faith dwell in the forest,
peaceful and wise, living a mendicants" life,
free from passion depart through the door of the sun
to the place of the immortal person, the imperishable Self.

Both *aranya* and *tapovana* are not impenetrable; they abound in flora and fauna. These places were *abhayaranya*, sanctuaries, which kings and commoners visited not for hunting games but to seek wisdom, blessings and guidance of the sages and the seers.

The Mahavana, A Great Forest

Unlike the *aranya* and the *tapovana*, a *mahavana* is impenetrable and thus it preserves flora and fauna without much interference. Its dense forest, spread over a vast area, envelops the mystery of natural and supernatural forces, but not a fearsome place. In this forest Lord Shiva, the God of fearlessness, resides.

The Srivana, a Place of Prosperity

Tradition holds that the *vana* should not be within the village, but the village should be within the *vana*. In ancient times most Indian villages were located within the boundaries of *srivana*. This can be seen even today in the so-called "tribal" region. The prosperity of man depends on his ability to conserve natural resources. The *vana* brings every spiritual and material treasure. Hence religious men have for the forest a sense of sacredness. For the ancient sages none was more important, or full of power, than Vanaspati, the Lord of the Forest.

The Devavana, a God's Forest

By linking the forest to Gods, preservation of the forest is assured. But the significance of *devavana* cannot be determined in terms of material loss or profit. Nor can it be said that the original intention of preserving the *vana* was to maintain ecological balance. Configurations of trees in a *devavana* are linked to the presiding Gods and Goddesses. The tradition of *devavana*, today, is preserved largely among the "tribes" of India. The Santhals, for instance, have a strong tradition of *jaher*, the sacred grove of the *sal* trees, *Shorea robusta*, where Lords, the "Five", the Lady of the Sacred Grove, the Pargana, Maran Buru, Gosai Era, and Manjhi Haram are invoked (Bodding, 1942). As one of their songs reveal, the sacred grove is not man-made (Hembram, 1988):

> Who erected a jaher around laterite soil?
> Who turn it into a place of sal-grove?
> Marangburu erected a jaher around laterite soil.
> Jaher era turned it into a place of sal-grove.
> Moreko enliven a jaher amidst of stony land.
> Turuiko took care to make it a green place.

The Lady of the Sacred Grove, or the Mother Goddess, is called *Jaher Era* in Santhali and *Sarana Buria* in Munda and Oraon dialects. The Santhals give importance to three kinds of trees among many trees, namely, the *sal*, the *peepal*, and the *vata* (Hembram, 1988).

The World of the sacred grove makes a sense to the "believers", not to the "agonistics". What has so far sustained this World are the people's "thinking faith", "acting faith" and "loving faith". To those who have lost the sense of the sacred may have rational awareness of the value of the sacred grove but, obviously, they have no means to control the machinery of its destruction.

The Sacred Grove Not of This Age

The *Matsya Purana* describes (Agarwala, 1963) Earth as the Goddess Lotus which Narayana-Vishnu created from the depth of the Ocean. From the churning of the Ocean appeared among several others the *Kalpavrksa*, the wish-fulfilling Tree. The *Markandeya Purana* gives a curious and interesting account of the primeval human race (Saraswati, 1988): "Social and moral evolution of this primitive humanity has been attributed to some Trees called *kalpa* which, until the beginning of the Treta Age, were producing houses and honey (not made by bees) and providing every kind of enjoyment and subsistence, clothing and ornament to those people; but afterward, in course of time when those people grew covetous, their mind being filled with selfishness they fenced the trees, and those trees perished by reason of that wrong conduct on their part".

The legend of the *Kalpa* Tree, as amplified in the Puranas, is purposive. It shows that the four Cosmic Ages—Krta, Treta, Dwapara, and Kali—represent a gradual decay in human life and morals. Of these Kali, the present Age, is the degenerated Age. Since the Ages are repeated in the same order, Kali will be followed by Krta, or Satya, the Age of Truthfulness.

The celebration of the Trees in the present Age has declined, as expected, to the extent that no sacred tree is favoured any longer. Cultural (moral) degradation leads to environmental degeneration. The sacred groves are fast vanishing. Here are a few examples:

SACRED GROVES TURN INTO SACRED CITIES

The Kali Age speaks for itself. Shiva's *anandavana*, the Forest of Bliss, is a name for Varanasi. The Puranas have described the sacred landscape of the *anadavana* (Eck, 1982) "as a garden paradise, sprinkled with the waters of the heavenly Ganges, abundant with flowers and blossoms, filled with the songs of birds, the buzzing of bees, and the tinkling of anklets of lovely women. In its blessed groves even animals who are natural enemies dwell in peace with one another. The mouse nibbles the ear of the cat, the cat sleeps peacefully in the tail feathers of the peacock, the crane leaves the fish alone, the hawk pays no attention to the quail and the jackal befriend the antelope". This description does not tally with Varanasi, the Forest of Bliss today. "Even in the late eighteenth century, when William Hodges and William Daniell sketched the riverfront of Banaras, it was a long spectacular bluff crowned with trees and a few prominent temples. By that time, however, the urban centre of the city had already shifted southward and the dense area around Chaukhamba and Thatheri Bazar called the *pakka muhalla*, the "well-built quarter", was completed. Even so, the people in that neighbourhood still refer to their quarter as the *ban kati*, the "cut-down forest", for the memory of the time when this was indeed a forest is not many generations past. Further south and west, the Forest of Bliss remained a forest until still more recently" (Eck, 1982). Today, that place is no longer a forest paradise. It is transformed into one of the most thickly populated cities of India. Varanasi, the holy land between the Varuna and the Asi, the two rivers set there by the Gods along the Ganga, was the place of the "eye of wisdom". Of these three rivers, the Ganga today is one of the most polluted rivers; the Varuna hardly merits the status of a river except during the rainy season; and the Asi which was the Dried-up river, even during the Pauranic Age, does not exist any more.

Naimisaranya, the *aranya* of the seers, is reduced to a temple town, a pilgrimage centre. Today, there is hardly any *mahavana* unapproachable to man; its association with Mahakala Shiva in Ujjain city is evidently at the symbolic level. The *tapovana* around Haridwar, a pilgrim city, is not a place of penance. Vrindavana is a city, no longer the *vrindavana,* the grove of basil plants and a garden bearing juicy fruits. Pancavati is a part of Chitrakut city, not the *pancavati*, the sacred grove of five kinds of trees, of the Tulsidas" epic *Ramayana*. Nasik is a city, not the sacred grove where Lord Rama lived during his exile. The forest of *peepal* trees where the Buddha performed penance and attained enlightenment is transformed into a large city of Gaya. Only one *bodhi* tree which stands there is said to be of the Buddha's time.

SACRED GROVES MEET RELIGIOUS ADVERSARIES

The sacred groves of the aborigines, the indigenous people, lasted as places of worship so long as they remained true to their natural religion.

Christianization and Islamization have provided such a view of nature that the converts themselves are guiltlessly causing destruction to the sacred groves which were once an essential part of their life and culture. There are numerous examples:

In Meghalaya, Shillong Peak (6445 ft) was considered sacred. Shillong is the name of a Khasi God dwelling in that peak. Another sacred place of the Khasi is Sohpetbneng Peak (4400 ft), literally means a "navel of the heavens". It is believed that (Pugh, 1976) "at one time this peak was like an umbilical cord connecting the heavens and the earth, along which denizens of the Khasi land (16 families in all) used to go up to the heavens and come down to earth again". Around these peaks there were magnificent sacred groves providing the unifying resonance between people and nature. Christianity brought an end to this. Sacred groves have been replaced by the Christian churches and missionary establishments. Shillong today is the capital town of Meghalaya.

According to Chandran and Gadgil (1996) sacred groves and sacred trees had a prominent role in the early religion of the indigenous people of the Western Ghats in South India. This practice suffered a setback on the arrival of Christianity and Islam which professed faith in one God and were against the paganism of the local people. The amalgamation of the primitive deities into the Hindu pantheon, often followed by temple construction, was another early threat to sacred groves all over India, including the Kans sacred forests.

Before the advent of colonization and Christianity and Islam, the African lived in harmony with nature. According to Anane (1997) missionaries who trooped to Africa alongside the colonial masters discouraged traditional practices, such as the worshipping of rivers, mountains and trees, which they described as idolatry and heathen.

Conquerors Gave the Final Death Blow

Five hundred years ago, European colonization of North and South America and Australia began. Indeed, it was the quest for India that led Columbus to stumble on to America. However, the experience of the conquerors and the colonial pattern of expansion remains the same everywhere:

"The acts of mass destruction are glorified as colonial heroism, as in the case of the decimination of the Native American and Australian Aboriginal cultures. The famous American historian, Francis Parkman, wrote of the American Indian: He will not learn the arts of civilization, and he and his forest must perish together" (Lawlor, 1991).

Such views were founded amid the realities of racism in Australia. "Even such a liberal observer as the famous English novelist Anthony

Trollope could conclude in 1873 in his book *Australia* and *New Zealand*: It is their (Aboriginals) fate to be abolished and they are already vanishing... Nothing short of abstaining from encroaching on their lands-abstaining, that is, from taking possession of Australia, could be of any service to them. Of the Australian black man we may certainly say that he has to go. That he should perish without any unnecessary suffering should be the aim of all concerned in the matter" (Whitelock, 1985).

"The Aboriginal Australia which adhered to its ancient Dreamtime Law, was a primeval garden rich with a variety of beautiful flora and fauna. The Aborigines enjoyed hunting and foraging, free and naked in the open air of primal forests. European conquest brought to them wholesale enslavement, displacement, and near cultural genocide" (Lawlor, 1991).

With the arrival of the white man in Africa (Anane, 1997), virgin forests that had been preserved for their sacredness were raped by the colonial masters, and the trees that were felled were exported abroad.

The impact of the British rule on the forests of India was equally disruptive. British colonial forest management was the principal factor leading to the destabilisation of India's traditional relations between man and environment: "One hundred years of British presence, by creating a market for wood and by introducing capital and increasingly sophisticated and efficient techniques, had a dramatic effect on the expanding development of forest exploitation. These effects on environment were much more disruptive than the activities of the rural populations of the region during the same period" (Buchy, 1996).

Regretably, the forest policy inherited from European colonisation remains nearly unaltered. A study of Western Ghats has shown that "between 1954 and 1985, 50,158 hectares of forest were given over to agriculture, whereas another 55,000 hectares were allocated to mining, various hydro projects, the rehabilitation of displaced populations as well as to diverse infrastructures. The bamboo stands around Dandeli have been destroyed by over exploitation during the last twenty years for the production of wood pulp (a considerable amount of which was entirely illegal). The remaining stands of Junglewood were destroyed by the plywood industry located in the region" (Buchy, 1996).

The root of the problem is not the growth in population, nor the use of the forest for sustainable development, nor even making a city out of the sacred grove. The seed of destruction lies in the *greed*, not the need.

THE GLOBAL CONCERN

The contemporary concern for environmental future is good for the future of man and the Earth. The challenges posed by the Machine Age have to be responded creatively. The wanton use of earth"s resources has raised many questions about our lifestyles and about how long humankind expects to continue its current rate of using resources before the earth is exhausted. The Amazon forest, the largest rainforest area in the world, will become extinct as a rainforest by 2020 unless the process of deforestation is stopped. Some of the trends in ecological destruction are almost irreversible. It is estimated that at the present rate of the death of species, twenty-five per cent of all the earth"s species will become extinct (Treston, 1991) by 2010.

The modern Western World is the product of the philosophy of the industrial era. It has created a "technocentric civilization" which gives primacy to technology, seeks solution to all problems in terms of technology, and believes in technocracy that lashes out at other civilizations. In sharp contrast, the traditional Eastern World has constructed a "cosmocentric civilization" which gives primacy to nature, seeks solution to all problems in consultations with nature, considers all trees and plants sacred-effective, and believes that the sacred complex contains the entire Universe. At a higher level of thinking, the dichotomy between East and West, tradition and modernity, sacred and secular, is false.

The Question of Human Survival

Must man plant trees, protect sacred groves, and save tigers, simply because his survival is threatened? Why the widespread concern about environmental catastrophe? "Saving man", is the answer. But, is this not a very narrow anthropocentric approach to life? Moreover, is it possible to halt the cosmic catastrophe? Modern scientific theorists such as Lovelock, Sheldvake, Kuhn and Capra have proposed paradigms of the Universe which emphasize interdependence and mutual dependence on all other life forms. If the living forms on Earth are components of a single more complex system, called the "biosphere" or "Gaia", how can "modern" man hope to live eternally while he destroys all other life-forms ruthlessly?

A "Selfless Vision" to Act

Modern man, inflamed with information technology, has a fantastic confidence in the "secular" science which has a body of knowledge of the material world but no "selfless vision" to act truthfully. For the ancient seers conservation was a sacred act. How can the sacred groves be protected without the "sacred heart"? How can one act toward conservation without the "selfless vision"? How can one protect a monkey sanctuary without having good faith in Hanuman, the monkey God? How can one protect wildlife without believing that animals have the Cosmic Power, and that they are the vehicles of Gods and Goddesses? How can all this be done without the "sacred" science of nature?

The New Experience

The sacred science of nature awakens man to "see" all that is and is-not in utmost simplicity. It's metaphors are different from the metaphors of the modern "secular" science; its *mythos* carry the ultimate mystery. The *logos* of the myth and the meaning of the metaphor, derived from a transcendental vision of the whole, wake up the larger self to the religion of the Earth.

The question about the nature and value of the sacred grove arises at the very moment one thinks about its new context, new experience. The *negative* way should read: If the Cosmic Tree is rooted in the ultimate, how can sacred grove vanish for ever? If there is a mystical tie between trees and men, how can men survive and trees die? If all trees are sacred and all have curative power, how can the sacred grove become worthy of conservation and all other forests subject to environmental deterioration? If the Old World sacred groves have survived today, how can it make no sense to the New World people? The *positive* formulation of the problem will refer to the traditional attitude, an attitude that is reflective, understandable and meaningful. this is more clear in the following religiuos poetry of the saint Kabir (Vaudeville, 1974) :

> Where there is "sacred heart", there is "sacred grove".
> Where there is greed, there is doom.
> Attending to the "sacred grove", all is obtained.
> Attending to "forest" of greed, all is lost.
> Within the "sacred heart", the sacred grove is at Peace.
>
> In that forest where no lion roars, no bird takes to flight.
> Where there is neither *peepal* nor *kadamba*, even an ant cannot hold its footing.
> Where no sandal fragrance enters, the green shrub knows no water.
> Where the heart is so hard, dead wood does not know.
> The body is forest, mind an elephant gone mad.

The Optimistic Vision of the Reality

Traditions die hard. As has been reported: "In Meghalaya, many sacred groves are still well-protected, in spite of a rapid decline in the traditional value system with the advent of Christianity. The traditional religious belief is that the Gods and the spirits of the ancestors live in these groves. The Mawphlang grove close to Shillong town is one of the best preserved, set in a degraded landscape all around. Indeed, the Mawswai grove in Cherrapunji of about 6 km of protected mixed broad-leaved rain forest, though subject to some disturbance along the peripheral region, is an island in a bleak desertified landscape. Though ceremonies used to be performed regularly in this grove and others to propitiate the ruling deity, they have been stopped in many of them for the last few years. Removal of plants or plant parts is considered to offend the ruling deity leading to local calamities. There are in the Cherrapunji region, 21 sacred groves with varied degrees of human disturbance (Ramakrishnan, 1994). In Meghalaya today sacred groves, totalling about 1000 km^2, are scattered in small pockets all over Khasi and Jaintia Hills. These groves remain the last refugia for 700 rare plant species.

Sacred groves contribute greatly towards conservation of biodiversity. Originally, the groves were based on religious and cultural beliefs, but they have since made significant contributions to the protection of wildlife and other biological resources.... Over 80% of the sacred groves in Ghana serve as watersheds for catchment areas where they protect sources of drinking water. So far about 1.5% of Ghana"s land is covered by some 2000 fetish groves (Anane, 1997), and most taboos and beliefs surrounding many of these groves are conservationist in nature and approach.... The luxuriant green abundance of trees of different species and thick undergrowth in some parts of Ghana are living examples of what religion can do for conservation.

Alas! the surviving sacred groves were enough to sustain the ecosystem of the Earth. Sacred groves are not man-made. Man plunders them at his own peril. How he can reconcile with the values and symbols of the sacred is a complex question.

THE FOURFOLD NOBLE PATH

1. Defining

By "sacred grove" we understand that forests have been formed by the "sacrifice" of the first Absolute, so that the Cosmic Tree, the Tree of Life, stands at the Centre of the Universe. That the Divine Nature manifests itself as a "sacred grove" where superior men and women sacrifice for protection, prosperity and pleasure for all.

2. Identifying

The sacred grove is that where sacrifices are made to Nature. Where every tree, whether upright or fallen, is worshipped. Where the keystone Trees—the Five, the Three, the One—keep a close watch on themselves. Where ants and elephants, even poisonous insects and reptiles, roam about freely, fearlessly. Where the human guardians venture not to enter the sanctum sanctorum of the Divine Forest. Where the memory of creation is cherished by the creator's descendents. Where man nourishes his faith and gives all things their due.

3. Renewing

Man brings back to his original. The forest is like a human body. Scorched by the fire of human neglect, the trees of the sacred grove cry aloud. The Forest is the seed of faith. He who waters the seed of Forest, has flowers and fruits in abundance. There can be no Moonlight in the house where the sacred grove is not. Set out for that land where there is *pancavata*, the Physician. Stay near the sandal Tree where even *nim* becomes sandal. Take your repose under the Tree which bears fruits twelve months a year. On earth there are Forests of many kinds with endless fruits and flowers; pick the sweet and fragrant ones, avoid the poisonous. Gather honey from every flower, finding in them the supreme Soul.

4. Conserving

The modern World is a blind cow. The calf (of materialism) is dead, yet it keeps licking its skin. Man has thrown away the precious Gem and taken a peeble in his hand. Find a worthy buyer of the Gem. People mistake the sandal for a *palas*. Throw the sandal (of faith) into the fire (of the sacred grove), its fragrance will increase. The wretched people burn the forest, they have not found wisdom. Their heart is so hard that nothing can wound them. The green shrubs know the boon of the water, but the dead wood does not know. Conservation of the Forest is a sacred act. The "sacred grove" is sacred to the entire humanity. The World Tree is the "World Heritage". Bring the sacred back to its original purity. Renew the people. Renew the sacred grove. Renew the sacred grove, the tangible cultural heritage. Renew the true identity interexistent with all the sentient beings. Let the government of the state declare *all* the sacred groves as "National Heritage Groves", all the Ficus trees as "National Trees". Let everyone remember: the sacred groves are not for riches, but for justice to the faithful, the noble people. Let UNESCO unite all nations in one principle of "World Heritage Grove".

Rugged is the path, inscrutable is the mystery of the sacred grove. Complex is the human mind, formed by the Sky of ambitions and the Earth of limitations. Let us follow our true intuitions: carefully and reverentially step forward, and we reach the "sacred grove".

REFERENCES

Agrawala, S.V. 1963. *Matsya Purana—A Study: An Exposition of the Ancient Purana-Vidya*. All India Kashiraj Trust, Ramnagar-Varanasi.

Anane, M. 1997. Religion and Conservation in Ghana. In: *Leyla Alvanak and Adrienne Cruz* (eds.). *Implementating Agenda 21: NGO Experiences from Around the World.* United Nations Non-Governmental Liaison Service, Geneva. Pp. 99-107.

Bodding, P.O. 1942. *Traditions and Institutions of the Santals: Horkoren Mare Hapramko Reak Katha*. Oslo Ethnografiske Museum, Oslo.

Buchy, M. 1996. *Teak and Arecanut: Colonial State, Forest and People in the Western Ghats (South India) 1800-1947*. Institut Francais de Pondichery-Indira Gandhi National Centre for the Arts, New Delhi.

Chandran, M.D.S. and Gadgil, M. 1996. *Sacred Groves and Sacred Treed of Uttara Kannada.* Pilot Project Report, IGNCA, New Delhi.

Eck, D.L. 1982. *Banaras City of Light.* Alfred A. Knopf, New York.

Hembram, P.C. 1988. *Sari-Sarna (Santhal Religion)*. Mittal Publications, Delhi.

Lawlor, R. 1991. *Voices of the First Day: Awakening in the Aboriginal Dreamtime.* Inner Traditions International, Rochester Vermont.

Prem, S.K. 1982. *The Yoga of the Kathopanishad.* The New Order Book Co., Ahmedabad.

Pugh, B.M. 1976. *The Story of a Tribal: An Autobiography.* Orient Longman, Calcutta.

Ramakrishnan, P.S. 1995. *Ecology and Traditional Wisdom.* Paper presented at the UNESCO-sponsored Conference "The Cultural Dimension of Education and Ecology", Indira Gandhi National Centre for the Arts, New Delhi.

Saraswati, B. 1985. *Rituals in Relation to Social Life and the Lifestyle of Individuals in Maithil Indian Society.* UNESCO Project Report (mim). N.K. Bose Memorial Foundation, Varanasi.

Saraswati, B. 1988. *Thinking About Tradition: the Indian Vision.* N.K. Bose Memorial Foundation, Varanasi.

Treston, K. 1991. Living in Unitary Age. In: *Creation Spirituality and the Dreamtime.* Catherine Hammond (ed.). Millennium Books, Newtown NSW.

Vaudeville, Ch. 1974. *Kabir, vol 1.* Oxford University Books, London.

Whitelock, D. 1985. *Conquest to Conservation.* Wakefield Press, Netley South Australia.

4

Natural Sacred Sites—A Comparative Approach to Their Cultural and Biological Significance

T. Hay-Edie and M. Hadley

Division of Ecological Sciences, UNESCO, Paris

INTRODUCTION

One result of unchecked globalisation has been that many forms of local idiosyncrasy including dwelt-in sacred geographies are inexorably challenged by an economicist planetary perspective which propounds a uniform conceptualisation of land and resources according to standardised topographic or statistical co-ordinates.[1] Local knowledge is constantly assailed by modernist political and bureaucratic power structures which treat the cognitive maps of small-scale societies as parochial and incomplete. However, in many small-scale societies identity remains rooted within a dwelt-in experience of landscape where divisions between the discourses of science, religion and art may not exist. Alexander von Humbolt the great nineteenth century amazonian explorer and scientist first recognised this in his discussion of historical geography as 'cosmography':

> *"In classical antiquity the earliest historians made little attempt to separate the description of lands from the narration of events, the scene of which was in the areas described. For a long time physical geography and historians appear attractively interminded." (von Humbolt 1845)*[2]

'Cosmography' of this kind may be what is necessary to revalidate local perspectives of landscape. Distanced, clinical depictions of space culminating in remote-sensed imagery fed into geographical information

---

1 The Ecologist (1992) 'Whose Common Future' Vol. 22, No. 4 (July/Aug.) p. 182

2 Sauer (1925) p. 318: Alexander von Humbolt 1845 *'Kosmos'* vol. 1, Stuttgart & Tubingen

systems (GIS) are important heuristic devices for modelling and prediction of physical hazards but reach a threshold of usefulness when faced with the arduous reincorporation of the 'human dimensions' of decision-making in land management. A number of growing fields of enquiry inlcuding ethnobiology, ecological anthropology and historical ecology criticise this alienation of 'scientific' technical analysis from local knowledge for the maintenance of healthy ecosystems.

As one part of this movement to revalorise the traditional ecological knowledge (TEK) of local peoples, many natural landscape features such as forests, groves, mountains, islands, springs and caves are becoming the object of conservation interest owing to their status as sacred places. Environmentalists see in such sites a capacity by traditional societies to conserve biodiversity as a result of their local symbolic systems while indigenous peoples themselves are mobilising their efforts to have these sites, central to the perpetuation of their cultures, *legally* recognised and protected at the national and international level while remaining under their own local *administrative* control.[3]

It is argued here that it is not sufficient to simply take note of such 'vernacular conservation' nor to analyse the systems as intriguing ecological and cultural case studies. In each individual sacred site there exists a complex inter-relationship between the cultural continuity and integrity of a people's knowledge and practices and the material biological diversity made manifest in the landscape. Alongside an elucidation of the functioning of such ecological-cultural relations, efforts need to be focused on creative advocacy techniques and ways to valorise attitudes of spiritual reverence.

THE SACRED

As a eurocentric analytical concept, the 'sacred' first appeared in the mid-nineteenth century when theories of 'deep time' and biological evolution disrupted the Church orthodoxy concerning the age and creation of life on Earth, and social scientists began to treat 'religion' as a subject of enquiry. According to Durkheim, a dialectic between the sacred and the profane existed in all 'primitive societies' made increasingly ambiguous in an irreversible trend towards modernity, while Weber predicted a secular desacralisation or 'disenchantment' with the world. In this classic sense,

---

[3] The term 'traditional' is problematic and is not used here to imply that indigenous and other small-scale societies are "unchanging" in a linear Eurocentric progression of history as defined by current industrialised countries (Redford & Padoch 1992 (eds.) (1992) 'Conservation of Neotropical Forests-Working from Traditional Use' Columbia University Press p.9)

the sacred originated as a category of study to be applied in what we may today refer to as 'traditional' societies.

As a reaction against these intellectualist definitions, religion came to be redefined in the 1960s as the construction of a subjective social reality. In 'The Sacred Canopy' Berger stressed a universal human craving to impose a meaningful order upon reality, while Geertz analysed 'worldview' as a cognitive sense of order in the "really real"—a response to the threat of a "tumult of events" lacking both interpretation and interpretability. Worldview and the sacred in this sense exist therefore at a level of 'deep-embeddedness' where the line between moral and natural law are blurred in a unified perspective of a causal universe.[4]

As embedded worldview, the sacred can be defined as the human perception of an encompassing cosmos rather than the apprehension of an external 'view-of-the-world'. This cognitive sense of order is therefore also a deeply personal experience of the sacred as described by Otto and James in the early twentieth century in the individual feeling of the 'numinous' or "wholly other".[5] Natural landscape features or buildings considered sacred therefore become 'markers' for individuals to ground an idiosyncratic claim of the sacred in their own life-histories.[6] The sacredness of natural sites such as forests, springs or mountains is thus imbued with ultimate levels of reality unique to the spiritual integrity of a local place as set-apart.

What is sacred cannot however be reduced to personal experience. What is sacred can only be expressed according to wider collective forces and relations in both historical and mythological time.[7] Reference or memory places act as organic markers to register past events in individuals' lives as well as to maintain the cultural continuity and integrity of an entire peoples'

---

[4] Geertz contrasts comprehensive 'worldview' with doctrine or 'ethos': *"A people's ethos is the tone, character, and quality of their life, its moral and aesthetic style and mood; it is the underlying attitude toward themselves and their world that life reflects. Their worldview is their picture of the way things in sheer quality are, their concept of nature, of self, of society. It contains their most comprehensive ideas of order"* (In Banton 1966 p. 3)

[5] Rudolf Otto (1923) 'The Idea of the Holy'

[6] The practice of planting a tree for a deceased person is a good example of how such a 'marker' can be created to commemorate a person or ancestor.

[7] David Bellamy, for example, feels that the sacred can be reduced to a matter of personal whim: *"Each and every one of us knows what the word sacred means to us, that tingle of something special, an aura of peace and tranquillity, a sense of being safe. The Welsh have a world for it-cynefin-which means the place in which I am at ease, the place I want to be, perhaps even the place of my being. Many of us have special places where we feel totally at ease; mine is a chair in exactly the spot from which I can see out of one window. That is my sacred spot, where I feel completely at ease, even after the most frustrating day at work."* Foreword to Martin Palmer and Nigel Palmer 'Sacred Britain—A guide to the Sacred Sites and Pilgrim Routes to England, Scotland & Wales' (1977) p. 12 Piatkus.

identity. Ramakrishnan (1996) discusses such marker species and sites as social 'keystone' symbols whose enduring presence helps in the process of cultural continuity. How then, if each site is locally-specific, can one sacred site be compared to another sacred site? The sacred may depend on whose point of view-the actions and representations of local people, or the theories of social scientists and academics?

If we follow Geertz's original use of the term worldview, utilitarian scientists, developers and conservationists only have an objectivist 'view-of-the-world' which fails to grasp the sacred. Reductionist science may contain and comment but it cannot encompass. Interdisciplinary fields of enquiry such as ethnobiology and cultural ecology are therefore moving back towards an earlier cosmography. Yet there is a real danger that, as anthropology considers that all cultures can be read as multiple 'texts', the key concepts of worldview and the sacred come to be casually applied to almost anything. Paradoxically perhaps, in the context of natural landscape features considered sacred, the original Durkheimian sense of the sacred could be legitimately focused once more on the manifestation on the numinous in the dwelt-in universe of traditional societies.

SACRED SITES AND LANDSCAPE PERCEPTION

A number of questions have been raised concerning the relationship of sacred sites to wider landscape ecology issues. What causes some features of the landscape to be selected as sacred sites while others may be considered taboo? Why do particular landmarks such as rocks, mountain peaks and caves attract more symbolic attention than others? Do sacred groves and forests come to reflect a different set of symbolic values to those inorganic places? Environmental topography and constituency it is argued below represent critical variables to interpret memory inscription work and sacred associations in a dwelt-in landscape.

In indigenous, small-scale and subsistence societies innumerable natural features in the landscape are used for a variety of daily needs such as the collection of water, cutting firewood and finding medicinal plants, but also as symbolic 'markers' to demarcate territory, ownership of resources and ancestral connections to land. Impermanence, transience and renewal are therefore central themes to all local and traditional resource users for both functional as well as symbolic and spiritual reasons. As scientific disciplines, ecological anthropology, ethnobiology and historical ecology explore these links between environmental stability and cultural change by examining the relationship between ecological change and associative shifts in spiritual attachment invested in landscape (Crumley,1996).

Features in the landscape may appear in a number of inter-linked symbolic forms: either as biographical memories-"that waterfall was where I first caught a fish with my father"; historical markers orienting a sense of temporal depth and social continuity in a locality-"that clearing in the forest-or grove in the savannah-was cleared or planted by our great-grandparents when they arrived in this valley"; or as mythological power places imbued with numinous qualities owing to the presence of deities or spirits-"that mountain peak was transported across the ocean and dropped in our land during the time of the creation of the world and is still inhabited by the presence of a goddess". Each of these levels of memory-work are involved in describing natural sacred sites.

Recognising the key role of ecological memory inscription in defining the 'sacred', Bernbaum (1996) presents an important seven-part descriptive typology for sacred mountains as: (a) places of 'power'; (b) abodes of deity(ies); (c) as a garden, 'paradise' or 'wilderness'; (d) as connected to ancestors; (e) symbols of communal identity; (f) symbols of purity and longevity; (g) loci of inspiration and revelation. Such a list of attributions however offers little comparability and does not distinguish between the temporal depth of memory inscription and features in the landscape which are more likely to attract such 'memory magnetism'. While biographical and historical marker places (which need not necessarily be visual and can be lingering smells or particular sounds) often evoke a strong sense of belonging in a place, the realm of the sacred as the set-apart and the transcendental has, by definition, to remain associated with ultimate levels of reality.

The elemental "composition" of a sacred site is also critical in determining the position of such a place in human perception (Box 1). Sacred sites made up of inorganic matter, stone or rock formations prominent in a landscape horizon offer powerful functional reference points and navigational aids for journeys through a territory and may be visible to all visitors to an area. The Inuit, for example, use 'inuksuk' to orient themselves in a largely flat tundra landscape (Hallendy ,1996). In Nepal and Tibet 'chortens' are such references, and in the Andes 'apashektas' fulfill a similar role. Yet while the shape and location of such markers give rise to innumerable legends and stories which can be "grafted on" and interpreted by all those in contact with such sites, inert rock must be "brought alive" by symbolism and the presence of divinities. In a human life-span, withstanding sculpted and displaced rock formations such as Stonehenge or Carnach, weathering of rock is invisible to the human eye (Fig. 1).

In an 'Atlas of Sacred Sites and Holy Places of the World' Colin Wilson (1996) notices that many of the earliest religious monuments in the world such as the pyramids in Giza and Mexico may have been built to resemble mountains thus reproducing the permanence of a human-made structure as substitute 'natural' mountains. In this way, many sacred edifices such as the symbolism of Borobudur and other Buddhist stupas in the shape of a mandala of Mount Meru-the mythological axis mundi of the world-reflect an "ecological concern" with the transience of matter and energy in material forms and constructions. Compositions of elements thus display varying levels of durability and symbolic continuity through time: rock as the hardest; water as the most unchanging yet flowing (i.e symbolism of the holy river Ganges); and trees and forests as symbolic of organic renewal yet vulnerable to ecological change.

Many traditions describe elemental forces and the renewal of life in poetic and symbolic terms which echo the change of the seasons, the coming of the rain,. the immutability of the mountains. Features of the natural world thus inevitably creep into all religious traditions by way of analogy, allegory and metaphor, but it is in the so-called animist or 'traditional' peoples where this reverence of the cycles and laws of the natural order are most apparent and the numinous presence of vital life forces, deities and spirits most present. Persons who live closer to nature end up feeling a deeper sentiment of attachment. Indigenous people who have done so for hundred of years may take such 'indivisibility' for granted.[8]

By indivisibility we do not imply a crude depiction of the ecologically noble savage, but rather an index of both traditional ecological knowledge as well as 'environmental envelopment'. Different environmental envelopes offer contrasting perspectives of the world[9]: mountain communities have semi-aerial vistas of plains and valleys marked by high-low binary categories; coastal groups must negotiate the boundary between land and sea;[10] island populations live on potentially demarcated units incorporating forests and mountains; forest peoples live in dense, visually limited lifeworlds where smell and sound may be the primary senses. Caves, once the habitat of our human ancestors continue to attract attention as sacred places for pilgrimage

---

[8] Proximity to nature and animism are not necessarily confined to 'remote' conrners of the planet. In Japan, Shinto shrines remain honoured today across an apparently secular society: small woodland shrines and *'torii'*, red wooden gates, continue to mark the presence of *'kami'* conceived as deities which represent natural qualities such as growth and fertility, ancestral spirits or national heroes. In large cities such as Osaka, buildings continue to be planned around such sites with reference trees poking through railway platforms.

[9] Appendix 1 presents a cursory overview of a spread of diverse natural sacred sites from a broad range of biogeographic situations.

[10] Cape York Aborigines' mental maps of underwater coral reef cognitive markers eliminate this distinction between water and soil for example.

Box 1 : Glimpses of Natural Sacred Sites from around the World*

Africa

Kenya : *Mount Kenya*—local legends claim that Ngai the creator of all things, dwells on *Kirinyaga*, one of the highest points of Mount Kenya. It is believed by the *Kiembe* that humans were created at the summit of the mountain.

Ethiopia : Source of the Blue Nile—considered a holy site by the Ethiopian orthodox Church. From the source, the stream called the Abay flows down to Lake Tana, another sacred site.

Egypt : *Jebel Musa*—possible location of Mount Sinai where Moses saw God in a burning bush and received the tablets of the law containing the ten commandments. At the bottom of the mountain now lies the Greek Orthodox Monastery of St. Catherine, founded in the fourteenth century A.D.

Ghana : *Lake Bosumtwi*—sacred to the *Ashanti* who believe that their dead souls come here to bid farewell to the God Twi.

Nigeria : *Oshogbo*—forest containing shrine complexes sacred to the *Yoruba*. Its main shrine is to the river Goddess *Oshuno*, with a sacred river running through its centre.

Madagascar : (a) *Mont Passot*, Nosy-Bé-mountain site of sacred volcano lakes said to be the home to the spirits of *Antakarana* and *Sakalava* princes; (b) *Lac Anivorano* -sacred lake believed to have once been a semi-arid desert until a mythical traveller was refused drink by the locals and, in revenge, flooded the area.

Australia and Pacific

Australia : (a) *Yarnda Thalu*, W. Australia—mound of stone placed across the ridge of a hill used by elders of the *Ngarluma* people to regulate the temperature of the local area; (b) *Laura*, Queensland-Home of the *Quinkan* spirit-beings and an area rich in rock art; (c) *Wullunggnari*, sacred to the Kimberley peoples as the resting place of the ancestors. *A. Walguna* (tree of wisdom) stands in front of the sacred cave.

Fiji : (a) *Labasa*, Vanua Levu—small town surrounded by sacred sites such as the 'Floating Island' and the 'Growing Stone' beneath a Hindu temple attracting many pilgrims each year; (b) *Korolamalama*, Vatuele-a small cave considered sacred for its holy red prawns; (c) *Navatu Rock*, Rakiraki-a steep cliff and small island said to be the departure point for the afterlife. Excavations reveal Navatu to be one of the first Fijian places of habitation.

Hawaii : (a) *Kailau*, Oahu—once a giant, the first chief of Oahu turned himself into this mountain ridge; (b) *Kilauea*, Hawaii-site of *Halema'uma'u*, a cooled lake of lava said to be the body of *Pele*, the goddess who created Hawaii. Devotees leave Pele offerings of hibiscus flowers and tropical fruit.

Micronesia : *Wichon Falls*, Chuuk—waterfalls site of ancient petroglyphs which play a major part in the island's legend of a spirit which took part of a nearby mountain and dropped it into the sea to form the atoll.

Northeast Asia

Tibet : *Mount Kailas*—with its four faÛades, Kailas resembles an enormous diamond. One of the most revered peaks of the Himalaya, Buddhists compare Kailas to the legendary mount Meru and consider it to be a mandala from which the sacred rivers the Indus, Sutlej, Bramaputra and Ganges flow like spokes from an eternal wheel.

China : (a) *Mount Jizu*, Yunnan—sacred Buddhist mountain and pilgrimage site. During the Qing dynasty, 100 temples were found on the mountain with around 5,000 monks in residence; (b) *Emeishan*, Sichuan-'Lofty eyebrow mountain', western and highest of the four Buddhist sacred mountains (*Wutaishan*, North; *Putuoshan*, East, and *Jihuashan*, South). The mountain is thickly wooded with cool bamboo thickets full of fast-flowing streams, butterflies, monkeys and pandas; (c) *Hengshan*, Datong-The 'Northern Guardian' was named by the Ming emperors in the 14th century A.D. after they overthrew the Mongol rulers and regained the territory; (d) *Hengshan*, Hunan-The most southern of the Taoist holy mountains [In Chinese the pronunciation differs from the northern Hengshan]; (e) *Taishan*, Shandong-Most venerated of the five sacred mountains of China, revered by Taoists, Buddhists, and Confucianists.

South Korea: (a) Mt. *T'aebaeksan*—one of the three most sacred peaks and popular pilgrimage sites in Korea. Site of the *Ch'onjedan* altar for the performance of religious ceremonies. In the Tanngol valley below is a shrine to the mythical

Contd.

Box 1 : Contd.

progenitor of the Korean people, *Tan'gun*; (b) *Mt. Hallasan*, Cheju-sacred mountain considered to the counterpart of *Paektu* in North Korea. Both have a lake in the center and a rich history of shamanistic ritual practice.

Japan: (a) *Ibusuki sacred wood*, Kyushu-sacred woods in Japan have since early times been places of veneration and evocation of *Kami* (nature spirits) found in trees, rocks, and streams, whose shape are often imparted with animistic significance; (b) *Saitobaru*, Kyushu-sacred Shinto mountain where legendary figure *Ninigi no Mikoto* allegedly "landed" with a shrine dedicated to him; (c) *Ama No Iwato-Jinja Shrine and Sacred Caves*, Kyushu-shrine split into two by a river: on the east bank is a cave-the boulder door of heaven-where the sun goddess was lured out to bring sunlight back to the world.

Southeast Asia

Myanmar: *Mt. Popa*—mountain centre for worship of the nats, a group of spirits propitiated with offerings to protect from their harmful actions by a folk cult extending into Thailand.

Vietnam : (a) *Sam Mountain*, Mekong Delta—holy mountain surrounded by dozens of pagodas and temples; (b) *Nga Hanh Son*, Da Nang - five stone hills and Buddhist sanctuaries each representing one of the five elements. *Thuy Son* (water), *Moc Son* (wood), *Hoa Son* (fire), *Kim Son* (metal or gold), and *Tho Son* (earth).

Indonesia : (a) *Karang Tretes* Cave, Java—cave devoted to the virgin queen of the Southern ocean, who lives in a splendid palace on the seabed surrounded by many spirits. The Javanese make elaborate sacrifices to her before venturing along the dangerous coastal zone; (b) *Gunung Penanggungan*, East Java-sacred Hindu Mountain and major site of pilgrimage with over 81 temples, said to be the peak of the holy Mount Mahameru, which broke off and landed when the holy mountain was being transported from India to Indonesia.

South Asia

Nepal/Sikkim: *Mt. Kanchenjunga*—third highest mountain in the world deeply venerated by both Buddhists and Hindus. Padmasambhava, the Indian saint revered throughout the Himalaya blessed the *Yoksum* area and the surrounding landscape of 'Demojong' in the district of West Sikkim by placing within it a number of hidden treasures (*'ter'*) to be slowly revealed to enlightened Lamas at appropriate times.

Sri Lanka : *Mihintale Mountain*—site of the "couch of Mahinda", carved in the rock. Mahinda brought Buddhism to Sri Lanka from India by air and this mountain is the place where he landed.

India : (a) *Trimula, Tirupathu*—holy hill, one of the largest pilgrimage sites in India, and site of the ancient Vaishnavaite Temple of Sri Balaji, a god whose eyes are covered in case they might scorch the world; (b) *Amarnath*, Kashmir-ancient site of pilgrimage situated in a narrow gorge at the farthest end of the Sind valley. Favourite haunt of Shiva and his divine consort Parvati symbolised by rare pigeons who live in the cave nearly 4,000 metres above sea-level.

South & Latin America

Guatemala : (a) *K'umarcaaj*, Western Guatemala—partly excavated capital of the *Quiché* Maya people, for whom it remains a ritual center for Mayan practices; (b) *El baul*, Cotzumalhuapa—sacred site with buildings and monuments in the Izapan style which still have an active place in animist devotion and worship.

Venezuela : *Sorte Mountain*—sacred mountain used a as healing center for the Maria Lionza religion, a recent shamanistic group whose devotees worship the jaguar.

Colombia : *Lake Guatavita*—sacred site of the Muisca people where upon appointment the new king was adorned in gold dust and put on a raft covered in torches to set sail to the center of the lake and offer gold and emeralds to the water and mountain gods.

North America

Canada: *Moose Mountain*—2,000 year-old 'medicine wheel' with five spokes and cairns at the end of each. Central cairn is surrounded by a stone ring for former ritual purposes.

United States: (a) *Bear Bute*—high ridge sacred to native Indians of South Dakota used for vision quests, the search for spiritual power and self-knowledge. Around 4,000 native Americans come to pray here every year; (b) *Multnmoah Falls*—waterfalls on the Columbia river sacred to the Multnomah people. Legend tells of a chief's daughter who, to end an epidemic, threw herself from the clifftop as a sacrifice to the Great Spirit who created the falls in her memory.

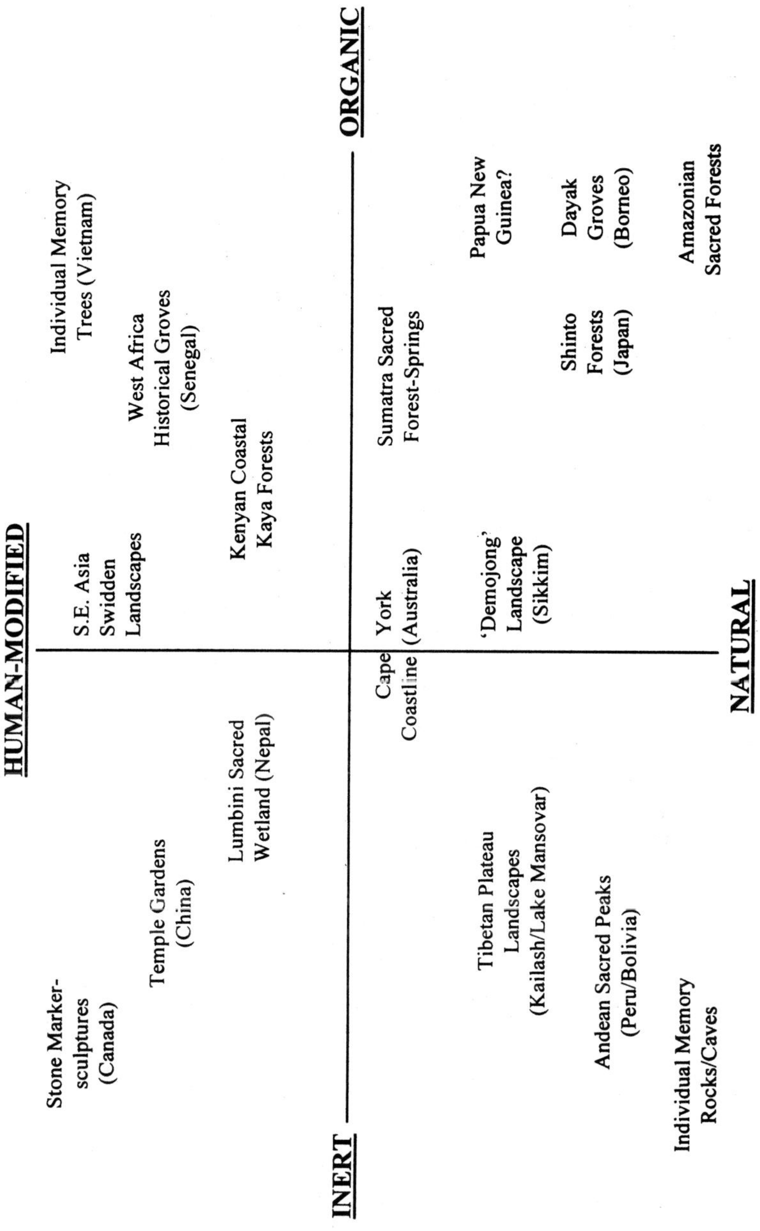

HUMAN-MODIFIED

Stone Marker-sculptures (Canada)

Temple Gardens (China)

Lumbini Sacred Wetland (Nepal)

S.E. Asia Swidden Landscapes

Individual Memory Trees (Vietnam)

West Africa Historical Groves (Senegal)

Kenyan Coastal Kaya Forests

INERT — **ORGANIC**

Cape York Coastline (Australia)

Sumatra Sacred Forest-Springs

Tibetan Plateau Landscapes (Kailash/Lake Mansovar)

'Demojong' Landscape (Sikkim)

Papua New Guinea?

Andean Sacred Peaks (Peru/Bolivia)

Shinto Forests (Japan)

Dayak Groves (Borneo)

Individual Memory Rocks/Caves

Amazonian Sacred Forests

NATURAL

Fig. 1: Human Impact on Natural Sacred Sites (suggested examples only)

and meditation.[11] In the light of richer ethnographic description heuristic categories may be dispensed with, but a few basic characteristics can nonetheless be useful to compare sacred sites within an analytical framework of ecological memory assessment:

(1) **Degree of organic/inorganic composition**
- rock (natural/carved)
- water (river/lake/spring)
- vegetation (continuous forest/grove)
- soil (anthropogenic or 'natural')

(2) **Sensory perception of landscape**
- sight (hidden or conspicuous)
- smells ('natural'/introduced)
- touch/contact? (unvisited/limited access)
- sounds (animals/winds/waters)

(3) **Boundary identification**
- oral/written? (elders knowledge or texts)
- markers? (organic/inorganic)
- size and shape
- scale (single tree or place/network of sites)
- hierarchy of sacred? (centre/periphery)
- link to monument?

Indigenous people living in dense tropical forests in lowland South America, Southeast Asia, Papua New Guinea and parts of Africa are quite literally re-enfolded within an acoustic, visual and olfactory organic landscape quite unlike the open fractured landscapes inhabited by the majority of the world's population. Such forest-dwelling people live in enveloping environments where sounds and smells provide the guiding sensory instruments to situate oneself in the world. In the Bosavi region of Papua New Guinea, Feld (1996) describes Kaluli inter-sensuality as a "perfume of hearing" which recognises a diffuse 'synesthesia' where sound, sight and smell are intertwined within the same being-in-the-world.[12]

---

[11] Amarnath cave, a major Hindu pilgrimage site in northern Kashmir is considered to be an abode of the gods Shiva and his consort Parvati.

[12] The Kaluli verb *dabuna* signifies absorption of sensory information through both ear and nose as muti-sensory 'sago-places' emit a smelling 'presence' of the aroma of fresh or rotting sago pith accompanied by the social history and mythological memory inscribed in the place activities of the ancestors.

Concerning the issue of scale and hierarchy of a sacred site, Palmer (1996) depicts an 'interconnectedness' of sacred mountains in China by including both Taoist and Buddhist mountains and caves into a 'network' of pilgrimages associated with a hierarchy of gods presiding over the most holy mountain, Tai'shan, through to lesser mountains situated in the four cardinal points of the country. Pei (1991) also describes the 'holy hills' of the Dai ethnic minority in Yunnan demonstrating how the perceptions of set-aside, sacred forests are intimately intermeshed with a regional system of natural resource management. In both cases, the authors argue that an emic awareness of the interlinkages tie these dispersed points at the regional and national scale in a large country such as China into a single referential network(Fig. 2).

Cultural categorisations of features of the natural world will also depend on the stability of an ecological system. Environments are dynamic and change considerably over time as forests are cut and planted, soil is leached and erodes, and long term cycles in the weather can alter patterns of rain and sun. Vernacular linguistic categories of vision, sound and smell attempt to address these ecological patterns. As the analysis of the ongoing dialectical relations between human acts and acts of nature may manifest in the landscape, historical ecology can thus serve to examine structural inversals in associations of the 'sacred' with features of the landscape undergoing fragmentation and ecological change. In formerly extensive forests being reduced to vestigial groves, a series of crucial inversals may be taking place in relation to the boundaries of the sacred and its phenomenological emplacement (Hay-Edie In Press).

As Walpole (1991) points out, 'visibility' of natural sacred sites is both a literal concept referring to the material presence of trees and vegetation but also a metaphysical relationship to land which cannot be disassociated from such a perception of vital energies. As environments are altered over time and populations migrate or are (sometimes forcibly) resettled, forest areas and landscapes may become disconnected from their former inhabitants and become spiritually "empty", while remnant forests may become spiritually "alive" while scientifically too small to regenerate and "standing dead". Ecological destruction leads to a form of 'cultural blindness' where the markers and memory places which once "attached the thoughts" and structured the existence of indigenous people are suddenly and brutally swept away:

> *"The old datu was sitting at the edge of a deserted logging road beyond Opis high in the mountains of Bukidon, Mindanaoà. He was looking out over the valley, but he could not see; he was nearly blind. He explained the demise of his people: 'We had our*

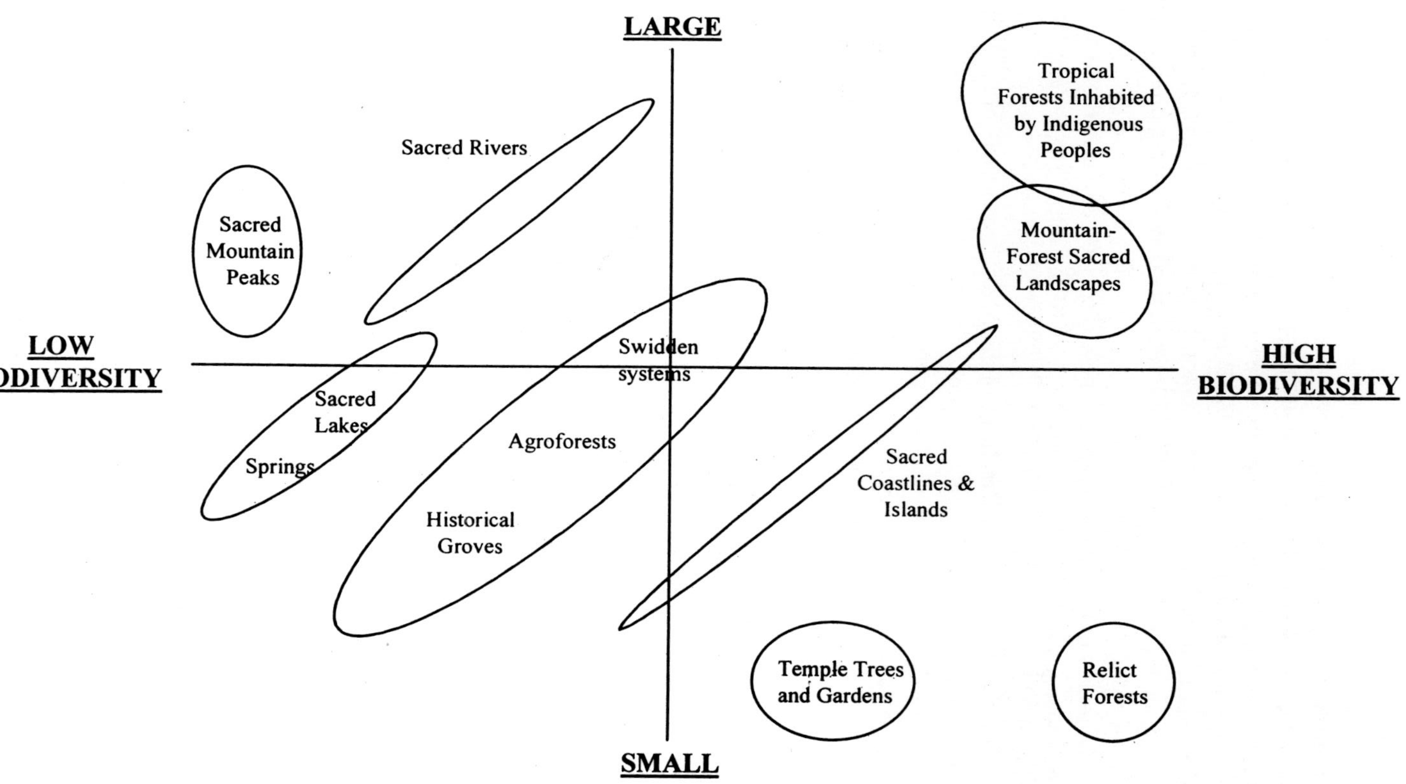

Fig. 2 : Scale and Type of Natural Sacred Sites (large UNIT size on vertical axis, estimated abundance in POLYGON size)

> *own way of seeing the life of the forest. But some time ago others came and they only saw logs, not life. They were more powerful than the spirits of the forest, so we were forced to see what they saw. When the logs had gone, the forest had gone, and life too. Everyone left us, even the spirits. We are blind.' (Walpole, 1991).*

All natural sacred sites display a spiritual interweaving of historical and mythological depth, yet in a rough scale of the vulnerability and fragility to disappearance through loss of respect and traditional knowledge, highly organic natural sacred sites are the most susceptible to destruction. The life-world of many forest people vividly reflects the encompassing totality of an environmental envelope engaging both the senses physically as well as in a metaphyscial sense. Highly organic natural sacred sites of this nature epitomise the symbiosis between the continuation of an unfragmented ecosystem and the maintenance of a people's cultural integrity.

CONSERVATION AND SACRED SITES

The role of natural sacred sites as aforementioned is attracting increasing interest in international organisations and conservation organisations such as UNESCO, the World Wide Fund for Nature (WWF), The Mountain Institute (TMI) and has significant relevance for the implementation of article 8j of the Convention on Biological Diversity (CBD) which sets out to:

> *"respect, preserve and maintain knowledge, innovations and practices of indigenous and local communities embodying traditional lifestyles relevant for the conservation and sustainable use of biological diversity and promote their wider application with the approval and involvement of the holders of such knowledge, innovations and practices and encourage the equitable sharing of the benefits arising from the utilisation of such knowledge, innovations and practices".*

Through an ongoing process of consultation and reflection within the CBD, UNESCO and other institutions, a series of questions are being formulated regarding the possible benefits and uses arising from natural sacred sites as 'vernacular protected areas'. Pei (1997) asks the question: can the existence of sacred sites be tapped by conservation organisations as a category of 'cultural resource'? Dinh (1996-97) also comments for *Ficus* species in Vietnam "l'arbre-repere est aussi l'arbre-repère" ("reference trees are also repair trees"). Through ongoing work in ethnobiology, UNESCO recognises that local peoples' semantic categories combine biological descriptions with socio-cultural qualities. Wherever possible it is necessary to examine the classification of sacred sites in vernacular languages and the position they hold in relation to other 'profane' areas before proceeding to a more generalisable comparative level of analysis. Nonethless it is possible to formulate a few of simple working hypotheses:[13]

1. Do sacred sites have similar sanctions on cutting wood or the collection of certain plants and are the location of tombs or ancestral shrines managed by 'sacred' specialists (Camara,1994; Aumeeruddy, 1995)
2. Do sacred sites occur equally in both resource-rich and resource-poor ecosystems and can the gene-pools of plants in sacred sites be used in programmes of restoration ecology? (Ramakrishnan et al., 1994; Schaaf,1995)
3. In what measure are sacred sites 'natural'? Does one find in them vestiges of primary forest or rather anthropogenic habitats where certain plants are encouraged to provide selected natural resources?
4. Can sacred sites be examined as indicators for assessing the potential natural vegetation of ecosystems degraded or modified by human impact? (Roussel, 1992)
5. What is the mutual overlap between cultural and spiritual 'values' of biodiversity and so-called scientific or economic 'valuation' techniques (Posey, 1998 forthcoming).

Conservation agendas however seldom address political issues and many descriptions of sacred groves and mountains by biologists do not consider the relations of religious or economic power inherent in reality claims of the sacred. Often, where conservationists do adopt an activist or advocacy role these tend towards a naive vision of an 'ecologically noble savage' in static unchanging societies (Ellen, 1985). Further, in the quest for commensurate cross-demonstration of ideological lessons, some conservationists adopt diffuse notions of the sacred. Palmer (1996) in presenting an idea of pan-Chinese spirituality does not consider the competing appropriations of state, confucianist, buddhist, taoist and shamanistic spirit-cults which may contest perceptions of sacred places.[14] Nakashima and Roué (1997) therefore pose the question for the Cree Indians of Quebec: can the selective protection of some sacred places become an alibi for powerful developers to destroy other sites?

[13] See UNSECO-CNRS Symposium September 22-25, 1998 Information Note

[14] Anthropologists such as Humphrey (1995) on the other hand show how many such 'power places' in Mongolia are sites where the perceptions of popular shamanship and Buddhist stately ritual have entered into intense conflict over long periods of time.

Large + Low Biodiversity *Ecological Awareness* Component? i.e. Mountain Pilgrimage	*Large + High Biodiversity* *International recognition* Associative Cultural Landscape Biosphere Reserve
Small + Low Medium Biodiversity Strengthening Ecological Memory Markers i.e. rocks, trees, sculptures	*Small + High Biodiversity* 'Strict Protection' :- National Monument Gazeteer/Inventory

Box 3 : Institutional Approach towards Natural Sacred Sites

SACRED SITES, CULTURAL INTEGRITY AND BIOLOGICAL DIVERSITY—SOME FIELD EXAMPLES

Reflecting the increasing interest in sacred sites from these numerous perspectives, plans are currently taking shape for a UNESCO-sponsored collaborative initiative which seeks to explore the interrelations between natural sacred sites, cultural integrity, and the conservation and management of biological diversity.[15] The stimulus for this initiative comes from several different programmes and interests, including concern with traditional ecological knowledge, so-called "vernacular conservation" and bottom-up approaches to conservation, the notion of 'cultural landscapes' and the cultural dimensions of natural resource use.

Questions raised concern the impact of existing or potential biosphere reserve or world heritage 'cultural landscape' status on local sacred sites. A major concern is whether international protected areas serve to revalorise and protect local systems of values or rather to dislocate local people from their symbolic ties to land? In Australia-Pacific and North America for example many indigenous groups now present ancestral toponyms and local place-names as arguments to seek assistance from international bodies and human rights NGOs in legal court cases to substantiate claims for land and territorial ownership.

[15] Concerning guidelines on 'cultural integrity' the 1966 UNESCO *Declaration of the Principles of International Cultural Co-operation* proclaims in its first article: 1. Each culture has dignity and value which must be respected and presereved. 2. Every people has the right to develop its culture. 3. In their rich variety and diversity, and in reciprocal influence they exert on one another, all cultures from part of the common heritage belonging to all mankind.

Within the UNESCO international protected area framework which seeks to positively reinforce associative cultural values of landscape, Engel (1985; 1989) considers that biosphere reserves and world heritage sites thus emerge as entirely new types of sacred areas. A number of recent globalisation theorists also argue that within the growing transnational discourse of conservation and development, numerous hybrid representations of place emerge through a process of involution known as 'glocalisation' where the global and the local perpetually re-encompass each other. In this light, international protected areas come thus to represent globally-enchanted areas lifted beyond the confines of national legislation into a transcendent arena of global symbolic value. A number of field projects have at the local level already addressed aspects of the 'sacredness-culture-biodiversity' triptych as an ensemble engaging a wide range of the perspectives and features mentioned in the discussion above. Insights from three field situations follow:

Sacred Forests in Casamance, Senegal

In a study of biodiversity and sacred forests in the Ziquinchor region of Casamance, Senegal, a team of collaborators recorded the perceptions of sacred forests at various levels (individual, family, village, community, young people) and the role of these forests in the life of Baynouk and Jiola-speaking populations in both a traditional and modern political context. The study showed that a number of taboos and prohibitions serve to conserve biological diversity within sacred forests. Some taboos were found to be internal to the forests such as prohibitions on tree-felling, burning, poaching crop production and entry, while others were external to the forest, such as the formal interdiction to divulge secrets of the sacred forest (i.e. on medicinal plants) to the non-initiated. One of the basic rules of the sacred forest is that *"tout ce qui est vue àl'interieur ne peut etre divulgué á l'exterieur"* ("all that is seen on the inside cannot be divulged outside"). Other 'outside' regulations relate to the setting of bush fires and tree-cutting in the vicinity of a sacred forest (Camara,1994).

Sacred Village Forests in Kerinci, Sumatra

In the Kerinci valley in Sumatra, sacred village forests fulfil a range of functions-economic, religious, social and environmental. According to oral history, the first people who arrived in Kerinci were seeking a land where springs never ran dry, even in the dry season. To this day, the inhabitants of Kerinci attach great importance to the preservation of springs which are considered to be sacred places. Following the foundation legend, heard in many permutations in most of the villages throughout the region, marriages took place between women described as deities (those posessing supernatural powers) who lived on the forested mountain summits near-by springs, and the first people who arrived in the region (Aumeeruddy,1994).

Symbolic access is therefore required to the forest, to the land and to the water as the vital element for rice production. Set-aside forest is perceived as the dwelling place of ancestors and many forest spirits. Through this foundation myth, a close relationship is established between real and mythical ancestors, forests, springs, rivers and rice paddies in the local representations of nature. In practice, there are many indications of this concern to preserve the springs and, more generally, the zones upstream. Taboos prohibit cutting forests around streams and along river banks. Village forests are thus preserved amidst the agricultural land, generally in catchment areas that feed the principal streams and rivers.

The village forest (*temedak*) at Keluru village is an example of a sacred place, where one of the ancestors who founded the village lived. The village shaman or healer fulfils the role of customary chief, secondary only to the village chief who represents the Indonesian administration. The customary chiefs control the gathering of forest products which include condiments and spices, construction materials and timber, and plants with medicinal, magical-ritual, or technological purposes. Trees cannot be felled without the permission of the customary chiefs. No product can be sold. Encroachment on the forest can lead to a fine and the farmer responsible has to replace the trees cleared (Aumeeruddy, 1995).

The forest thus fulfils an economic function by providing multiple products; a religious function through the link maintained with ancestral spirits; a social function through the intermediary of the shaman who wields considerable power over the inhabitants; and an environmental function reflected in the inhabitants' conservation of forest cover in order to safeguard springs and rivers. These integrated functions associated with the forest remain at the root of the local farmers' indigenous rationale for conserving and managing their own natural resources.

Tongariro—First 'Cultural Landscape' on the World Heritage List

Tongariro national park in New Zealand became the first cultural landscape on the world heritage list in July 1994 when the natural property was renominated as an 'associative cultural landscape' (ACL) under new criteria within the World Heritage Convention. A revised set of operational guidelines introduced three categories of cultural landscape of outstanding value as: (1) 'designed cultural landscapes' such as gardens and parks; (2) 'organically evolved continuing or fossil landscape'; and (3) 'associative cultural landscapes' identified as:-

"justifiable by virtue of the powerful religious, artistic or cultural associations of the natural element rather than material cultural evidence which may be insignificant or even absent."[16]

In the case of Tongariro, Maori mythology holds that all life forms come from the sky and the earth and humans are related to the mountains. The identity of the people and their chief is related to the volcanic landscape expressed in the words "te ha o taku maungu, ko taku manawa" ("the breath of my mountain is my heart"). For the Maoris, the mountains are sacred and travelling into them was forbidden-it is said that long ago travellers wore special hats to prevent them from seeing the mountains.[17] In 1887 paramount chief Te Heuheu donated the sacred mountain peaks of Tongariro, Ruapehu and Ngauruhoe to the British crown to ensure their protection thus making the park the first in the world to be donated by an indigenous people to a state (World Heritage Newsletter No. 2, 1995). However while Mt. Tongariro has today been nominated as a 'cultural landscape' of an international order, to the Maori a small area of land hidden somewhere within the Park, remains "fenced off".[18]

Uluru Kata-Tjuta in Australia then became in 1994 the second ACL to recognise the ancestral relationship of aborigines as traditional custodians of their environment, while the inscription of the Ifugao rice terraces in the Philippines as a continuing cultural landscape further engaged the need for participatory co-management.[19] A workshop in Sydney on ACLs also emphasised "mental images embedded in a people's spirituality, cultural tradition and practice" as well as intangible "acoustic, kinetic (air movements) and olfactory" elements. Participants felt that 'inspirational' ACLs central to indigenous peoples' sense of a dwelt-in sacred landscape could come to be expressed in the world heritage nomination process through words in poetry and song, or through photography and paintings as 'landscapes of memory' (ICOMOS Australia, 1995).

CONCLUSION

Attempts by modern institutions to broach the sacred are critically linked to the form by which the set-apart comes to be expressed and well-intentioned attempts to rationalise the sacred may, paradoxically, reveal or even

---

[16] Operational Guidelines Paragraph 39 (iii)

[17] The volcanoes in the area of Tongariro are linked to the foundation story of the Polynesian ancestor *Ngati Tuwharetoa* who arrived in the region around 1300 AD and continues to provide a powerful sense of identity to the Maori people.

[18] Sarah Titchen, personal communication November 1996.

[19] World Heritage Centre (1995) 'Asian Rice Culture and its Terraced Landscapes—Report on the Regional Thermatic Study Meeting' Manila, Philippines March-April 1995, UNESCO.

exacerbate a secular alienation from land. For most local people, symbolic imagination and scientific observation of the natural world remain profoundly intermingled. Once explained scientifically, the apparently arbitrary existence of natural sacred sites as objects of veneration may become "visible" for interpretation by conservationists yet such an interpretability may not in fact be desirable in terms of the cultural integrity of local and indigenous people.[20] For Geertz, interpretability represents a defining aspect of worldview, but how far is the explanatory power of Cartesian reductionism compatible with a dwelt-in sense of place inherent in holistic ecological knowledge?[21]

What is revealed and what remains concealed remain important hallmarks of the mysterious quality inherent in the sacred and undue rationality or surface visibility to sacred objects, knowledge or sites may be injurious or a serious affront towards cultural sensibilities. Abstract ecological categories of sacred sites thus rapidly appear hollow if dissociated from a detailed ethnography of a cultural system under consideration. Biological and cultural conservation discourses therefore need to ensure that any discussion of sacred sites (or networks of sacred sites) remain rooted in a grounded and numinous sense of a real sacred place evident in the words of Lama Anagarika Govinda who describes a sacred mountain in this way:

> *"Nobody has conferred the title of sacredness on such a mountain, and yet everybody recognises it, nobody has to defend its claim, because nobody doubts it; nobody has to organise its worship, because people are overwhelmed by the mere presence of such a mountain"*[22]

Alongside the above analytical framework to compare natural sacred sites it is therefore also essential that site-specific protocols and guidelines be "tailor-made" to accord respect and pay tribute to local and indigenous people's sense of spiritual intimacy. Box 3 presents a proposed set of insititutional appraoches for the valorisation and protection of natural sacred sites. As part of this process a number of innovative media of an artistic, poetic or audio-visual nature faithful to emic perceptions of sacred sites or associative cultural landscapes will need to be èxplored involving

---

[20] This said, sacred knowledge by its very nature is difficult to access. Hallendy (1996) for example describes how Inuit elders effectively circumvent inappropriate questions concerning secret knowledge with the simply reply *"who knows"* avoiding to reveal the sacred while not denying its existence.

[21] Ernest Gellner once referred to this cartesian 'cogito' process as one of "rational enlightenment fundamentalism".

[22] Bernbaum, E. (1990) 'Sacred Mountains of the World' p. xiii Introduction, Sierra Club Books, San Fancisco.

collaboration between indigenous people, cultural promoters and specialists from ethnoscience, anthropology and environmental conservation. Detailed scientific case studies focusing on boundary demarcation and community mapping of the spatial dimensions of traditionally venerated sites should thus be joined with culturally-sensitive ethnographies to provide a clearer voice for the traditions and innovations of indigenous and local people actively maintaining diverse worldviews and ecosystems.

REFERENCES

Aumeeruddy, Y. 1995. Local Representations and Management of Agroforests on the Periphery of Kerinci Seblat National Park Sumatra, Indonesia In: *People and Plants*. WWF-UNESCO-Kew Initiative Working Paper No.3, Paris.

Aumeeruddy, Y. 1994. Perceiving and managing natural resources in Kerinci, Sumatra. *Nature and Resources*, 31: 3-5.

Bernbaum, E. 1990. *Sacred Mountains of the World*. Sierra Club Books, San Francisco.

Bernbaum, E. 1996. Sacred Mountains: implications for protected area management. *Parks*, 6: 41-48.

Camara, T. 1994. *Biodiversité et forets sacrées en Casamance, région de Ziquichor*. Afrinet Report 10, UNESCO-ROSTA, Dakar.

Crumley, C. 1996. *Historical Ecology—Cultural Knowledge and Changing Landscapes*. School of American Research Press, New Mexico

Dinh T. H. 1996-97. Signes-Nature, Signatures, Biodiversité (1): Les bosquets cultuels" au Vietnam-pour un concept de "vestiges verts. *Cahiers d'Etudes Vietnamiennes*, 12: 13.

Engel, J.R. 1985. Renewing the bond of mankind and nature: biosphere reserves as Sacred Space. *Orion*, 4: 52-59.

Engel, J.R. 1989. The Symbolic and Ethical Dimensions of the Biosphere Reserve Concept In: *4th World Wilderness Congress*, pp.21-32.

Ellen, R.F. 1986. What Black Elk left unsaid-On the illusory images of Green primitivism. *Current Anthropology*, 6: 9-13.

Feld, S. 1996. Waterfalls of song: An Acoustemology of Place Resounding in Bosavi, Papua New Guinea. In: S. Feld, and K. Basso.(Eds.) *Senses of Place*. School of American Research Press, New Mexico.

Hallendy, N. 1996. The Silent Messengers. *Canada Journal, Jan/Feb 1996*: 40.

Humphrey, C. 1995. Chiefly and Shamanist Landscapes in Mongolia. In: E.Hirsch and M. O'Hanlon (Eds.), *The Anthropology of Landscape—Perspectives on Place and Space*. Clarendon Publishing, Oxford

ICOMOS Australia, 1995. The Asia-Pacific Regional Workshop on Associative Cultural Landscapes. A report by Australia ICOMOS to the World Heritage Committee.

Palmer, M. 1996 . *Travels through Sacred China*. Thorsons, London

Pei, S. 1991. Conservation of Biological Diversity in Temple-yards and Holy Hills by the Dai Ethnic Minorities of China. *Ethnobotany*, 3: .27-35.

Pei, S. 1997. *Report from UNESCO-ICIMOD National Training Workshop Application of Ethnobotany to Conservation and Community Development*. Chitwan, Nepal .

Ramakrishnan, P.S., Campbell, J., Demierre, L., Gyi, A., Malhotra, K.C., Mehndiratta, S., Rai, S.N. and Sashidharan, E.M. 1994. *Ecosystem Rehabilitation of the Rural Landscape in South and Central Asia: An Analysis of Issues*. Special Publication. UNESCO, New Delhi.

Ramakrishnan, P.S. 1996. Conserving the Sacred: from Species to Landscapes. *Nature and Resources*, 32: 11–19.

Roussel, B. 1992. A propos de l'Historicité des forets Sacrées de l'ancienne cote des esclaves' In: *Plantes, paysages et histoire en Afrique Sud Saharienne,* Karthalo-ORSTOM-CRA

Schaaf, T. 1995. Sacred Groves: Environmental Conservation based on Traditional Beliefs. In: *World Decade for Cultural Development 1988-1997: Culture and Agriculture, Orientation Texts,* UNESCO, Paris, pp.43-45.

Walpole, P. 1991. In the Shade of the Forest. In *Asian Pacific Culture 'Forest'*-UNESCO No.45.

Wilson C. 1996. *The Atlas of Holy Places and Sacred Sites.* Dorling and Kindersley, New York

5

Sacred Groves Around the Earth: An Overview

J.D. Hughes and M.D.S. Chandran***

*University of Denver, Denver, Colorado 80208, USA
**Department of Botany, Dr Baliga College of Arts and Science
Kumta 581343, India

INTRODUCTION

The groves were God's first temples. Ere man learned
To hew the shaft, and lay the architrave...
in the darkling wood,
Amidst the cool and silence, he knelt down.

(Bryant, 1848)

Sacred groves are segments of the landscape, containing trees and other forms of life and geographical features, that are delimited and protected by human societies because it is believed that to keep them in a relatively undisturbed state is an expression of important relationship to the divine or to nature. Diverse cultures perceive this relationship in different ways, and establish various rules of behavior in regard to sacred space and its denizens. But wherever they occur, sacred groves are of ecological and cultural interest.

Ecologically, sacred groves tend to be fragments of the original ecosystem in a given region, although in some cases local people plant trees in groves, and there were, and are, groves consisting entirely of planted trees. Because they are protected, the groves that are derived from natural forests often shelter plant and animal species that may have disappeared elsewhere in the region. Thus they serve as refugia, and possible centers of dispersal and restoration. They are not immune to ecological change caused by several factors, however. As a rule, sacred groves are relatively small, from a fraction of a hectare to a few square kilometers in size, and are therefore island-like, or like individual tesserae in a landscape mosaic. Studies of island biogeography have shown that such fragments may suffer extinctions of species. Also, sacred groves may be vulnerable to invasions of aggressive introduced exotic weed-like species, and to natural or human-caused disasters such as fire, windstorms, diseases, predation by hunters and poachers, encroachment, and deliberate destruction.

Culturally, sacred groves are of considerable interest because they exemplify phases of social interaction with the local ecosystem. The practices permitted or forbidden in them often reveal much about the attitudes of various societies to nature. In early times, and often down to the present, their use was and is regulated by local communities. These may be villages or cities, religious organizations, families, or other affinity groups. In a few cases, they are recognized and granted protection by modern states or nations. Any of these entities may also sometimes diminish or destroy them. Sacred groves may serve as examples of local ownership and autonomy, and may serve as rallying points for local people when these are threatened.

DISTRIBUTION IN SPACE AND TIME

There is evidence of sacred groves in past and present in many parts of the world, and among people with many different religions and forms of social and economic organization. While previously no coordinated, exhaustive search has been made to find all the localities in which they occur, examples are numerous and extend to every inhabited continent, and to many oceanic islands.

Africa

To begin with Africa, possibly the original home of humankind, sacred groves still exist widely in almost every sub-Saharan section of the continent. In East Africa among the Kikuyu, Frazer (1935) reports, "Groves of this tree [the *mugumu*] are sacred. In them no axe may be laid to any tree, no branch broken, no firewood gathered, no grass burnt; and wild animals which have taken refuge there may not be molested. In these sacred groves sheep and goats are sacrificed and prayers are offered for rain or fine weather or in behalf of sick children." In West Africa as well, the aura of holiness has kept botanically rich groves from falling prey to lumber companies (Bachmann, 1992). Sacred groves of Sierra Leone are repositories of medicinal plants of value to traditional medical practitioners (Lebbie and Guries, 1995). Sacred groves still exist in Ghana. For nearly three centuries, the community of Malshegu in Northern Ghana has preserved a small forest (0.8 ha) that they believe houses a local spirit. Although threatened by various nearby activities such as mining and the building of roads and power lines, this grove is being protected by the strong traditional religious beliefs of the community. The fact that local people guard it against fires, farming, grazing, gathering of forest resources, and hunting has allowed this grove to grow from a once open forest into a partially closed forest (Dorm-Adzobu and Veit, 1991). The government of Ghana has recognized the sociocultural and religious importance of sacred groves. The religion of the Senufo, inhabitants of the Ivory Coast, is "centered inside a sacred grove, where ritual initiations are held and objects of power are stored" (Spindel, 1989). Among the many groves in Nigeria is one in Oshogoo dedicated to the goddess Oshun (Molyneaux, 1995). In Old Calabar, the village folk worshipped a grove of trees in a ravine with a clear stream (Frazer, 1935). There are many sacred groves associated with villages in the southern nations of Africa, such as Zimbabwe and South Africa (Farieda Khan, pers. comm.). In rural Africa one often finds a "King's Grove," where permission for use is entirely the responsibility of the local tribal leader. In Egypt, a North African country, it was the practice in ancient times to have a sacred grove, along with a sacred lake, in the precincts of every sizable temple (Erman, 1894). These groves were planted with many species of trees including sacred species such as palm and persea (*Mimusops laurifolius* or *schimperi*; called *ished* in Egyptian) (Manniche, 1989). Egyptian gods and goddesses were believed often to reside in trees (Budge, 1911).

Western Asia

There are, or were, sacred groves in most regions of Asia, and they are associated with many different traditions. In Western Asia, one of the earliest literary traditions deriving from ancient Sumeria (fourth millennium BC) reveals their existence. Gilgamesh, King of Uruk, and his companion Enkidu, a formerly wild man, are said to have made a journey to distant mountains where a cedar grove stood surrounded by a palisade and guarded by a semidivine giant named Humbaba (Sandars, 1960). The defeat of Humbaba and the logging of the grove represent a historical process of deforestation on the margins of Mesopotamia. Later Babylonians and Assyrians planted groves; the "Hanging Gardens," one of the seven wonders of the world, was possibly a grove high on a ziggurat; archaeologists have found provisions for water drainage in these temple mounds. The Roman writer Diodorus of Sicily (see Oldfather, 1963) reported a palm forest in Arabia, with "an altar very old in years, bearing an inscription in ancient letters of unknown tongue," adding that the inhabitants are long-lived. A sacred oak grove was venerated in Asia Minor by the Galatians. Hebrew tradition honors groves associated with the patriarchs: Abraham built an altar among the holy oaks of Mamre, and planted a grove in Beersheba (Genesis 13:18, 21:13). Similarly Christians, who often zealously cut down groves where others worshipped, sanctified the Garden of Gethsemane, an olive grove where Jesus prayed, building a church beside it, and preserved groves of Lebanon cedars in monastery precincts (Mikesell, 1969). Christian cemeteries in the Holy Land, planted with cypresses and other trees, often resemble ancient sanctuaries. To the north, in Armenia, Christians to this day make offerings by tying cloths in sacred trees, and sacrificing sheep in groves. In pre-Christian times, their neighbors in the Caucasus, the Tzanoi, according to the Byzantine historian *Procopius* (see Dewing, 1961), lived independently, "regarding as gods the groves and birds and various creatures besides, and worshipping them." Among the Ingushes of the Caucasus, each community had its own sacred grove.

North, East and Southeast Asia

In north Asia, many Siberian people honored sacred groves, and used them for the rites of shamanism. The Ostyaks and Voguls, nomadic people of the Ob River basin, had sacred groves in which nothing could be touched, "no grass or wood might be cut, no game hunted, no fish caught, nor even a drop of water drunk" (Frazer, 1935). Eagles which alighted in a grove were also sacred. Among the Buryats, each tribe has a grove of firs where its esteemed shamans are buried. When in open country, these are visible from afar. There are also village groves in Korea (Kim, 1994). In Japan, Shinto shrines are as a rule surrounded by trees, and often by extensive thick forests including sacred trees. Indeed, the grove itself is one of the reasons for originally locating the shrine there (Ono, 1962). Deer are often kept in these precincts, and one of the ritual acts of worship is to release captive birds and animals. Many Buddhist temples both in Japan and China have carefully tended gardens, including trees that have the aspect of groves. Traditional Chinese religion honored sacred mountains which were tree-covered, and Taoism is well-known for its respect for nature; its shrines and hermitages were often located in forests. Monastery groves preserved species such as the Ginkgo (*Ginkgo biloba*), a curious archaic gymnosperm (Bachmann, 1992). In Yunnan province, China, members of hill-farming communities designated sacred woodlands where the gods were believed to dwell, and strictly forbade anyone from entering, except to make annual sacrifices to the mountain and earth gods. Since villagers felt that trespassing for tree felling, cultivation, hunting or gathering would bring disaster, these forests generally remained pristine. Remnants of them exist as patches of primary forest among villages in Lancang County and the Xishuangbanna Region. The idea of sacred forests was not limited to swidden cultivators such as the Dai minority, but the wet rice cultivators in the lowland basins also venerated sacred hills, preserving the local fauna and flora, including medicinal plants, from destruction (Yin, 1991, 1994; Wachtel, 1993). Southeast Asia, Thailand in particular, has many sacred groves associated with Buddhist monasteries and temples. The Balinese, in Indonesia, have "monkey forests," fragments of the ancient rainforest where animals sacred to the monkey-god Hanuman are protected.

South Asia

South Asia is particularly rich both in the historical existence of sacred groves and in their modern survival as compared to other regions, even though many have disappeared or been degraded. They are found from the Himalayas in the north to Kerala in the south. D. Brandis, the first Inspector General of Forests in India, wrote as follows on the sacred groves:

> Very little has been published regarding sacred groves in India, but they are, or rather were, very numerous. I have found them in nearly all provinces.... I may mention the Garo and Khasia hills which I visited in 1879, the Devara Kadus or sacred groves of Coorg with which I became acquainted in 1868, and the hill ranges of the Salem district in the Madras Presidency examined by me in 1882. Well known are the Swami Shola on the Yelagiris, the sacred grove at Pudur on the Javadis and several sacred forests on the Shevaroys. These are situated in the moister parts of the country. In the dry region sacred groves are particularly numerous in Rajputana. In Mewar they usually consist of *Anogeissus penduala*.... in Partabgarh and Banswara, in a somewhat moister climate, the sacred groves, here called *malwan*, consist of a variety of trees, teak among them. These sacred forests, as a rule, are never touched by the axe, except when wood is wanted for the repair of the religious buildings (Brandis, 1897).

Gadgil and Vartak (1976) recorded four important regions for extant sacred groves in India that had been reported to them. They were the Khasi and Jaintia Hills of the Northeast, the Western Ghats, Aravalli Hills of Rajasthan and the Sarguja, Chanda and Bastar areas in Central India. Subsequent reports have added other sites.

To begin with the north, Burman (1992) found sacred groves all along the Himalayas from Northwest to Northeast. The Kalash Kafirs of Hindu Kush have sacred areas that must be kept permanently pure and unpolluted, such as the grove of Sajigorr (Hewitt, 1986). There are sacred groves in Nepal. In West Himalayan Kumaon and Garhwal, many temples have small groves, some of which shelter monkeys. In the Darjeeling hills groves are known as *deorali:* these are at the meeting points of bridle paths from various directions.

In Meghalaya, the sacred groves known as *lakyntang* are present in the Khasia and Jaintia hills. The Jaintia hills may have over 200 groves, each no more than 3–4 ha (Burman, 1992). Balkrishnan (1981) found six major groves in these hills, each measuring 1–2 km^2. Groves protected by local gods and ancestral spirits stand out as green islands in a devastated "wet desert" landscape in the Cherrapunji district of Meghalaya in the northeastern hill region (Ramakrishnan, 1996). In the olden days, every village had at least one sacred grove, ranging in area from 0.04 km^2 to 12 km^2, but at present many of these have vanished. Rituals and ceremonies have often been discontinued, with the exception of some groves. The advent of Christianity and modernization have contributed to this. In many villages, though sacred groves have lost their significance, the headman with the concurrence of the village council has protected patches of forests. Limited extraction of timber is permitted. Further, the timber so extracted is shared by all the families in the village. Removal of dead biomass is strictly prohibited (Khiewtam and Ramakrishnan, 1989).The people of the Mizo Hills keep "supply forests" managed for conservation, and sacred groves called "safety forests" from which no biomass is removed.

In 1864, Brandis (1864) visited a sacred forest in the compound of a building inhabited by a Muslim saint, Mian Sahib, in Gorakhpur (Uttar Pradesh). This forest, dominated by sal *(Shorea robusta)* was in good condition. Nothing was allowed to be cut except the wood required for the perpetually burning sacred fire. This required annually the cutting of a small number of trees from among those showing signs of age and decay.

The groves of Dudh and Dhelki Kherias of Central India are known as *jankor* or *sarna,* and to the Santal tribals the groves are known as *jaher.* It is said that the Gonds always saved a grove when clearing a jungle, and called them *pen geda* (Burman, 1992). The tradition of the Munda people of Bihar was that every village should have a grove, and that permission must be gained through sacrifice before cutting a tree, or the tree gods would withhold rain (Frazer, 1935; Ramakrishnan, 1996). The Mundas call their groves *sarna;* evidently Sarnath, the "deer park" pilgrimage center where the Buddha gave his first sermon, is named for a sacred grove. Buddha upheld the right to live of every creature and incorporated many elements of the nature worship which already existed among the Indian people. Buddha found the *chaityas,* or cult spots with groves of trees or single trees, dedicated to the *yakshas* or other nature gods, to be ideal places to preach his sermons. For instance, in Vaisali, Buddhist worship places came into existence in the Mahavana (Great Forest), Gotamaka Chaitya, Chapala Chaitya, Ananda Chaitya, Sarandada Chaitya, etc. These were already pre-Buddhist worship places dedicated to sylvan deities (Mishra, 1962; Chandran, 1997a).

The tribals of Chotanagpur in Central India generally worship in groves of sal trees *(Shorea robusta)* located at the outskirts of the villages. All over Chotanagpur were, in the beginning of this century,

> Many jungle-clad hills, the favourite haunts of bears, which the beaters of the animistic races steadfastly refuse to approach until the mysterious power which pervades them has been conciliated by the sacrifice of a fowl. Everywhere we find sacred groves, the abode of equally indeterminate beings, who are represented by no symbols and of whose form and function no one can give an intelligent account, which have not yet been clothed with individual attributes; they linger as survivals of the impersonal stage of early religion (Risley, 1908).

These groves served as a religious rallying point for a tribal rights movement in the 1930s and 1940s (Hembram, 1982).

The groves of Rajasthan provide biodiversity refuges in an arid region; there the religious law of the Bishnois protects all *khejari* trees*(Prosopis cineraria)* and the blackbuck *(Antilope cervicapra).* A grove of 363 *khejari* was planted in 1978 in a village in honor of Amrita Devi and the other Bishnois who sacrificed their lives defending their trees in 1731 against the axemen of the Maharaja of Jodhpur, who wanted the timber for a palace (Weber, 1987). Elsewhere in Rajasthan are fewer groves, known as *jogmaya*..

Sacred groves are particularly noted all along the Western Ghats, mountains which border the west coast of South India; Gadgil and Vartak (1976) surveyed a number of them in Maharashtra, the northern part of the range, and noted their preservation of biological diversity. The central Western Ghats, in Karnataka, have hundreds of surviving groves, and are the subject of an accompanying paper by Chandran (Chandran, et al., this volume) along with the southern portion in Kerala, where there are village groves, temple groves, and groves on private property, some of which are refuges where snakes receive worship.

Europe

Europe had thousands of sacred groves in ancient times, although most of them have disappeared. There, as elsewhere, groves were the original temples. Ancient Greece was as full of them as Bacchus' leopard skin was full of spots, according to mythology, and *als-*, as they were called, were dedicated to gods major and minor (Hughes, 1984). Their management and protection was in the hands of the *poleis*, or cities. The situation was similar in Italy, the homeland of the Romans. Frazer (1935) based his monumental work, *The Golden Bough*, on the practices of succession to the priesthood of Diana in the grove of Aricia, near the lake and village of Nemi (*Nemus* is a Latin word for grove). The Celts, Slavs, and Germans, the Romans' northern barbarian neighbors, customarily worshipped in groves, and regarded the oak as the most divine tree. The Celtic word for sacred grove is *nemeton,* a wilderness sanctuary where nature worship occurred and earth-festivals were celebrated (Vest, 1985). There the Druids, the Celtic shamans, learned their lore and practiced their spirit-journeys. Many sacred groves existed in the British Isles; Frazer (1935) gives numerous examples such as that of a sacred lake and grove on the Isle of Skye in Scotland where not a twig could be broken, and Belly parish in Ireland, where in the grove around the chapel of St. Ninian, it was considered a sacrilege to take branches or fruit. Often the sacred precincts were demarcated by banks or ditches. Holy Island, off Wales, has a placename, Llwyn-y-Berth, meaning "the beautiful grove," only one among many Welsh names including *llwyn* (grove). The Dumnonii had a grove of their tribal god by the Tamar River between Cornwall and Devon. Boudicca, the Briton queen, fought a battle against the Romans at Vernemeton, a sacred grove in the center of Britain.

On the continent, there were manifold groves in Gaul (France), and Tacitus says that the Semnones, a Germanic tribe, considered their sacred grove to be the "cradle of the race and the dwelling-place of the supreme god to whom all things are subject and obedient". There were groves in the Scandinavian countries; the Swedish sacred center at Uppsala had one in which every tree was divine. The Baltic-Slavic people honored powers called *siela* that guarded the groves and would not allow anyone to whistle or shout there (Vest, 1985). They also protected animals, particularly the bear (*Ursus arctos*). The Lithuanians maintained oracles in holy groves about their villages and houses, and believed anyone who injured a grove would die within a year (Frazer, 1935). The Finno-Ugric people of Finland and arctic Russia commonly fenced their sacred groves, permitted no wood to be hewn, and excluded women from certain ones, a common rule in many parts of the world, although in other cases it was men who were banned. A fir grove survived to 1650 in Kuopio, Finland. The artist-ethnographer Wassily Kandinsky found active sacred groves among the Zyrians (Komi) in 1889 (Weiss, 1995). Relics of sacred groves exist to this day for the Nenets of the Arctic. The holy grove at Kozmin Copse in the Kanin tundra is a notable survival in which offerings are still made including modern industrial products such as a thermos and an alarm clock. The trees in this grove, mostly birch and fir, are sacred (Ovsyannikov and Terebikhin, 1994). The Wotyaks, who live toward the Urals, believed that the violator of a grove would die the very next day.

Other Parts of the World

Sacred groves are widely known in the Austro-Pacific region. Australian aborigines find groves among the many natural spots made sacred by the experiences of ancestral spirits in the ancient "Dreamtime" when the landscape was shaped. They also have countless groves of different trees for different uses, including ritual and worship. The Yarralin people of the Northern Territory have a "Dreaming" grove for uninitiated males *(karu);* each tree represents one *karu,* and there is a huge tree called "boss *karu*" (Molyneaux, 1995). The Dieri tribe of Central Australia regard certain trees, believed to be their fathers transformed, as very sacred, and they take care to see they are not cut or burnt (Burman, 1992). For the Maoris of New Zealand, humans form an intimate part of the environment Maori sacred sites, *Waahi Tapu,* include trees and forests among many other natural features (Matunga, 1994). In Polynesia, the coconut, breadfruit, etc., play important parts in myths and some groves of them were sacred places.

In the Americas, a widespread view in pre-Columbian times was that Earth and all her creatures are sacred. Trees were seen as having spirits and the power to help or hurt; their permission would have to be sought before any part of them was taken or used, and if harmed unwillingly, they could cry out and seek revenge (Hughes, 1996) In light of this universal sacredness, the marking out of specific groves is problematic. Still, the native people believed that although the whole Earth is sacred, still in certain places the spirit power manifests itself more clearly and readily, "like a fawnskin" with its many white spots (Hughes and Swan, 1986). Many of these power places were groves of trees. Pedro Simen, an early Spanish explorer of Costa Rica, described "copses" of many species of trees decorated with small golden bells (National Museum, San Jose). Bishop Diego de Landa wrote in the Yucatan in 1566, "They have sacred groves where they [the Maya people] cultivate certain trees, like cacao" (Tozzer, 1941). Cacao *(Theobroma cacao)* produced a valuable drink for Mayan priests and royalty, and its seeds were widely used as currency in Mesoamerica. Its existence in arid Yucatan was unknown to botanists until 1977, when surviving specimens were found in sacred natural limestone sinkholes called *cenotes* (Gomez-Pompa et al., 1990). Tribes such as the Ojibwas and Utes reserved certain sections of forest where they would not hunt any animals except in case of great need. Americans of European descent also have sometimes hallowed groves of trees. Certain ones have been dedicated to early settlers, or to the war dead, or to noted leaders. Muir Woods National Monument, north of San Francisco, has a redwood grove *(Sequoia sempervirens)* named for the great conservationist, John Muir. Muir wrote that the "clusters" of

Sequoia trees *(Sequoia gigantea)* were places where he could worship "King Sequoia... Lord Sequoia" and take the sacrament with "Sequoia wine" (Cohen, 1984). There are also Sequoia groves in California dedicated to the memory of General-President Ulysses S. Grant and President William McKinley, of the Humboldt Pioneers, of Jedediah Smith, Julia Pfeiffer Burns, Harris Whittemore, Henry Cowell, Franklin K. Lane, and many others. It might also be claimed that some of the national parks are sacred groves of a "civic religion." In Golden Gate Park in San Francisco itself, a quiet, wooded grove of six hectares was designated as the AIDS National Memorial Grove in 1996. It is dedicated to those lost to AIDS, and in support of those living with HIV. It is funded through private donations, is cared for by volunteers and a city gardener, and local people visit it to plant, to "pour water into bowls carved into rocks for small flower bouquets... and each week, one man gently places eggs in rock crevices—a Hindu offering" (Associated Press, 6 October 1996). Other such groves are planned in San Luis Obispo and other cities.

ORIGIN

That the practice of honoring sacred groves is ancient is clear, but how far back in human history was it found? It seems that groups of hunter-gatherer people began the process in Palaeolithic times. The evidence for this statement is that many hunter-gatherer societies observed in modern times have sacred groves, and that the symbolism encountered in the rituals of people who worship in sacred groves includes wild animals such as bears, deer, tigers, snakes, etc., hunter-gods and goddesses, and weapons used by hunters, such as the bow and trident. Though hunting was usually forbidden in the holy precincts, it was sometimes permitted for the express purpose of making an offering to a resident god or goddess. The sacred groves, wherever they existed, belonged originally to traditional societies, who considered themselves as being linked in a web of spiritual relationships with their biophysical environment. Primal hunter-gatherers see the world as filled with spiritual power and populated by spirit beings. All beings are alive and sentient; including the Earth and Sky. Animals and plants are to be approached with reverence, killed only when needed, and treated with honor even after they are killed. In this primal perception, the human being is one with all nature. People stand within the natural world, not separate from it; and are dependent upon it.

Agriculture in forested regions began with "swidden" or slash-and-burn methods. The differentiation of sacred groves from surrounding land probably first became clearly visible with the appearance of shifting cultivation;

they were the areas spared from slashing and burning. Village farmers respected the same groves that the hunters and gatherers had reserved, along with others like them. Many of these societies came to believe that the groves sheltered gods of the growing crops, deities that could assure a good harvest but also could bring famine if they were angered by volations of their sacred space. In this phase, sacred groves became isolated patches of the original forest in a landscape mosaic of villages, cleared fields, and secondary forest in various stages of growth.

As agriculture became intensive, and larger towns and cities asserted their claims to resources, sacred groves were further isolated. The cult of city gods was linked to the growth of professional priesthoods and the rise of temples as centers of worship. Temple buildings were erected in the groves, and the association of temple and grove was so close that even temples inside the cities were flanked with trees. Many of these had to be planted, so urban groves became less distinguishable from gardens or plantations. Those in the countryside, however, often persisted as fragments of the original forest ecosystems in the area concerned. Loss of species, especially large animals,occurred in those fragments in spite of the protection the smaller refuges received. Preservation through long periods of time depended on cultual continuity and the persistence of belief in the resident powers and the sacredness of life and land in the groves.

CONSERVATION OF BIODIVERSITY

"The network of sacred groves in such countries as India has since time immemorial been the locus and symbol of a way of life in which humans are embedded in nature and in which the highest levels of biological diversity are found where humans interact with nature," according to Marglin and Mishra (1993). One of the facts that makes sacred groves such important factors in conservation is that the behavior enjoined in regard to them protected not just the trees, but every possible element of the habitat. This is true also of sacred species of plants and animals. As Ramakrishnan (1996) observed, "ecologically valuable species, which perform the function of keystone species in an ecosystem and contribute to the maintenance and enhancement of biodiversity, are also species that are socially valued by local communities for cultural or religious reasons." Although specific rules varied from place to place, and over time, the usual attitude concerning sacred groves was that nothing ought to be broken, killed, or removed unless there was special need. Animals could not be hunted, fish could not be caught, branches and leaves could not be carried away. At the same time, domestic animals and many people, including hunters, were kept out. Planting of crops and construction of roads and buildings were proscribed. Certain limited forms of use, such as gathering for medicinal purposes plants that grew only in the groves, were sometimes permitted. When these regulations were obeyed by people who understood and accepted them, or could be enforced by the local authorities, they were effective in providing refugia for species that otherwise would have disappeared from the ecosystem. The grove would contain trees and understory plants that had been destroyed in neighboring tracts, and these would in turn provide food and shelter for other plants, small animals, and microorganisms. There would be loss of species due to the small size of most groves, and studies show that this loss might be quite serious, but the relic would still be of great importance in the total pattern of the landscape (Harris, 1984). Certainly the losses would be less catastrophic than if the grove, too, were removed.

A further factor that amplifies the significance of sacred groves is the fact that almost all of them contain water in the form of springs, ponds, lakes, streams, or rivers. Not only that, but the vegetative mass of the grove itself conserves water, soaking it up like a sponge during wet periods and releasing it slowly in times of drought. Plato noted that when the trees were cleared from around the shrines of springs on the hills of Attica, his homeland, the springs themselves dried up (see Bury,1929). A similar phenomenon has been noted in Uttara Kannada and elsewhere. Conceptualization of ecosystems as watershed-based units existed in several Amerindian, European and Asian cultures (Gadgil and Berkes, 1991). It is evident that one of the functions of groves in preserving biodiversity is to provide a more dependable source of water for the organisms inside them and in the neighborhood. Another function may be to reduce the incidence and intensity of fire, at least in some climates. In addition, transpiration will increase the humidity in the immediate vicinity and produce a more favorable microclimate for many organisms. When forest litter is allowed to accumulate, organic material builds the soil and is returned to the biomass of the standing forest. In the process, many microorganisms, invertebrates, fungi, etc., will flourish, and a vast array of species not indigenous to plowed fields and secondary forests will be present in the groves.

All of these factors indicate that the preservation of sacred groves is of key importance in maintaining biodiversity and the comprehensive health of a landscape. They provide greater benefits than their small size would indicate. Generally speaking, if they were larger and more numerous, these benefits would be even greater.

DESTRUCTION OF SACRED GROVES

It is ironic that the Epic of Gilgamesh (Sander, 1960), perhaps the oldest surviving piece of literature to refer to a sacred grove, describes its destruction. Gilgamesh's clearcutting of the cedar forest and use of its timber to build a palace did not go unpunished; his companion Enkidu died as an indirect result. The constant repetition of rules to protect groves, and a number of legends that tell of the terrible things that happened to those who violated them, indicate that transgressions occurred in early times.

One of the motives for incursions into the groves was desire to use their resources: timber for building, as in the case of Gilgamesh; hunting, when Agamemnon slew the sacred deer of Artemis; or use of vegetation for browsing of domestic animals, as with the Roman shepherds who were required to ask forgiveness for this trespass at the festival of Parilia (Hughes, 1994). Another was warfare. Battling armies often honored the gods and spared the sanctuaries of defeated enemies, but sometimes they burned sacred groves to flush out or to kill foes who had sought shelter in them, or simply to show their contempt for defeated adversaries. During the Persian invasions of Greece, in the fifth century BC, the aggressors burned some sacred forests as well as temples. In the Peloponnesian War, the Athenians cut wood from the sanctuary of Apollo at Delium to build fortifications. At the end of the third century BC, Philip V cut down the entire grove of the Nikephorion at Pergamum. In the same period Turullius built ships with wood cut in the sanctuary of Asklepios on Cos. Sulla cut the trees of the grove of Academus, the site of Plato's school, to build siege engines, and Julius Caesar was accused of felling a Gallic grove during the Roman Civil War.

A religious reason was often given for using the groves' resources. Wood from the groves was used for construction and repairs of temples in various countries including Greece and India. A 5th century inscription from Paros prohibits all cutting of trees in sacred groves except as required for the sanctuaries' own buildings (Jordan and Perlin, 1984). Sometimes a shrine or temple building might usurp most or all of the space formerly occupied by a sacred grove. Religious change, or conflict, might result in the destruction of sacred groves. The Israelites, during their occupation of Canaan, were instructed to cut down and burn the "groves" of the local inhabitants Modern scholars maintain that the Hebrew word *Õasher‰h*, translated "grove" in the Authorized Version, actually refers to a large sacred pole of wood which was an emblem of the old tree-worship. However, destruction of sacred groves is still quite probable, given passages such as the following: "You must demolish completely all the places where the nations whom you are about to dispossess served their gods, on the mountain heights, on the hills, and under every leafy tree" (Deuteronomy 12:2). In the areas where they became dominant, Christians also eliminated most of the groves and sacred areas of people that they considered "pagan" (the word refers to rural folk) or "heathen" (that is, dwellers in the wild). A few were preserved in monastery gardens and churchyards. Christian missionaries exhibited the same intolerance for sacred groves almost everywhere they went. For example, St. Boniface destroyed the great oak of Jupiter (Thor) at Geismar in Hesse around 722 AD. The grove of Irminsul, the most sacred place of the Saxons, was leveled at the order of Charlemagne (Heer, 1975). The Council of Nantes in AD 895 specifically commanded the destruction of trees that were consecrated to "demons," i.e. local gods. Stephen of Perm confronted the Zyrian (Komi) shaman, Pam. The saint cut down a fir grove that "was the site of many deities. Stephen was known for the fervor with which he destroyed the sacred groves of the heathens" (Weiss, 1995). All out Christianization of the Russian tundra from the early nineteenth century included physical destruction of much that was associated with the sacred tradition of the Nenets, including the Kozmin Copse monument that was destroyed in March 1837 by Christian missionaries led by Archimandrite Veniamin (Ovsyannikov and Terebikhin, 1994). Similar scenes were enacted in many nations when missionaries arrived.

Since the beginning of the Industrial Revolution, however, the greatest threat to sacred groves has been economic exploitation. In New Zealand, for instance, the Europeans began a flurry of economic activities, including logging and commercial intrusions into the landscapes, housing, high rise buildings, roadways, etc., which destroyed many Maori sacred sites (Matunga, 1994). In the United States, the American Indian Religious Freedom Act of 1978 was enacted to inventory and preserve the traditional religious values and practices associated with the sacred places of native people, but it has proved completely ineffective in the courts (Hughes and Swan, 1986). In India, in colonial times and afterwards, sacred groves were included in forests managed by the Forest Department, and subjected to selection fellings, clearcut to be replaced by plantations, or leased to private concerns. The remaining sacred groves in many parts of the world are falling victim to the demands of the market economy for timber, woodchips, and pulp.

IN DEFENSE OF SACRED GROVES

One of the most significant contributions that can be made by a consideration of sacred groves to the worldwide concern for the preservation of biodiversity is the fact that these groves are places where local communities have been involved in the conservation of nature in their own immediate environments. This could continue to be the case if local communities are given ownership, the responsibility for management, the economic resources to survive and to protect their reserves, and the support of governments for traditional forms of organization. If they had the power to act in their own interests, they would provide an impetus for conservation and sustained use. Unfortunately, they are seldom allowed to participate in the management of their forests by governments, wood products businesses, and multinational banks and corporations. Even international conservation organizations tend to ignore them. Where a village might still control its own groves, the appropriation by outsiders of other resources outside the sacred groves, and the incremental loss of access by villagers to fields and forests, may force the local people to resort to the groves for daily biomass needs, to occupy them for housing, or to sell them for short-term economic profit.

It is a misfortune that local communities are generally small, poor, and often inadequately educated. They face powerful foes who can summon up huge amounts of money, numbers of employees, and sometimes weaponry. Violence, however, is seldom necessary; large national and multinational corporations can promise jobs and other rewards. Agencies set up by governments to protect local people sometimes prove amenable to bribes and other forms of pressure.

A further obstacle lies in the rapid social changes which are eroding traditional attitudes and practices everywhere around the world. Rising material expectations; access to motion pictures, television and other media; mobility and the strains on family structure; and the waning of belief not only in the traditional religions that hallowed the groves, but also in perhaps any image of the sacred, have lessened the chance that local people will find the time and energy to protect biodiversity in refugia near their homes.

In the face of these difficulties, the maintenance and extension of systems like the sacred groves would not be easy. For both people and forest ecosystems, however, it would be good to try. Some positive developments may give cause for guarded optimism. One is the fact that sacred groves have, in a few cases, served as rallying points for movements to restore local rights. Another is that the revolution of rising expectations has already, in a number of places, led to a demand for a healthy and pleasing environment. Still another is the growth in and desire for education, which can include teaching young people the value of biodiversity and the importance of its conservation. Sacred groves would provide wonderful outdoor illustrations for the classroom. A recognition of the fact that humans depend on functioning ecosystems for survival may supplement the ethical structures provided by a variety of religions. Communities that have been given a chance, and the resources, to increase the protection of their sacred groves have elected to do so. It is possible that other communities might be willing to undertake the restoration of ecosystems in enlarged, or entirely new groves, such as the AIDS groves mentioned above. Whether the sacredness of these groves will be expressed in terms of traditional religions, or memorials, or in terms of other values such as reverence for all forms of life, or conservation, or perhaps several of these, will be up to the communities themselves to decide.

REFERENCES

Bachmann, V.K. 1992. *Mythen*. Geo. Wissen., Hamburg.

Brandis, D. 1897. *Indian Forestry, Working*. Oriental Institute, Woking.

Bryant, W.C. 1848. *Poems*. Carey and Hart, Philadelphia.

Budge, E.A.W. 1911. *Osiris and the Egyptian Resurrection*. 2 vols. The Medici Society, London.

Burman, R.J.J. 1992. The institution of sacred grove. *Journal of Indian Anthropological Society 27:* 219-238.

Bury, R.G. 1929. *Critias* of Plato (4th Century B.C.). Loeb Classical Library, Harvard-Heinemann, Combridge.

Chandran, M.D.S. 1993. *Vegetational Changes in the Evergreen Forest Belt of the Uttara Kannada District of Karnataka State*. Ph.D. thesis, Karnataka University, Dharwad.

Chandran, M.D.S. 1997a. Transitions in Post-Indus Ecological Traditions of India. Paper presented at *the Conference on Ecological History and Traditional Sciences of India, 27-29 March, 1997*. Centre for Science and Environment, New Delhi.

Chandran, M.D.S. 1997b. On the ecological history of the Western Ghats. *Current Science* 73: 146-155.

Cohen, M.P. 1984. *The Pathless Way: John Muir and American Wilderness*. University of Wisconsin Press, Madison, Wisconsin.

Dewing, H.B. 1961 *Buildings* of Procopius (6th Century A.D.). Loeb Classical Library, Harvard-Heinemann, Combridge.

Dorm-Adzobou, Clement, Ampadu-Agyei, Okyeame, and Veit, Peter G. 1991. *Religious Beliefs and Environmental Protection: The Malshegu Sacred Grove in Northern Ghana*. From the Ground Up Case Study No. 4. Center for International Development and Environment, World Resources Institute, USA, Nairobi and New York.

Erman, A. 1894. *Life in Ancient Egypt*. Macmillan, London.

Frazer, J.G. 1935. *The Golden Bough, Part I: The Magic Art and the Evolution of Kings*, 2 vols. Macmillan, New York.

Gadgil, M. and Berkes, F. 1991. Traditional resource management systems. *Resource Management and Optimization* 18: 127-141.

Gadgil, M. and Vartak, V.D. 1976. The Sacred Groves of Western Ghats in India. *Economic Botany 30:* 152-60.

Gomez-Pompa, Arturo, Salvador Flores, Jose, and Aliphat Fernandez, Mario. 1990. The Sacred Cacao Groves of the Maya. *Latin American Antiquity* 1: 247-257.

Harris, L. D. 1984. *The Fragmented Forest: Island Biogeography Theory and the Preservation of Biotic Diversity*. University of Chicago Press, Chicago.

Heer, F. 1975. *Charlemagne and His World.* Macmillan, New York.

Hembram, P.C. 1982. Return to the Sacred Grove. In: K. S. Singh (Ed.) *Tribal Movements in India*, Manohar Publications, New Delhi, Vol. 2, pp. 87-91.

Hewitt, K. 1986. Axis Mundi: Living Space, Custom, and Identity among the Kalash Kafirs of the Hindu Kush. In: M. Tobias (Ed.) *Mountain People*. University of Oklahoma Press, Norman, pp. 82-87.

Hughes, J.D. 1996. *North American Indian Ecology*. Texas Western Press, El Paso.

Hughes, J.D. 1994. *Pan's Travail: Environmental Problems of the Ancient Greeks and Romans.* Johns Hopkins University Press, Baltimore, Maryland.

Hughes, J.D. and Swan, J. 1986. How Much of the Earth is Sacred Space? *Environmental Review 10 (4):* 247-260.

Hughes, J.D. 1991. Spirit of Place in the Western World. In: James A. Swan (Ed.) *The Power of Place*. Quest Books, Wheaton, Illinois, pp. 15-27.

Hughes, J.D. 1984. Sacred Groves: The Gods, Forest Protection, and Sustained Yield in the Ancient World. In: Harold K. Steen (Ed.) *History of Sustained-Yield Forestry: A Symposium.* Forest History Society, Durham, North Carolina, pp. 331-343.

Jordan, B. and Perlin, J. 1984. On the Protection of Sacred Groves. In: A.L. Boegehold, et al. (Eds.) *Studies Presented to Sterling Dow on his Eightieth Birthday*. Greek, Roman, and Byzantine Monograph 10. Duke University Press, Durham, North Carolina.

Khiewtam, R.S. and Ramakrishnan, P.S. 1989. Sociocultural studies of the sacred groves at Cherrapunji and adjoining areas in northeastern India. *Man in India* 69: 64-71.

Kim, H. 1994. *Maul sup: Han'guk chont'ong purak ui tangsup kwa sugu magi = (Maulsup: the Korean village grove).* Yorhwadang, Seoul.

Lebbie, A.R. and Guries, R.P. 1995. Ethnobotanical value and conservation of sacred groves of the Kpaa Mende in Sierra Leone. *Economic Botany* 49: 297-308.

Manniche, L. 1989. *An Ancient Egyptian Herbal.* University of Texas Press, Austin.

Marglin, F.A. and Mishra, P. C. 1993. Sacred Groves Regenerating the Body, the Land, the Community. In: W. Sachs (Ed.) *Global Ecology: A New Arena of Political Conflict.* Zed Books, London.

Matunga, H. 1994. Waahi Tapu : Maori sacred sites. In: D. L. Carmichael, J. Hubert, B. Reeves, and A. Schanche (Eds.) *Sacred Sites, Sacred Places*. Routledge, London, pp. 217-226.

Mikesell, M.W. 1969. The Deforestation of Mt. Lebanon. *Geographical Review* 59: 1-28.

Mishra, Y. 1962. *An Early History of Vaisali.* Motilal Banarasidas, Delhi.

Molyneaux, B.L. 1995. *The Sacred Earth.* Little, Brown and Company, Boston.

Oldfather, C.H. 1933. *Bibliotheke* of Diodorus of Sicily (1st Century BC). Loeb Classical Library, Harvard-Heinemann, Combridge Mass., and London.

Ono, S. 1962. *Shinto: The Kami Way*. Bridgeway Press, Tokyo.

Ovsyannikov, O.V. and Terebikhin, N.M. 1994. Sacred groves in the culture of the Arctic regions. In: D.L. Carmichael, J.H.B. Reeves, and A. Schanche (Eds.) *Sacred Sites, Sacred Places*. Routledge, London, pp. 44-81.

Plato. [4th century BC]. *Critias*. Translation by R.G. Bury. 1929. Loeb Classical Library, Harvard-Heinemann, Cambridge.

Procopius. [6th century AD]. *Buildings*. Translation by H.B. Dewing. 1961. Loeb Classical Library, Harvard-Heinemann, Cambridge.

Ramakrishnan, P.S. 1996. Conserving the Sacred: from Species to Landscapes. *Nature and Resources 32:* 11-19.

Risley, H. 1908. *The People of India*. Thacker, Spink & Co.

Sandars, N.K. 1960. *The Epic of Gilgamesh*. Penguin Books, Harmondsworth. Translated.

Spindel, C. 1989. *In the Shadow of the Sacred Grove*. Vintage Books, New York.

Tozzer, A.M. (Ed. and transl.) 1941. *Diego de Landa. Relaci—n de las cosas de Yucatn.* Manuscrito en la Real Academia de la Historia, Madrid. Papers of the Peabody Museum of Archaeology and Ethnology Vol. 18. Harvard University, Cambridge, Massachusetts.

Vest, J.H.C. 1985. Will-of-the-Land: Wilderness Among Primal Indo-Europeans. *Environmental Review* 9: 323-329.

Wachtel, P.S. 1993. Asia's Sacred Groves. *International Wildlife* 23: 24-27.

Weber, T. 1987. *Hugging the Trees: The Story of the Chipko Movement*. Penguin Books, New Delhi.

Weiss, P. 1995. *Kandinsky and Old Russia: The Artist as Ethnographer and Shaman*. Yale University Press, New Haven.

Yin S. 1991. *Yige chongman zhengyi de wenhua shengtai tixi: Yunnan daogeng huozhong yanjiu [A Highly Controversial Culturo-Ecological System: Slash-and-Burn Cultivation in Yunnan]*. Yunnan Renmin Chubanshe, Kunming.

Yin S. 1994. *Senlin yunyu de nonggeng wenhua: Yunnan daogeng huozhong zhi [A. Farming Culture Formed by Forest: Slash-and Burn Cultivation in Yunnan]*. Yunnan Renmin Chubanshe, Kunming.

6

Where Nothing is Sacred

B. Burrows

The Edmonds Institute, Edmonds, Washington 98020, USA

INTRODUCTION

In the United States of the 1960's it was the style of some scholars to divulge anything that might have influenced their research procedures and conclusions. Whatever one's aspirations for objectivity, they told their students, it was necessary to state "where you are coming from." Here at the end of the twentieth century, in the midst of a flourishing corporate culture brimming with confidence in the ultimate victory of capitalism and the market, there is little place for the affirmation, delineation, or acknowledgement of the sacred. "Nothing is sacred at Kvoerner. Anything can be sold," one learns from an airline magazine story (Richards, 1997) about a successful business tycoon. There is little intellectual or physical or natural space for the sacred here. Even the sense of the sacred dulls and atrophies. One marks oneself as old-fashioned and quaint even to suggest that the sacred is anything but conceptual and historical. It is not so much that the sacred is absent. It is simply not noticed. Put another way, it may seem non-existent because it is rarely looked for. Unseen and unreported, the sacred is nevertheless still here.

SNAPSHOTS FROM THE COMMODIFICATION OF THE SACRED

In 1993, I attended a conference in Seattle, Washington on "The Future of Intellectual Property Protection for Biotechnology in the United States, Europe, and Japan". For three days eminent speakers discussed patents. The last day of the conference, one of the panelists-I forget his name-bemoaned the situation in Europe (where at that time it was nearly impossible to obtain a patent on any form of life). The panelist hoped his colleagues in other places would never have to face the situation he faced with "... environmentalists and those who would bring ethics and other irrational considerations to the table." Those were his words: "ethics and other irrational considerations." No member of the audience challenged his pairing of "ethics" and "other irrational considerations." Not one lawyer. Not one official. Not one academic. Not any one.

That same year, a draft proposal for the work of the Human Genome Diversity Project (HGDP) circulated by the Rural Advancement Foundation International, Canada came into my hands. It arrived two days before I was scheduled to go to San Francisco to address a group of indigenous people on the intellectual property implications of the (then-proposed) North American Free Trade Agreement and (the then latest version of) the General Agreement on Tariffs and Trade. The event was a confernce entitled "Commodification of the Sacred", sponsored by the Cultural Conservancy, a U.S. non-governmental organisation headquartered in California. The HGDP appeared to be a study of high intellectual purpose headed by scientists and scholars with extraordinary credentials. Proposing to collect human tissue from 722 human populations, including indigenous people on the verge of extinction, the project aimed to use, to quote one admiring article, "new genetic techniques... (to solve) the ancient mysteries of mankind's origins and migrations." (Goodheart, 1996). Nevertheless, I found the ethical implications of the HGDP troubling and when I spoke in San Francisco, I used my allotted time to share what I knew of the project.

I remember the looks on people's faces and the great silence and sadness that followed my remarks about collecting samples of hair and blood and cheek scrapings. The silence lasted a long time and when finally someone spoke, it was Jeanette Armstrong, an indigenous woman from Canada. She said to me, "You people. We thought you folks had taken everything you could. You took our land, you took our homes. You stole our pottery and our songs and our blankets and our designs. You took our language and in some places you even took our children. You snatched at our religion and at our women. You destroyed our history and now, now it seems you come to suck the marrow from our bones."

Other indigenous people would later call the HGDP "the Vampire Project"—not because it intended collection of blood samples but because its only apparent interest in the people from whom it would collect-some of whom it labelled "Isolates of Historic Interest" and considered people on the verge of extinction-was interest in their genetic material.

In 1994 the World Council of Indigenous People, Canada issued a pamphlet about the HGDP entitled, "Presumed dead...but still useful as a human by-product." The pamphlet charged that the proposed collections of

DNA samples "will supposedly help preserve indigenous gene cultures for generations to come...(but the) real agenda is the future development of pharmaceuticals that will generate huge corporate profits, long after indigenous people have been left to disappear."

The pamphlet went on to assert, "The assumption that indigenous people are doomed adds insult to the indignity of being used as human guinea pigs. The millions of dollars to be spent on the Human Genome Diversity Project could fund healing and community development for those indigenous people that are considered at risk... .The Human Genome Diversity Project has already begun dehumanizing us by labelling us as 'Isolates of Historic Interest' (IHI)."

While admitting that a "human harvest" of blood samples, cell scrapings, and hair roots could have potential scientific value—"The unique characteristics of some genetic samples could support the development of drugs to fight cancer or prevent Alzheimer's disease"—the pamphlet also warned that : "At least two countries, the United States and Japan, allow companies to claim patent rights over human genes and DNA segments. This means that the bloods and living cells of indigenous people will become the property of a private research group."

Long before the HGDP controversy began, the extension of patents-time-limited grants of monopoly ownership—to living organisms and materials derived from living organisms was well-advanced in the United States. The beginning of what Andrew Kimbrell would call the "slippery slope" of life patenting is generally thought to be a patent given to Ananda Mohan Chakrabarty, a microbiologist who worked at the time for General Electric (see Kimbrell, 1993). According to a publication of the Office of Technology Assessment of the Congress of the United States, that patent, for an oil-eating bacterium, "helped to precipitate increased research and development, assuring commercialization of biotechnology in the United States" (US Congress, OTA, 1989).

Originally the Chakrabarty patent application was turned down by the U.S. Patent and Trademark Office (PTO) but on appeal to the Court of Customs and Patent Appeals, the rejection was reversed. In a path-breaking decision, the court held that, "The fact that microorganisms...are alive...(is) without legal significance."

The PTO subsequently appealed the reversal to the U.S. Supreme Court. In an amicus brief to the court, Ted Howard argued that, "...a ruling in favor of patenting genetically engineered living organisms (would mean that) manufactured life-high and low-will have been categorized as less than life, as nothing but common chemicals." (Howard, 1980). In 1980, the Court affirmed that "a live, human-made microorganism is patentable subject matter... as a 'manufacture' or 'composition of matter" (US Congress, OTA, 1989).

Ethicist and biologist Leon Kass saw in the court ruling "deep and problematic truths about our (Western) culture":

"...the implicit teaching about ownership of life in the ...decision is indeed problematic... What is the principled limit to this beginning extension of the domain of private ownership and dominion over living nature? Is it not clear, if life is a continuum, that there are no visible or clear limits once we admit living species under the principle of ownership? The principle used in *Chakrabarty says there is nothing in the nature of a being*, no, not even in the human patenter himself, that makes him immune to being patented: not what he is, but only the accident of his non-man-made origin renders man himself a nonpatentable organism. If a genetically engineered organism may be owned because it was genetically engineered, what can we conclude about a genetically altered or engineered human being? To be sure, in general it makes sense to allow people to own what they made, because they have artfully made it. But to respect art without respect for life is finally self-contradictory." (Kass, 1985).

Perhaps no case concerning the patenting of life has been more analyzed than the case of John Moore versus the Regents of the University of California (Burrows, 1997; Annas, 1993).

While Moore was under treatment for hairy cell leukemia, the doctor supervising his care noticed that Moore's blood cells produced an unusual blood protein that showed promise as an anti-cancer agent. Without Moore's knowledge or consent, the doctor obtained a patent on Moore's cell-line, listing himself and his research assistant as "inventors". In 1984, Moore became aware of the patent and sued the doctor and the doctor's university and the pharmaceutical company to which the doctor and the university had licensed the cell-line. The case raised novel questions: Did Moore's doctor wrongfully take and profit from parts of Moore's body? Did Moore deserve a fair share of the profits from the products resulting from his cell-line? What would happen to science if the court recognized John Moore's property rights? The case took years to be resolved and in 1990, the California Supreme Court ruled that Moore's doctor had breached his "fiduciary duty" to Moore by not revealing his research and financial interest in Moore's cells. Nevertheless, the court denied Moore's claim to ownership of cells removed from his body, explaining:

"Research on human cells plays a critical role in medical research. This is so because researchers are increasingly able to isolate naturally occurring, medically useful, biological substances and to produce useful quantities of such substances through genetic engineering...The extension of conversion law into this area will hinder research by restricting access to the necessary materials."

Ethicist and lawyer George Annas, commenting on the court decision, noted, "California's courts decided who could reap profits from a cell line... as is clear from the text, the majority simply accepts the Chicken Little argument that if John Moore's property interest in his cells is upheld, the biotechnology industry's sky will fall on them and medical progress will suffer a major setback. In this regard the justices seem to have been blinded by science and unable or unwilling to distinguish it from commerce. The court essentially concluded that the biotechnology industry is both wonderful and fragile. Since it is wonderful, we must do our part to foster it; since it is fragile, we must protect it from harm.... In the court's flowery words, recognizing (Moore's property claim) would threaten 'to destroy the economic incentive to conduct important medical research..." (Annas, 1993).

Finding the decision unsatisfactory, Annas noted that effectively,"... the majority opinion concluded that 'only the little people can't sell cells.'"

Years after that court decision, I met John Moore. In the hope of gaining insight into the ethical dilemmas surrounding the patenting of life, I had arranged an interview. I began the discussion by asking John, "What did it feel like to be patented?"

He said an odd thing to me in reply: "Where have you been? Where have you been?"

I was startled and it took some time for me to understand his meaning. In the years since his case, John Moore had been characterized as everything from a greedy villain out to destroy the progress of medical science to an ungrateful biological freak. No one seemed to remember that others had patented his cell line for profit. Few, if any, had noticed his deepest wound:

"How does it feel to be patented? To learn all of a sudden, I was just a piece of material... It's so beyond anything you can conceive of. There were so many issues involved... There was a sense of betrayal... I was told that, in a dinner conversation with a colleague, (my doctor) had said that, 'John Moore is my gold mine.' This was certainly true; he had discovered genetic gold in my cells." (Burrows, 1997)

In 1996 John Moore appeared at a hearing before a committee of the United States National Academy of Science. He was there to speak against the Human Genome Diversity Project and he began his testimony to a distinguished panel of scientists saying, "I am known as Patent #4,438,032. Some of you may be familiar with pieces of me in your laboratories." Recounting the story of his cell-line, he reminded the panel, "Ultimately, everyone was protected and rewarded—the researcher, the physician, the entrepreneur, even Science. Everyone was protected and rewarded...but the patient, the subject, the source. Yes, me. Didn't I have a right to know? They were exploiting the uniqueness of my genetic material, growing my cells in a Petri dish, patenting my genetic essence. Shouldn't I have been told? Shouldn't someone have explained to me, the one from whom the gold was mined, the person who did not yet know anything about cell lines, genetic codes, blood proteins, GMCSF, patents, contracts, or stock options? Because I knew nothing, was I to be treated as nothing?"

Moore continued, "You may at this point be asking yourselves, 'What does this have to do with the HGDP and why is John Moore here telling us his personal story? How does this relate?"

Then he concluded, "Here are my concerns... I am concerned because the dehumanization of having one's cells conveyed to places and for purposes that one does not know of can be very, very painful. Why should I or any individual or group of individuals have their unique genetic materials borrowed, stolen, or bought for some fraction of their value for some project of others?...How can anyone else set and get a price on what may be priceless or sacred to someone else? ...I am concerned because even in this country where the rights of the individual are supposed to be protected from the grasp of institutions and certainly from the greed of private corporations and researchers, they are not. ... Do you think a system that could not protect me will protect the rights of people and individuals that live in other countries? ...I don't... I have had a very sad experience. I was violated by a legal system and a set of values in my own country. I know what to expect and I have no reason to believe that the environment which allowed the theft of my genetic material has changed in any positive or material way."

Ironically, at the HGDP hearing there was another John Moore, an anthropologist similar in physical appearance to the patented John Moore. The anthropologist was there to testify on behalf of the HGDP; he was, it turned out, one of the heads of the North American Human Genome Diversity Project.

In November 1996, during an afternoon break at the third Conference of the Parties (COP 3) to the Convention on Biological Diversity, (the patented) John Moore shared his experiences once again, this time with a group of indigenous people. He talked about his sense that it was his case that paved the way for the collection and patenting of indigenous cell-lines: "The first thing that flashed through my mind when I heard about the attempt (of the U.S. government to patent the cell-line of a Guaymi Indian woman) was that here was the consequence of a decision made by the California State Supreme Court."

Among those listening to Moore was Columbian Senator Lorenzo Muelas, an eloquent campaigner in his own country for the return of all genetic material taken from indigenous Columbian people. When Moore noted that the press in the U.S. had branded him a man "against the progress of science," Muelas laughed.

Referring to the press reaction in Columbia, he noted, "They said the same thing to me." And then Muelas added, "I got very angry."

Moore retorted, "I got pretty angry myself."

In 1997, seventeen years after the Chakrabarty decision, a large group of U.S. citizens and public interest organizations wrote to members of a European Parliament on the verge of considering legislation that would extend life patents in Europe:

"Once the U.S. Supreme Court decided to allow the patenting of a microorganism..., other patents for higher life forms followed quickly. Currently, the United States Patent and Trademark Office has granted approximately 35 patents that actually claim animals, including animals that have NOT been genetically engineered. We are now in the midst of massive attempts by patent seekers to divide up the entire biotic community for private interest.

"The adverse effects of patenting life daily become more and more apparent (and burdensome) here. Several years ago the California State Supreme Court effectively sacrificed individual civil and property rights for the good of an industry (Moore v. Regents of the University of California). Later our leaders thought it wise to pursue patents on the cells of far-off indigenous people; when that proved the source of international embarrassment, the patent claims were dropped but the precedent was established. In the meantime we passed technology transfer laws that resulted in great universities corrupting themselves by the pursuit of patents and profits and becoming indentured to industry. Increasingly, we found that patents constricted innovation and distorted the pursuit of knowledge itself: our peer (expert) review system has been put at risk by fear of theft of patentable ideas; universities have sued graduate students over patents resulting from ideas they had while students; graduate students have organized to prevent theft of their intellectual property by their universities and professors; our researchers have postponed publication of medical breakthroughs while they and their corporate sponsors await issuance of patents; and fruitful lines of research have been abandoned altogether to avoid the expense of licensing, the possibility of patent litigation, and the limitations of doing research on the terms of licensing agreements. In the United States researchers boycotted the use of the Oncomouse (the first patented animal), for example, because of the patent holder's licensing agreement sought rights to derivative inventions. Similarly, farmers have balked at taking up new innovations that carry the burden of licensing agreements that specify in exact and sometimes onerous terms the method of farming that must be used with the patented product. Finally, those who looked on the mire of our patenting regime and despaired of its ability to offer them protection sought to expand our trade secrets law and find solace there. In doing so, they in some cases foreclosed the sharing of innovation altogether. Thus, instead of being a stimulus to creativity, patents on life became the catalyst that narrowed and discouraged research in all directions.

"Is this situation what you wish for Europe?" (Mendelson, et al.,1997).

SOME THOUGHTS

In his book, *In the Absence of the Sacred*, Jerry Mander (1991) noted of his culture (and mine), "We have entered into a universe that has been reformed by machines; we are a species that lives its life within mechanistic creations; our environment is a product of our minds. Locked inside our cities and suburbs, working in our offices, controlling and conceptualizing nature as a raw material for our consumption, and now even including ourselves as raw material suitable for redevelopment, we are at one with the process" (Mander, 1991).

The "universe", Mander was describing, was to him a creature of a failed technology, a technology that promised to "expand the human potential, and to create a world of abundance for human enjoyment" but was also a technology engaged in the "utter conversion of nature into commodity form even that part of nature that remains wild within human genes and molecular structures." The "failure" of the failed technology, according to Mander, consisted not simply of an inability to deliver its promises but also of the "awful sacrifices" it demanded along the way to promise fulfillment—sacrifices "now becoming visible in oil spills, global warming, ozone depletion, toxic pollution, and deforestation, all of which affect our sense of well-being in everyday life."

Perhaps no sacrifice made in a technological society is greater than that which accompanies the adoption of the patterns of thought and the methods of problem-solving inherent in the new "high" technologies. Where everything must become a resource, all must be made fit for trespass. Where nothing is sacred, not even life, all may be commodified and sold. In such an environment, we can treat living beings as inputs, industrial resources, objects to be examined and owned, patented and manufactured, leased and even lost. There is simply no use for the sacred (except perhaps for the sacred Market itself).

It may make us queasy to ignore what our ancestors or we ourselves once considered sacred. We may regret having to violate the civil rights of our own citizens and to steal from the commons and the dignity of other people in order to serve Science and our own particular needs and technological, market-based view of the universe. But ultimately, where nothing is worthy of being called sacred, it will seem glorious to say, as Craig Rosen, the vice president of research and development for a U.S. genomics firm, once boasted to the magazine *Popular Mechanics*: "What you see here is the beginning of industrial-scale biology. . . We are changing biology the way Henry Ford changed automobile production" (Wilson, 1996).

With nothing deemed sacred, including the natural world, "the world out there is basically just resources for humans to use or abuse as they see fit," according to Father Sean McDonagh. With "the spiritual dimension out of the natural world," we may also remove the natural from the realm of personal responsibility.

"I have heard confessions for twenty-seven years," notes McDonagh, "(but) I have never heard anyone coming in and saying, 'I sinned against the planet, I poured poison into the rivers.' Now we know that there is major sin against the planet... against our soils and our water and our air but people don't see it as part of their moral conscience."

Biologist Martha Crouch has talked about the importance of culture in guiding individual behavior. In 1992 she gave a presentation to a conference to the National Agricultural Biotechnology Council in the U.S. Speaking of the religion and cosmology that supported indigenous systems of agriculture in Liberia and Mexico, Crouch noted, "In all these cases, people have direct incentive to use the land wisely, they have the power to do so, and they have the knowledge to be able to participate in feedback loops." Then Crouch went on to ask, "What underlying rituals and cosmologies guide technological agriculture, and who has the power and knowledge to respond to feedback? In technological societies such as ours, the world revolves around global industrialism. Our view of the world does not imbue the land with sacred qualities; all things are profane. Most food production in our societies can be described as agribusiness. Culture has been uprooted as the sense of place has been taken out of the picture; the varieties and techniques are deemed successful to the extent they can be applied to vast areas, rather than specific locales. Less than 5% of the population is involved directly in growing food, so very few people have the opportunity to see what happens on the land. Thus the knowledge to respond to changes is in the hands of a small fraction of the community.

The rural people who see the water, soil, and organisms on a daily basis are often not the same ones who have the power to respond, because urban-generated global market forces drive the adoption of genotypes and practices... Monocultures are a logical outcome of the business worldview" (Crouch, 1992).

Looking at modern biotechnology, Crouch outlined how the transformation to monocultures takes place, "By turning everything it touches into commodities, biotechnology also has the effect of making products and processes that fit more easily into the global market. Seeds that used to be saved by the farmer now must be purchased each year, for example. Genotypes that used to be specific to a slope, soil type, and rainfall amount in a particular valley are replaced with a genotype that will grow in a whole region. Markets that respond to short term increases in production replace subsistence or local markets that respond to the need for a secure food supply in unpredictable conditions. Diversity is lost."

CONCLUSION

In a technological world that requires the divorce of the sacred from the natural, we may come to have no sense at all of the sacred or the natural. We may take a fundamentally anti-ecological view of the universe and become comfortable in exchanging social contracts and *ordre publique* for business contracts and corporate efficiency. Together with the chief technology officer of Microsoft, we then may celebrate cloning as "the only predictable way to reproduce," finding sexual reproduction "a crap-shoot by comparison" (Newman, 1997). And, having subverted or rejected our natural embodied selves, we finally may join with those admirers of nanotechnology and robotics who welcome the intelligent machines of a post-biological age in which "human beings are kind of out of date" (see Mander, 1991).

REFERENCES

Annas, G.J. 1993. Outrageous Fortune: Selling Other People's Cells. In: *Standard of Care: The Law of American Bioethics*. Oxford Univ. Press, New York. pp. 167-177.

Burrows, B. 1997. Second Thoughts about U.S. Patent #4,438,032. *Bull. Med. Eth.* 124: 11-14.

Crouch, M. 1992. Biotechnology and Sustainable Agriculture. In: *Ethics and Patenting of Transgenic Organisms*, NABC Occasional Papers #1. National Agricultural Biotechnology Council. Ithaca, New York. pp. 94-98.

Goodheart, A. 1996. Mapping the Past. *Civilization* (the Magazine of the Library of Congress), 3: 40-47.

Howard, T. 1980. *The Case Against Patenting Life: brief on behalf of the People's Business Commission*. Amicus Curiae, in the Supreme Court of the United States. No. 79-136.

Kass, L.R. 1985. Patenting Life: Science, Politics, and the Limits of Mastering Nature. *In: Towards a More Natural Science, Biology and Human Affairs*. Free Press. New York. pp. 128-153.

Mander, J. 1991. In the Absence of the Sacred, The Failure of Technology and the Survival of the Indian Nations. Sierra Club Books, San Francisco. pp. 189-190.

Mendelson, J., Burrows, B., Flynn, E. 1997. *Letter to the Members of the European Parliament.* Brussels, Belgium.

Newman, S.A. 1997. Commentary: Embryo Technology:Cloning our way to "the next level". *Nature Biotechnology*, 15: 488.

Richards, H. 1997. *Cut and Run*. Scanorama. June: 44-46.

U.S. Congress, Office of Technology Assessment (OTA), 1989. *New Developments in Biotechnology: Patenting Life*. Special Report OTA-BA- 370 U.S. Government Printing Office. Washington, D.C.

Wilson, J. 1996. Finding New Wonder Drugs. *Popular Mechanics*. p. 30.

PART 2
CASE STUDIES

7

Sacred Groves of the Ancient Mediterranean Area: Early Conservation of Biological Diversity

J.D. Hughes

University of Denver, Denver, Colorado 80208, USA

INTRODUCTION

Everywhere in the world where trees grew, the people of early times regarded particular groves as especially sacred, and these were protected from damage and everyday use. The ancient Mediterranean area provides an excellent example of this practice. In the classical world of the Greeks and Romans, areas of forest land that had been set aside to be sacred to the gods dotted the landscape. Their boundaries, often marked with stone walls, had been surveyed carefully by public or private authority. They were an important institution for which we have evidence in the form of inscriptions and written leases, in addition to passages in many ancient writers referring to them. Pausanias, the author of a guide for travelers in Greece in the second century AD, noted them wherever he found them, and they were very numerous.

These outdoor sanctuaries were the first temples of the gods, as the Roman natural historian Pliny the Elder (1st century AD) tells: "Trees were the first temples of the gods, and even now simple country people dedicate a tree of exceptional height to a god with the ritual of olden times, and we...worship forests and the very silences they contain." In Greek, an *alsos* or grove that had been consecrated to a deity was called a *temenos*, a "cut-off" or demarcated place. The Latin equivalents were *lucus* or *nemus*, grove, and *templum*, a space marked out. The latter word is, of course, the origin of the English word "temple".

A RELIGIOUS MOTIVE FOR CONSERVATION

The ancients never forgot the original motive that had created the groves: religious awe. The areas were considered to be the precincts of the gods. It was not the act of setting them aside or surrounding them with walls that consecrated them; rather, the act of dedication recognized an original sacred character of the places. As the philosopher Seneca (2nd century AD) remarked in *Epistles*, "If you come upon a grove of old trees that have lifted their crowns up above the common height and shut out the light of the sky by the darkness of their interlacing boughs, you feel that there is a spirit in the place, so lofty is the wood, so lone the spot, so wondrous the thick unbroken shade." The poet Virgil (1st century BC) in *Aeneid* made his character Evander remarked that when they saw the old tree-covered Capitoline Hill, the rural folk who lived around it exclaimed, "Some god has this grove for dwelling!"

In archaic times sections of the primeval forest were dedicated (Thirgood, 1981), and an effort was made to keep them untouched: wilderness areas, one might have called them, and this would have been true in the sense of a forest kept as near its original state as possible. Virgil (1st century BC) noted in the *Georgics* that the gods favor wild trees "unsown by mortal hand." But they were not wilderness in the ancient Greek or Roman sense of a region which lacks human beings (*eremos, deserta, solitudo*). The groves were in use for worship, supervised by local authorities, and priests or guards often lived in them. Management policies were developed as time went on.

In the earliest period the grove itself was the temple, and sacrifices were offered there without benefit of statues or buildings. Indeed, the Greeks and Romans continued to conduct most of their public rituals and sacrifices out-of-doors throughout their history. Notable exceptions to this rule were the mystery religions, open only to initiates, which demanded secrecy and therefore met for their most sacred ceremonies in halls that were sometimes partially underground. But the mysteries also consecrated groves for their use; the cult of the wine-god Dionysus was famous for its "mountain-dancing" (*oreibasia*), in which the devotees roamed forests high on the slopes of sacred mountains such as Mount Parnassus.

Later, statues of gods or goddesses were erected in the groves, and then shelters were built over these images to protect them and to provide a place to put votive offerings. It is this shelter, originally of wood but later of stone with fluted columns and carved reliefs, that took the name, "temple," that formerly belonged to the whole grove. The association of grove with temple was never lost. Every Greek or Roman temple, it was felt, needed to have trees around it, and where there were none, they were planted. When the Athenians chose the barren limestone outcropping called the Acropolis as the site for the Parthenon, a temple dedicated to the virgin goddess Athena, they excavated two rows of pits in the rock, filled them with soil, and planted cypresses in them. Similar holes have been found beside other temples, such as the Hephaesteum (Theseum) near the Athenian Agora.

SIZE OF SACRED GROVES

The groves varied in size. Some were only a few trees around a temple that must have looked from a distance like the cypresses that fill many walled cemeteries next to Orthodox churches in modern Greece. Travelers, it is said, prayed under the trees on a little sacred hill fenced all around. But the historian Arrian (2nd century AD) declared that an entire island, called Icarus, was dedicated to the goddess Artemis, who was the protectress of wild animals. The grove of Daphne was ten miles in circumference (Strabo, 1st century AD). The sacred land of Crisa, near Delphi, was a plain reserved from cultivation that covered many square kilometers and a grove near Lerna stretched all the way down a mountainside to the sea (Pausanias, 2nd century AD). Pausanias (2nd century AD) also says that the sacred grove of Asclepius, god of healing, at Epidaurus was "surrounded by mountains on every side." Judging from the lay of the land there it must have been extensive. Since the number of sacred groves was many hundreds, the area included in them was considerable, perhaps as much as ten percent of the total forest, and the practices used within them must have been of importance in regard to the conservation of the biological diversity for which they served as refugia.

BIODIVERSITY IN SACRED GROVES: EVIDENCE

Because many sacred groves were tracts of the original forest that had been protected from time immemorial, it would be expected that at least a proportion of the original biodiversity would have survived. For evidence of this, it is possible to list species mentioned in ancient Greek and Latin texts that are associated with groves (Table 1). These texts, a large selection of which were listed by Birge (1982), are of varying quality for this purpose. Most dependable are inscriptions that were placed in the groves by the authorities, including cities, communities, or private owners, who protected and managed them. Next would come mentions by natural historians such as Aristotle, and agricultural writers, Cato, for example. A third group comprises the prose historical writers, for instance Herodotus. The least

Table 1. List of species mentioned in Greek and Latin Texts as present in/near sacred groves.

English Name	Latin (Scientific) Name	Greek Name	Loci[1]
PLANTS:			
Trees and Shrubs: Wild or Timber-producing:			
Acacia	*Acacia arabica*	êkanya	1
Acantha	*Atractylis gummifera*	êkanya ² YhbaÛkā	2
Agnus-castus (Chaste-tree)	*Vitex agnus-castus*	ígnow	3
Alder	*Alnus glutinosa*	klÆyrh	1
Arbutus	*Arbutus unedo*	kÒmarow	1
Ash	*Fraxinus ornus*	mel¤a	3
Cassia	*Cinnamomum iners*	kas¤a	2
Cornel	*Cornus mas*	krāneia	4
Cypress	*Cupressus sempervirens*	kupārittow	24
Elm	*Ulmus glabra*	ptel°a	15
Fir	*Abies cephalonica*	§lāth	2
Frankincense	*Boswellia carteri*	l¤banow	4
Hazel	*Corylus avellana*	karÊa	3
Ivy	*Hedera helix*	kittÒw	12
Juniper	*Juniperus* sp.	êrkeuyow, k°drow	2
Laurel	*Laurus nobilis*	dāfnh	32
Mistletoe	*Loranthus europaeus*	Þj¤a	1
Myrtle	*Myrtus communis*	murr¤nh	12
Oak	*Quercus robur*	drËw	16
Oak, kermes	*Quercus coccifera*	pr›now	5
Oak, valonia	*Quercus aegilops*	fhgÒw	8
Oleander	*Nerium oleander*	nÆrion	1
Olive, wild	*Olea oleaster*	agri°laiow, kÒtinow	15
Pear, wild	*Pyrus amygdaliformis*	éxrāw, êxerdow	3
Palm, wild Cretan	*Phoenix theophrasti*	fo›nij	2
Pine, Aleppo	*Pinus halepensis*	p¤tuw	6
Pine, umbrella or stone	*Pinus pinea*	peÊkh	1
Plane	*Platanus orientalis*	plātanow	23
Poplar, black	*Populus nigra*	a‡gerow	17
Poplar, white	*Populus alba*	leÊkh	6
Sandarac[2]	*Tetraclinis articulata*	yÊon	1
Willow	*Salix* spp.	Þt°a	6
Yew	*Taxus baccata*	m¤low	1
Trees and Shrubs: Cultivated or Fruit-bearing:			
Almond	*Prunus amygdalus*	amugdal°	1
Apple	*Pyrus malus*	mhl°a	4
Fig	*Ficus carica*	suk°	6
Grapevine	*Vitis vinifera*	êmpelow, ²mer¤w	18
Mulberry	*Morus nigra*	sukāminow	1
Olive	*Olea europea*	§la¤a, mor¤a	50
Palm, date	*Phoenix dactylifera*	fo›nij	12
Pear	*Pyrus communis*	êpiow	1
Persea	*Mimusops schimperi*	pers°a	2
Pomegranate	*Punica granatum*	Ñroā	1
Herbs:			
Alsine (Mouse-ears)	*Parietaria cretica*	éls¤nh	1
Flowers	*(unspecified)*	ênyh	14
Grass	*(unspecified)*	pÒa	4
Lily	*Lilium candidum*	kr¤non	1
Lotus	*(unspecified)*	lvtÒw	1
Reed	*(unspecified)*	kālamow	3
Rose	*Rosa* spp.	ÑrÒdon	2

Contd.

Table 1. Contd.

English Name	Latin (Scientific) Name	Greek Name	Loci[1]
Smilax	*Smilax aspera*	m›laj	1
Thistle	*Scolymus hispanicus*	skÒlumow	1
Cultivated Crops:			
Barley	*Hordeum sativum*	kriyÆ	2
Pulse	*(unspecified)*	ˆsprion	1
Wheat	*Triticum vulgare*	purÒw	1
ANIMALS:			
Domesticated:			
Cattle	*Bos taurus*	boËw, etc.	11
Dog	*Canis familiaris*	kÊvn	5
Fowl (Cock)	*Gallus gallus*	él°ktvr	1
Goat	*Capra hircus*	a‡j	3
Horse	*Equus caballus*	·ppow	4
Sheep	*Ovis aries*	prÒbaton	5
Mammals:			
Bear	*Ursus arctos*	írktow	1
Deer, red (Stag)	*Cervus elaphus*	¶lafow	8
Deer, roe	*Capreolus capreolus*	dÒrkaw	1
Elephant	*Loxodonta africana*	§l°faw	1
Goat, wild	*Capra ibex*	trãgow	1
Hare	*Lepus capensis*	lagÅw	1
Lion	*Panthera leo*	l°vn	2
Wolf	*Canis lupus*	lÊkow	1
Birds:			
Birds	*(unspecified)*	ˆrniyew	9
Blackbird	*Turdus merula*	kÒssufow	1
Crow	*Corvus corax*	kÒraj	1
Eagle	*Aquila heliaca*	pÊgargow	1
Falcon	*Falco* spp.	ß°raj	1
Jackdaw	*Corvus monedula*	koloiÒw	1
Kingfisher	*Alcedo atthis*	élkuÅn	1
Nightingale	*Luscinia megarhynchos*	éhdÅn	5
Owl, little	*Athene noctua*	sk–c	2
Peacock	*Pavo cristatus*	ta«w	1
Sea-crow (Shearwater)	*Puffinus puffinus*	korÅnh	2
Swan	*Cygnus olor*	kÊknow	1
Vulture	*Neophron percnopterus?*	Ùreip°largow, Ípãetow	1
Reptiles, Amphibian:			
Frog	*Hyla arborea*	bãtraxow	1
Snake	*(unspecified)*	drãkvn, ¶xiw	12
Tortoise	*Testudo hermanni*	xelÅnh	
Fish:			
Eel	*Anguilla anguilla*	¶gxeluw	2
Fish	*(unspecified)*	Þxyu°w	4
Peacock fish	*(unknown; mythological?)*	ÞxyËw õ ta«w	1
Invertebrates:			
Ant	*(unspecified)*	mÊrmhj	1
Bee	*Apis mellifera*	m°lissa	1
Cicada	*Cicada plebeia*	t°ttij	5
Mussel	*Mytilus edulis*	kÒgxh	1
Scorpion	*Euscorpius flavicaudis*	skorp¤ow	1
Spider	*(unspecified)*	érãxnhw, falãggion	2

1. Number of references found by author in ancient texts and inscriptions.
2. Called, confusingly, *citrus* by Pliny the Elder and other Latin writers.

dependable are the poets, who might mention a species because of the metrical quality and resonance of its name rather than personal observation, although a work like Virgil's *Georgics*, which deals specifically with agriculture, could be expected to yield better information than, say, a writer of imaginative fiction like Apollonius of Rhodes. In a survey of this kind, it is important to note that negative proof is of no value at all; the fact that no ancient writer happens to mention a species as present in groves does not mean that it did not occur there. It must also be admitted that a writer might place a species in a grove when it did not occur there, but multiple mentions by several authors pretty well establish that a species existed in at least some groves. Sometimes archaeological evidence available; for example, in 1985 a tree in a grove sacred to Hera, goddess of women, on the island of Samos was identified as a juniper (Kyreilis, 1993).

DIVERSITY OF VEGETATION

Sacred groves existed in every part of the Mediterranean region, and at every elevation from the seashore to summits where snow persisted through the summer. Although it may seem a contradiction in terms, sacred groves also could include meadows, marshes, and bare mountain tops. This implies that samples of every major vegetation type in the Mediterranean basin probably were represented within sacred groves. Of course a great number of them conformed to the archetypal image of relatively undisturbed, mature, high forest.

Sacred groves consisted of a mixture of tree species, although one species might predominate in any given grove. The gods had favorite species which were considered to be sacred to them. The oak belonged to the king of gods, Zeus; the olive to his daughter Athena; the laurel to the archer-god Apollo; the apple to the love-goddess Aphrodite; the fir to Pan, the god of nature; the list could be extended indefinitely. Thus it might be expected that the groves sacred to a specific god would consist of trees considered to belong to that god. However, upon an examination of the texts, this idea proves to be misleading. Indeed, groves usually consisted of mixed species; the expression, "all sorts of trees" is often encountered. Where a species is mentioned as comprising a grove, it might belong to any god, goddess, or hero, not just to the one to whom the tree was thought sacred. There were oak groves dedicated to Demeter, goddess of crops; to the messenger-god Hermes; the war-god Ares, and the hero Lycus, and those portentous goddesses, the Fates and Furies, as well as to Zeus himself. Olive groves were dedicated to Zeus, Apollo, Artemis, and the heroine Ino, not just to Athena. While the laurel was undoubtedly Apollo's special tree, there were also laurel groves dedicated to his twin sister Artemis; to Dionysus; to the sun-god Helios; to the Hesperides, guardians of the golden apples; and to the Twins, or Dioscuri. Artemis was the "goddess of the wild forest" in general. One could go on in this vein for pages.

Of trees or large shrubs said to be found in groves, there are more than forty species named that generally existed in the wild state in the ecosystems of the Mediterranean. Among the most often mentioned are three kinds of oak, two kinds of pine, cypress, elm, laurel, myrtle, wild olive, plane, and both black and white poplar. A word of warning is necessary here. "Kinds" is used advisedly instead of "species," since the Greek and Latin vocabularies give plants and animals common names that do not necessarily equate with the species recognized by modern taxonomy. Usually, but not always, it is possible to identify the genus. Often one can identify the species, but there are frequent cases where one cannot be certain.

About a dozen kinds of cultivated or fruit-bearing trees and vines are noted in the sacred groves, sometimes with the specific statement that they were planted there. Sometimes it is impossible to say whether a tree in a grove was planted, since plants often have both wild and cultivated forms, and forest trees were sometimes planted as well. It might be maintained that these plantations within the groves enhanced biological diversity, but it could also be noted that they altered the natural ecological state of the forest fragments that remained in the groves. Of cultivated plants in the groves, the olive is by far the most often mentioned in ancient documents, and the grapevine is second. This is not surprising, since they were two of the most important Mediterranean food plants. Others include apple, fig, palm, and pear. Some groves must have taken on something of the aspect of parks, with planted or cultivated trees. Pausanias (2nd century AD) described a temenos at Gryneum as follows: "Apollo has a most beautiful grove of cultivated trees and of all trees which, without bearing fruit, are pleasant to smell or see". On the island of Lesbos there was a grove of apple trees dedicated to Aphrodite, with whom, as the recipient of the famous golden apple awarded by Paris, apple trees were associated.

Species of herbs in sacred groves are harder to identify. Although writers say that grass and many flowers were found there, names of kinds of understory plants are seldom mentioned. Those that can be tentatively identified are acanthus, alsine, lily, lotus, reed, rose, smilax, and thistle. Undoubtedly there were many more, and the protection accorded to sacred groves must have improved their survival. Planting of cultivated annuals such as wheat, barley, and pulses was not permitted in sacred groves where there were trees, but sometimes happened on a sacred field, or *chorion*, that was dedicated to a god or gods, and might adjoin a grove.

DIVERSITY OF ANIMALS

All kinds of animals visited sacred groves or made their homes there. The attention of Greek and Latin writers was attracted mainly by those that were large, or that were the usual game species, even though hunting or fishing was not permitted inside the groves. As with the plants, it must be noted that there were undoubtedly more kinds than can be found in the written sources.

Among mammals, the most often mentioned are deer of two kinds, red deer and roe deer. Hares were there but, it is said, usually avoided tree-covered places. Also called denizens of groves are bear, wild boar, wild goat, lion, and wolf. The lion still existed in Greece in the fifth century BC, but became extinct there early in the Roman period. It persisted in North Africa, where, Aelian (3rd century AD) also notes sacred and unmolested elephants frequented the groves at the foot of the Atlas Mountains. These elephants were members of a smaller North African race that later became extinct due to the use of the animal in warfare, and to the ivory trade.

Domesticated animals are often discussed in connection with sacred groves, usually to say that they were excluded from them, or sacrificed there. The kinds mentioned are cattle, dog, goat, horse, pig, sheep, and (in a late source) chicken. Domestic animals were forbidden, and therefore usually absent from sacred groves, but in a few cases some of them were kept in a grove and had sacred status. For example, at Etna in Sicily there was a grove of Hephaestus with a pack of sacred dogs that "wag[ged] their tails and fawn[ed] on prudent people," but bit criminals and chased away the debauched.

Birds were apparently very numerous. Again, many writers simply remark that there were lots of them in sacred groves without saying which kinds. The one most noted is, as might be expected, the nightingale with its beautiful song. Others are the crow, jackdaw and sea-crow, along with blackbird, eagle, falcon, kingfisher, owl, peacock, stork, and swan.

The animal most often named is the snake, without distinction of species. The only others of interest to herpetologists are tortoise and frog. Fish were protected in the waters of several groves; the kinds specified are eel and the elusive peacock-fish, although it must be assumed that there were many other species in those places. Mussels are noted in one river on sacred land.

The taxa most neglected by writers are the invertebrates, and here we can be sure that there were countless species. The only ones that could be found in a survey of likely sources are ant, bee, cicada, and spider. It was said that the grove of Clarus was purified by Apollo from "noxious creatures" including scorpion, spiders, and, to name a vertebrate, the viper as well.

MANAGEMENT FOR CONSERVATION

The principle of forest management applied in the groves appears to have been the strictest preservation (Hughes, 1994). This followed directly from the motive of religious consecration: the groves are the property of the gods and must not be damaged in any way. It is important to remember that these were areas with set boundaries. Within the lines certain regulations and practices of a variable but predictable type applied, which were different from those that governed private and public land outside, and these regulations and practices strongly shaped forest management.

The specific rules that protected groves varied in different places and in different historical periods, but the pattern is consistent. First of all was a definition of the boundary of the sacred area and a prohibition against trespass. To step over the line, which might be marked by a wall, was to pass from ordinary ground to holy ground, and this was allowed only for those who were prepared and would not, in a religious sense, pollute it by their presence. It is a sacred grove from which unworthy mortals were excluded by the sonorous words, *Procul o procul este profani* ("Keep out, profane ones!") (Virgil,1st century BC; in *Aeneid*). In Pellene, there was "a grove surrounded by a wall, sacred to Artemis the Savior... no man [was] allowed to enter it except the priests" (Pausanias, 2nd century AD). One might be banned from a sanctuary because one was not a man, or woman, or priest, or citizen. These are exceptions; usually ordinary persons could enter the groves if they were ritually clean. Women were considered unclean during menses and for a time during and after childbirth, and were excluded then. Any persons might be barred if they were carrying iron or weapons of any kind, or were guilty of serious unabsolved crimes, especially blood guilt. One had to be careful to know about any unmarked boundaries, because the penalties, religious and civil, still applied even if the transgressors were unaware that they had strayed onto consecrated ground.

A basic law found everywhere forbade the felling of trees or the cutting of branches. "Men call them the holy places of the immortals and never mortal lops them with the axe" (*Homeric Hymn* 5. 257-72) (Evelyn-White, 1914). If a tree was felled, it was believed that its spirit, the

Fig. 1. Brauron, Attica; Temple of Artemis.
The appearance of an ancient temple in a sacred grove is well illustrated by this archaeological site. Artemis, goddess of the wild wood and wild creatures, was worshipped here.

Fig. 2. Wall of temenos, Pythagoreion, Samos.
Protective stone walls often surrounded sacred groves in Greece. This example is from the ancient city of Samos, now called Pythagoreion in honor of the great mathematician and philosopher, Pythagoras.

nymph, died, and that the god might leave the sanctuary. Even the removal of dead and down timber, or of fallen leaves, was prohibited. In the grove of the heroine Hyrnetho, "of the olives and all the trees that grow there, no man might take home the broken boughs, or use them for any purpose whatsoever, but they leave the branches where they lie, because they are sacred" (Pausanias, 2nd century AD).

SACRED GROVES AS WILDLIFE REFUGES

All living denizens of the groves were protected, and hunting, fishing, and bird-catching were strictly forbidden. Protection within the groves was given to the animals that lived there, and enforced by laws of the local communities to which the shrines belonged. A hunter was not allowed to take his dogs into a sacred enclosure, and "if his quarry takes refuge in the precinct, the hunter will not follow it, but waits outside" (Pausanias, 2nd century AD). At Mt. Lycaeus, an area sacred to Zeus, it was believed that a hunter who violated this rule would die within a year. A cautionary tale said that the huntress Atalanta had been turned into a lioness for violating a sanctuary of Zeus.In some sanctuaries there were deer or wild goats sacred to Persephone, goddess of the underworld, or to Artemis, none of which could be hunted, although special permission might be given to capture a victim for sacrifice to the proper goddess. Sacrifices of wild animals were rare in Greek but not in Roman times; however, most sacrifices in both periods consisted of domestic animals.

As a rule, wild animals in the sanctuaries were preserved as sacred to the gods, and to kill them incurred punishment. Mythology, literature and art are full of examples, such as Artemis' destruction of the hunter Actaeon by the horribly appropriate method of having his own hounds tear him to bits. Although the usual version says this was because he saw the goddess naked, that story is not found until late in Greek history (Lloyd-Jones, 1983). If, as seems likely, there was an earlier version of the myth that said his offense had been to hunt a deer in the goddess' sacred demesne, then the form of his punishment truly fit the crime. He had, in any case, unadvisedly boasted to Artemis of his hunting prowess (Euripides, *Bacchae* 337-340). Bragging was also a fault of Agamemnon, the best-known literary figure to be punished for hunting in a holy place. As Sophocles says (*Electra* 563-572), "when taking pleasure in [Artemis'] sacred grove, he startled an antlered stag with dappled hide, shot it, and shooting made some careless boast." In vengeance Artemis caused winds that prevented the sailing of the fleet against Troy until Agamemnon sacrificed his daughter "in quittance for the wild creature's life."

The evidence of inscriptions shows that it was not only in literature that penalties were exacted for hunting in the groves. To provide another example of a site where animals were given the protection of a temenos, tortoises were preserved on a peak in Arcadia where "the men of the mountain fear to catch them, and will not allow strangers to do so either, for they hold that they are sacred to Pan" (Pausanias, 2nd century AD). Similarly, the Athenians would not let anyone harm the little owls of Athena that nested on the Acropolis, or the snake that had a den there, and to which offerings of honey-cakes were made (Herodotus, 5th century BC).

Most of the groves contained springs and streams, and some had lakes. Pollution of these places was strictly forbidden. The fish in them were protected, usually by a total ban on fishing, under penalty of death. Priests, exempt from the penalty, could fish in the Rheitoi, lakes sacred to Demeter and Persephone, but the priest of Poseidon at Lepcis abstained from fish because they were sacred to that very sea god. Artemis' eels were shielded by a no-fishing ordinance in the spring of Arethusa, and at Pharae fish sacred to Hermes could not be caught. At the lake of Poseidon, it was said, "they fear to fish in the lake for they say that anyone who catches fish in it will be turned into the fish called the Fisher" (Pausanias, 2nd century AD).

OTHER PROTECTIVE REGULATIONS

While wild animals were granted heaven, domestic ones were excluded. Specific penalties were set for herders who allowed cows, sheep, goats. swine, or horses to graze in the precincts, nor were animal-drawn vehicles allowed to enter. Other common rules prevented plowing, sowing, and the erection of unauthorized buildings. The Sacred War was fought over the issue of illegal cultivation of Apollo's sacred ground. Setting fire to a sacred grove was regarded as the most heinous of crimes, even in wartime, although it did happen. Cleomenes, the Spartan warrior who burned five thousand Argives to death in a god's forest, was driven mad by the thought of divine retribution.

Civil regulations governing sacred groves are recorded in many inscriptions, often from carved stones that were placed in the precincts themselves. The local magistrates who had jurisdiction over the sanctuaries were generally those whose duty it was to administer religious matters: there was no separation of church and state in ancient times. In Athens and Chios, the officer was called "king" (*basileus*), not the head of state but the democratic appointee who retained the old royal title for religious purposes (Frazer, 1935). Plato (4th century BC)suggested that the task be given to the Commissioners of the Marketplace (*agoranomoi*), and this was done in some cities. The priests who guarded the groves, as well as anyone else, were required to report acts of injury to these officials, who could then prosecute the miscreants. Penalties were severe. Inscriptions set fines in amounts ranging from 50 to 1000 drachmas for cutting down a tree or carrying away wood or leaves (Jordan and Perlin, 1984). When one realizes that an average workman was paid one-third to one drachma per day, it is evident that the fines were high enough to discourage attempts to profiteer by theft from the groves. Expensive mandatory sacrifices might also be assessed. Slaves could be whipped or imprisoned for the same offenses. As Russell Meiggs (1982) noted, "when one of Antony's generals cut timber for shipbuilding from a grove dedicated to Asklepios on the island of Cos he was executed by order of Octavian in the grove itself to emphasize his impiety."

To these not inconsiderable punishments, ritual curses were joined. The ancients never laughed at these, since they believed that the gods listened to them, and people in the past were said to have been smitten with insatiable hunger for felling a tree in spite of the entreaties of the dryad, or driven mad, or were themselves changed into trees, fish or other creatures. Modern readers might doubt such stories, but the common people of Greece, Rome, and elsewhere took them seriously enough. In all times and places, however, there seem to have been a few who disregard both the rules of state and the threats of religion, and therefore it must be assumed that poachers invaded the sacred groves from time to time, hunting illicitly and stealing biomass. Why else make laws and post them in public? The birds in Aristophanes' play berate humans for persecuting them, setting snares and lime twigs for them even on temples.

EXCEPTIONS

Thus far we have discussed the principle of preservation. But exceptions to these rules were made under certain conditions. They were not common enough to destroy the groves, but they did allow some changes to the habitats in the groves, and therefore possibly some diminution of biological diversity. The basic purpose of the exceptions was to assure religious use of the grove's resources. As much wood might be taken as was necessary for sacrifice and worshippers were allowed to bear branches and wear crowns made of the holy trees. As noted above, animals in the grove might be captured for sacrifice to the appropriate deities.

There is also evidence that the trees in a grove could be used in building a temple. Although classical temples were mostly of stone, their roofs required long timber rafters and roof poles. Since large trees survived in groves when they had been cut elsewhere, it is understandable that temple builders should have looked to the very resources they supervised, and even beyond them to other sacred groves. The magistrates of the island of Carpathos ordered a tall cypress to be felled in the precinct of Apollo and sent to Athens for use in rebuilding the temple of Athena. The Athenians, in raising an inscription of thanks to their benefactors, recognized that such use of a sacred tree was to be considered appropriate (Tod, 1933). On the other hand, groves sometimes outlasted the temples with which they were connected, so that ruins stood in flourishing groves where worship was still offered (Birge, 1994). In addition to temples, other buildings of quasi-religious and public character were sometimes erected in sacred groves: baths, stadiums, gymnasiums, schools, old people's homes, and the ancient equivalents of hospitals. Rarely, public meetings and elections were held there. Wood from sacred trees was believed to keep its magical powers when fashioned into other objects. It was therefore used to make statues of the gods, staffs for augurs, speakers, and generals, military standards, scepters of office, heralds' wands, policemen's nightsticks, divining rods, and lot-tokens for oracles.

With a number of uses for wood from the sacred groves, one would expect that there were foresters to oversee their management, and there is evidence that this was the case. Mythology gives Aristaeus the title of "caretaker of groves" (*cultor nemorum*) (Virgil,1st century BC). The municipal guild of woodmen (*dendrophoroi*) in Athens was charged to locate, cut, and bring to the city each year the sacred tree for the Great Mother. At Olympia there was an official woodman (*xyleus*) on the staff of the sanctuary of Zeus. "His duty is to supply states and private persons with wood for sacrifices at a fixed price" (Pausanias, 2nd centry AD). He also helped to superintend the sacrifices. So wood to be cut in the groves was selected carefully by experienced personnel.

This idea is supported by the practices of the Arval Brethren, a Roman priesthood who had the care of a large forest, dedicated to Dea Dia, near the city. Every year, they "offered two young pigs in order to expiate the unavoidable desecration of the sacred grove by the use of the axe in pruning and felling it.... Whenever iron was brought into the grove, as for cutting inscriptions... or the lopping and felling of the trees... there were sacrifices *ob ferrum illatum* ["for the bringing in of iron"], and, when the work was done, *ob ferrum elatum* ("for the taking out of iron"]" (Peck, 1923). As in most groves, the rules of the Arval Brethren required that every tree that was cut had to be replaced. Also, "when the trees fell from decay, or, worse still, were struck by lightning, and when replanting was undertaken, still more solemn sacrifices [of a boar, ram, and bull] (*Suovetaurilia maiora*) were offered on the spot" (Frazer, 1935). Similar sacrifices for use before thinning a grove or working the ground in one are prescribed, with an all-purpose prayer that smacks of cynicism, by Cato the Elder (2nd century BC).

The groves were used for many purposes other than their timber. In spite of the rules against grazing, Roman art and Latin literature abound in pictures of cattle, sheep, and goats ruminating in sacred groves. Virgil says that sheep graze where "a grove black with thick holm-oaks broods with holy shade" (Virgil, 1st century BC). Ovid (1st century BC) reports that at the annual festival of *Parilia*, in April, shepherds offered expiation to the gods for entering sacred groves, sitting in the shadows of the holy trees, and lopping leafy branches for their animals to eat. There would have been no use for such a ceremony unless herders often resorted to the groves. Of course such domestic animals would damage the vegetation and compete with the native species. A second Roman festival of groves, the *Lucaria*, in July, probably began as a rite of atonement in apology to forest gods for clearing woodlands for cultivation (Bodel,1994).

Legal documents for private management of sacred groves survive from ancient times. On the island of Chios, a family called the Clytidae rented out a sacred grove on their land. Terms of the lease are recorded on an inscription (Wilhelm, 1933). The Clytidae required their tenants to carry out re-afforestation of the grove by planting young trees (Rostovtzeff,1941). This is a particularly interesting point, since it marks an extension of the practice of planting in the groves that were managed by religious bodies to similar practices in groves that were in private hands. The older religious motives were joined with motives that were more economic as this occurred. Since the same people were often managers of public and private forests outside the consecrated lands, they may have been able to see the advantages of replanting and conservation of other forest land.

RESULTS OF CONSERVATION

Conservation might not have been threatened by management that combined the care with which trees were selected with the firm obligation to replant for every tree that fell for whatever reason. Indeed, the sacred groves tended to survive. Aristotle (4th century BC) recommended that more of them be established around the countryside as a means of forest conservation.

The groves were protected over many centuries, and the individual trees in them were often of remarkable dimensions. This resulted from the religious feeling against cutting large, notable trees in the groves, since they were believed to be uniquely cherished by the gods, as well as serving as the dwelling places for particularly venerable tree spirits, or dryads. These trees remained undamaged and were allowed to live out their life spans until wind, fire, or rot brought them down. At Pharae, the plane trees were hollow with age and big enough to sleep or picnic inside. Pliny the Elder (1st century AD) reported one hollow tree in a Lycian grove that was measured to be 81 feet in circumference, and provided Licinius Mucianus space for a banquet for eighteen. Eyewitnesses describe many historical specimens of arboreal gigantism, such as the "Maidens," cypress trees at Psophis that "overshadowed a mountain," although the size of the mountain is not mentioned. "These cypresses they deem sacred to Alcmaeon, and will not fell them" (Pausanias, 2nd century AD).

Outside the sacred groves, large trees were sought out for timber and fuel, and tall, straight trees were in demand for ships' masts. Descriptions by travelers like Pausanias leave the impression that the sacred groves often visibly stood out from the surrounding landscape and lesser forests as tall, dark, flourishing patches of high forest where many species of animals, especially birds, could find refuge. At the sanctuary of Apollo at Kourias, people observed that the deer would flee into the large grove whenever hunters and dogs approached, and "by some mysterious instinct trust[ed] to the god for their safety" (Aelian, 3rd century AD). It was not a mysterious instinct, of course: the animals had learned that the hunters would not follow them there. The nesting of birds in sacred groves is commonly mentioned.

REFUGIA IN A WIDER PATTERN OF REDUCTION AND EXTINCTION

The importance of groves as refugia for various species becomes clearer when the general pattern of local extinctions during ancient Greek and Roman times is examined. The surviving evidence gives the impression of declining populations of wildlife, and the gradual extinction of certain species in one area after another. Writers often note that animals are no longer to be found where they were once abundant. Lions, leopards and hyenas disappeared from Greece, and lynxes, wolves, and jackals were limited to the mountains, where they hold out today in small numbers. Bears could be found in the Peloponnesus up to AD 100; a few still exist in mountainous northern Greece. Hunting reduced wild cattle, sheep, and goats to remnant herds, and eliminated them from some islands.

Procurement of animals for the Roman arena cleared larger mammals, reptiles, and birds from the areas most accessible to professional hunters and trappers. They exhausted the hunting grounds of North Africa, where elephant, rhinoceros, and zebra became extinct. The hippopotamus and crocodile were banished from the lower Nile to upper Nubia. By the fourth century AD, a writer could lament that there were no elephants left in Libya, no lions in Thessaly, and no hippopotami in the Nile. The Romans were persistent, efficient, could pay well, and came to dominate the commerce in animals and animal products throughout the Mediterranean basin, so the major responsibility for extinctions was theirs.

Bird populations also diminished. The former richness of Mediterranean bird life can be sensed today in such relatively undisturbed areas as the French Camargue near the mouth of the Rhone River, with its flocks of flamingos; or the Coto Doñana reserve in Las Marismas on the lower Guadalquivir River in Spain.

There were complaints of the depletion of fisheries. It has been suggested that the disappearance of mosquito-eating fish from Italian marshes aided the spread of malaria in the second century BC and subsequently. Increasing prices for fish on the Roman market indicate that there were fewer left to catch than before; the finest rare fish sold for their weight in gold, or more (Pliny, 1st century AD; Radcliffe, 1921).

Introduction of domestic species that compete with wild ones must have happened countless times, particularly on islands. The devastating effects of goats, rats, and opportunistic birds like pigeons have been observed in newly discovered lands since the fifteenth century AD; there can be no doubt that it happened in isolated places as they were visited by ships or settled by new groups of people in Greek and Roman times.

The process with the most damaging effect on all forms of wildlife was, as it still is, habitat destruction. The clearing of forests, the spread of agriculture, the introduction of weeds and other exotic plant species, the overgrazing of grasslands, and the draining of lakes and wetlands all affected wildlife even more seriously than hunting. The result of all these factors was depletion and extinction of wildlife and an impoverishment of ecosystems. In the face of this pattern, the preservation of even relatively small areas of habitat in sacred groves is significant.

CONCLUSION

In conclusion, then, how important were sacred groves to the conservation of biological diversity in the classical Mediterranean world? As forest lands in their own right, they were of not inconsiderable extent, and they served as reservoirs of animal and plant populations, refugia which may have allowed species to survive and even to recolonize other areas. The presence of this resource is revealed by the many references to their survival in writers as late as the fourth century AD. The area outside the groves, as Plato attests concerning his homeland, Attica was largely deforested.

The reservation of sacred groves was probably the greatest single means of conservation in the ancient world; as Greek and Roman writers note, plants and animals survived within them when they had disappeared from surrounding areas. It is possible that some of the preserved areas served as resource pools providing natural resources on a sustainable basis, as in the case of groves managed for timber or to supply sacrifices.

Human greed made the protection of these reserves difficult, because although the gods were believed to want to keep them inviolate, there was no doubt that humans desired to use them in many ways. While penalties exacted by governments and threatened by divine sanctions were effective, religion and the state also allowed exceptions from the regulations, permitting use of the resources they were intended to preserve, the construction of buildings, and even at times the alienation of land. These invasions altered the sylvan qualities of the groves and negatively affected the operations of the ecosystems within them. Undoubtedly they caused local extinctions of species.

The practice of setting physical boundaries for sacred spaces consecrated and protected what was within, but by implication unhallowed the land outside. Beyond the bounds, the gods no longer protected the earth, and people were free to use it as they saw fit. Inside the temenos there might be glimpsed a holy light, but outside shone only the ordinary light of day. Thus an enormous step had been taken toward desacralizing nature. Pausanias, writing in the first century AD, gives the impression that over much of Greece, sacred groves were isolated islands of forest in a generally denuded landscape. Strabo (1st century BC) provides similar descriptions for most of the Roman Empire, and further points out that even some sanctuaries had lost their trees: "But the poets embellish things, calling all sacred precincts 'sacred groves,' even if they are bare of trees."

The groves lasted as places of economic and religious importance until the Christianization of the Roman Empire. As centers of the older worship, they became the object of destructive zeal. The emperor Theodosius I, in 391 AD, issued an edict directing the closing of all pagan temples and directing that the groves be cut down unless they had already been appropriated for some purpose compatible with the new order. A similar law was included in the Code of Theodosius II in the fifth century. A few of the groves subsequently became monastery gardens and churchyards, and some of these continued to function as refugia. For example, monastery precincts in the mountains of Lebanon are the only places where the famous cedars survive in that country. Here and there in the Mediterranean even today one will find a gigantic old tree, the fragment of a grove, that is supposed, however unlikely it might seem, to have sheltered Abraham, Plato, or Hippocrates, and has been spared for that reason. Other than these fragments and some meager archaeological evidence, nothing else survives of the ancient Mediterranean groves outside the written evidence. But that evidence may serve as an interesting point of comparison with practices and processes associated with sacred groves in the countries where they still exist.

REFERENCES

Aelian, 3rd century AD. *On Animals*. Translation by A.F. Scholfield. 1958-1959. 3 vols. Loeb Classical Library, Harvard-Heinemann, Cambridge.

Aristotle, 4th century BC. *Politics*. Translation by H. Rackham. 1932. Loeb Classical Library, Harvard-Heinemann, Cambridge.

Arrian, 2nd century AD. *History of Alexander*. Translation by P.A. Brunt. 1976-1983. 2 vols. Loeb Classical Library, Harvard-Heinemann, Cambridge.

Birge, D.E. 1982. *Sacred Groves in the Ancient World*. Ph.D. dissertation, University of California, Berkeley.

Birge, D. E. 1994. Trees in the Landscape of Pausanias' *Periegesis*. In: Susan E. Alcock and Robin Osborne (Eds.). *Placing the Gods: Sanctuaries and Sacred Space in Ancient Greece*, Clarendon Press, Oxford, pp.231-245.

Bodel, J. 1986. Graveyards and Groves: A Study of the *Lex Lucerina*. *American Journal of Ancient History* 11: 1-118.

Cato, 2nd century BC. *On Agriculture*. Translation by W.D. Hooper and H.B. Ash. 1934. Loeb Classical Library, Harvard-Heinemann, Cambridge.

Evelyn-White, H.G. 1914. *Hesoid: The Homeric Hymns and Homerica*. (Translation). Loeb Classical Library, Harvard-Heinemann, Cambridge.

Frazer, J.G. 1935. *The Golden Bough. Part I: The Magic Art and the Evolution of Kings*, 3rd ed., 2 vols. Macmillan, New York.

Herodotus, 5th century BC. *Histories*. Translation by A.D. Godley. 1920-1925. 4 vols. Loeb Classical Library, Harvard-Heinemann, Cambridge.

Hughes, J.D. 1994. *Pan's Travail: Environmental Problems of the Ancient Greeks and Romans*. Johns Hopkins University Press, Baltimore.

Jordan, B. and Perlin, J. 1984. On the Protection of Sacred Groves. In: A.L. Boegehold (Ed.). *Studies Presented to Sterling Dow on His Eightieth Birthday*. Duke University Press, Durham, North Carolina, pp.153-159.

Jost, M. 1994. The Distribution of Sanctuaries in Civic Space in Arkadia. In: S.E. Alcock and R.Osborne (Eds.). *Placing the gods: Sanctuaries and Sacred Space in Ancient Greece*. Clarendon Press, Oxford, pp.217-230.

Kyrielis, H. 1993. The Heraion at Samos. In: N. Marinatos and R.Haegg (Eds.), *Greek Sanctuaries: New Approaches*, pp. 125-153. Routledge, London, pp.125-153.

Lloyd-Jones, H. 1983. Artemis and Iphigenia. *Journal of Hellenic Studies* 103: 99.

Meiggs, R. 1982. *Trees and Timber in the Ancient Mediterranean World*. Clarendon Press, Oxford.

Ovid, 1st century BC. *Fasti*. Translation by J.G. Frazer. 1931. Loeb Classical Library, Harvard-Heinemann, Cambridge.

Pausanias, 2nd century AD. *Description of Greece*. Translation by W.H.S. Jones. 1918-1935. 5 vols. Loeb Classical Library, Harvard-Heinemann, Cambridge.

Peck, H.T. (ed.). 1923. *Harper's Dictionary of.Classical Literature and Antiquities*. American Book Company, New York.

Pliny the Elder, 1st century AD. *Natural History*. Translation by H. Rackham and W.H.S. Jones. 1938-1963. 10 vols. Loeb Classical Library, Harvard-Heinemann, Cambridge.

Plato, 4th century BC. *Critias*. Translation by R.G. Bury. 1929. Loeb Classical Library, Harvard-Heinemann, Cambridge.

Radcliffe, W.1921. *Fishing from the Earliest Times*. J. Murray, London.

Rostovtzeff, M. 1941. *Social and Economic History of the Hellenistic World*, 3 vols. Clarendon Press, Oxford.

Seneca, 2nd century AD. *Epistles*. Translation by R.M. Gummere. 1917-1925. 3 vols. Loeb Classical Library, Harvard-Heinemann, Cambridge.

Strabo, 1st century BC. *The Geography*. Translation by H.L. Jones. 1917-1932. 8 vols. Loeb Classical Library, Harvard-Heinemann, Cambridge.

Thirgood, J.V. 1981. *Man and the Mediterranean Forest: A History of Resource Depletion*. Academic Press, London.

Tod, M.N. 1933. *A Selection of Greek Historical Inscriptions*, 2 vols. Clarendon Press, Oxford.

Virgil, 1st century BC. *Eclogues, Georgics, Aeneid*. Translation by H.R. Fairclough. 1916-1918. 2 vols. Loeb Classical Library, Harvard-Heinemann, Cambridge.

Wilhelm, A. 1933. Die Pachturkunden der Klytiden. *Jahreshefte des Oesterreichischen Archaeologischen Institutes in Wien* 28: 197-221.

8

Social and Cultural Aspects of Sacred Trees in Iran

A.A. Khaneghah

Faculty of Social Sciences, University of Tehran, Iran

INTRODUCTION

Thinking of the existing world, has always occupied the mind of the man. Contemplating on the universe, creatures and all surroundings, led man to create the myths, and beliefs. Among all the human beings' belief, trees have a unique place. In Iran, Man's desire and whim have found itself reflected in the divine sacredness of the trees.

Beliefs of sacredness of these trees have changed in the past. Some of these beliefs could be attributed to the era before Islam and some to the epoch after Iranians adhered to Islam. Attaching pieces of textile, chains, locks and affixing clues and signs to the trees are the signs of sacredness.

GEOGRAPHICAL DISTRIBUTION OF THE SACRED TREES

Some 158 sacred trees belonging to species like walnut, plane-tree, willow, cypress, turpentine, soor, bench, tabrizi, aras, ors, fig, spruce-fir, wild plum, elm-tree, poplar, mountain-ash have been identified in 21 provinces in the Country. Sacred trees or most abundant in the Central and Khorasan provinces. 'Morad' and 'Plane-Tree' followed by cypress are the most distinguished. 'Morad' is a typical name for those trees with unknown name. "Berries", "Palms" and "figs" are also common.

HISTORICAL BACKGROUNDS OF THE SACREDNESS OF TREES

The reasons of the sacredness of the trees, could be analyzed in different time scale. We have tried to analyze the roots of this sacredness in the eras before and after Islam, with a regard to the beliefs of the Iranians.

The Era before Islam

In this period, the trees known to possess secrets and magic forces gained sacredness. According to that belief, sacred trees have the ability to overcome the devil forces. The human beings' inability to revert calamities may have led the man to submit to supernatural forces of the trees. As time elapsed, man established close relation with animals, plants and other natural elements of his environment and applied these relations to improve his own life by articulating the belief of sacredness of trees.

Many people believe that, trees and bushes are heavens of angels and spirits. Therefore, damaging, cutting or destroying the trees, specially green ones, cause death or sickness. Some people believe that just recalling the name of some trees is majestic and causes felicity. In other words, people believe that, trees own independent souls and spirits that confer in them the ability to struggle with the natural forces detrimental to the well being of humans. Some times mysterious events helped these beliefs to be reinforced. Coincidence of cutting the brunches of a tree with occurring pain in the stomach or death could guarantee.

It was an Iranian conviction that trees were symbolizing the honest people, who were transpiring into trees after the death to achieve lasting life. Accordingly, cutting trees was a wrong doing and might lead to the death of relatives.

Some species trees like Cypress, "Plane-Tree", "Pomegranate" and "Grape" had achieved sacredness for different reasons. People believed that, feeding by Cypress leaves would prolong life. In the Persian myth, cypress is believed to have originated in the heaven and is called as "Heaven Tree". It was brought from the heaven to the earth by Zoroaster. It is believed that "plane-Tree" changes its crust annually due to availability of adequate water. It has been symbolizing a fertile environment liked by the gods and spirits. Pomegranate was considered sacred for the color of leaves and flowers. Grape plant was a symbol of energizing the blood. Sacredness is thus attached to health of human beings.

The Era after Islam

In Islam plants were considered as endowments of God. In Holy Qoran, Trees and plants are repeatedly mentioned as signs of God. "He is who sends down water from the cloud for you; it gives drink, and by it (grow) the trees upon which you pasture. He causes to grow for you therby herbage, and the olives, and the palm trees, and the grapes, and of all the fruits; most surely there is a sign in this for a people who reflect" (Quran, Chapter 20, Verse 11). Conservation of natural resources, including trees, has been highlighted in Islam -"For the holy prophet of Islam, breaking a branch of tree was equal to breaking the angles' wing".

After Iranians adhered to Islam, most of the sacred trees indebted their divinity to prophet, Imams, their descendants and beneficent. Areas where these holy people are buried, turned into sacred sanctuaries. Animals and trees inside these heavens were protected against any destruction. Any harm to these animals or plants used to be regarded a "sin".

A sacred "plane-tree" located in the ward of "Imamzadeh Saleh" (brother of the shi-it 8th Imam) tomb, in Shemiran, in the vicinity of Tehran, or a cypress tree located in Kashmar (a city in Khorasan Province) where Zoroaster is said to have publicized his teachings, are examples of such trees. Another example is a tree in "Razmian" village in Qazvin province. Local people believe that "Ismaeel" the son of the 7th Shi-it Imam is burried there. Sometimes, Trees are believed to have been witnessed a doleful event to have happend to holy Imams. In "Zerabad" village in Qazvin province, a sacred tree is believed to have sheltered a pigeon witnessing the martyrdom of "Imam Hossein" the third Shi-it Imam. The branches of the tree are said to have been weltered into the sacred blood of the Imam. In such condition, the sacred tree acts as an intermediator between people, the Prophets and Imams.

Positive functioning of the trees might have promoted respect and sacredness for them. An example of such a tree is the "berry tree" in "Abbaskadeh" village in Gorgan province in northern Iran. This tree used to be a watch-tower for protecting the village against burglars and a base for the watchmen as well. In some cases sacred trees, inspired peace and caused reconciliation among rival tribes.

SACRED TREES: THEIR PRESENT ROLE AND MANAGEMENT

Divinity of sacred trees in different parts of Iran may vary depending on culture and living style of the natives. Different faiths and beliefs related to the sacred trees include curing sickness and swooning, fertilizing, matchmaking, increasing productivity. Natives usually pledge themselves through attaching pieces of textile, chains, locks to the trees. If wishes are realized, religious ceremonies such as slaughtering and sacrificing sheep, cooking food and feeding the poor, are to be performed in the vicinity of the sacred tree. Those trees with mysterious folklore attract people on superstitious basis and generally are used for fallacious wishes.

Sacred trees serve meeting points, facilitating exchanges between local communities and the wider regional community. This enhances cultural proximity, economic transactions and tourism.

Sacred trees located in holy places like "Imamzadeh" (the Tombs of Imams' ancestors) are automatically under protection of "Haj and Oghaf", the organization, which administers "Imamzadehs" all over the country. Other sacred trees are protected through local support, and the cost is covered from people's donations. The "Cultural Heritage Organization" which is responsible for registering and protecting historical monuments, screens the sacred trees as well.

In the course of socialization, local and popular institutions are also protecting the sacred trees and encouraging protective attitudes and values.

SACRED TREES AND CONSERVATION OF ENVIRONMENT

Some sacred trees have close linkage with local ecological values. Water scarcity in many parts of Iran has motivated people to respect self-grown trees. Thus, distinctive beliefs about protection of these trees have developed, that helped them escape from human beings' manipulation and other events. Conservation of trees has continued through ages by informal teachings of man to new generations and socializing trends as well. Sacredness of trees could rescue the life of even an indecent animal.

In addition to positive role which sacred trees play in establishing small and diverse living systems in different regions, they provide valuable sources of information on the vegetation recovery. Any specific type of plant has its own effect on the environment; therefore, old trees could provide us with valuable data on species specific environmental amelioration effects. Thus, managers and administrators who are responsible for rehabilitation of the natural resources, can use these data for identifying suitable species for a specific environmental situation.

CONCLUSION

Iran, with its own distinctive climate, has anchored a civilization which has been unique throughout history. With such a historical background, religious beliefs have developed and flourished to its extreme. Arid climate condition and Shi-ism had a distinctive role in this formation. In the Iranian history, trees acquired special respect and religious beliefs. These beliefs have incredibly associated themselves with religion.

Throughout history, these trees have either responded to the needs of the local population or maintained ecological morale of the society. Sacred sites grew into pilgrimage sites surrounded by local markets, leisure and accommodation facilities. Job opportunities for local people were also created around these places. Despite frequent visit of pilgrims, these places remained unharmed and function as sanctuaries of various animal and plant species.

9

Sacred Trees, Groves, Landscapes and Related Cultural Situations may Contribute to Conservation and Management in Africa

G. Michaloud and S. Dury***

*ISEM (UMR 5554), CNRS et Université de Montpellier II, Laboratoire de Botanique, 163 rue Auguste Broussonet, 34090 Montpellier, France
**Délégation CIRAD Cameroun, BP. 2572 Yaoundé, Cameroon

INTRODUCTION

The aim of the present paper is to present as representative a review as possible on African sacred trees, groves and landscapes, and related situations for which the culture and local history play as an important part for conservation as "effective" worshipping does nowadays. Indeed, we will see how historically sacredness may evolve both in a positive or negative way. In the present paper, we attempt to synthesize significant information on diverse ecological situations in Africa in order to provide readers from other continents with comparative documentation.

To approach this purpose the authors have often made a compilation of complete paragraphs and sentances which did not need be changed to be informative and linked them together with their own interpretation of the described situations in order to obtain a synthetic text. For this reason, at the beginning of each geographic situation, the full references of the source are given, independently of the bibliographic list.

It would be useful to have a basic definition of a "sacred grove" to serve as a reference when comparing different situations. The UNESCO-Ghanaian "Co-operative Integrated Project on Savanna Ecosystems in Ghana" (UNESCO, 1996) defined sacred grove as an area of "natural" vegetation preserved through local taboos and sanctions that entail spiritual and ecological values. The ecological values are found in the traditional association of the sacred groves with wildlife and physical landscape features such as stream sources and hills. The grove is the focus for common religious and sociocultural affinity for a number of surrounding village communities. The respective communities, together with their lands, constitute the Grove Community Area.

However, we will see through the different analysed cases, that several levels of sacredness may exist, and that each of them contributes accordingly to conservation of biodiversity.

IN THE MOUNTAIN FORESTS

The following description is based on the report from Malawi National Commission for UNESCO (Majiga et al.,1995). The Mulanje Mountain of Malawi is the highest one in South and Central Africa. It lies close to the Mozambique border, and most of the higher parts are under forest reserve to protect the catchment area. The flora is very rich, making this place one of the centres of high biodiversity which, due to different local climatic conditions, ranges from tropical to afro-montane forest and moorland. This mountain harbours several endemic and rare plant and animal species. One of the most notable tree species of this mountain is a cypress called Mulanje Cedar (*Widdringtonia whytei*) which has a restricted distribution in the region and grows as much as 40 m high and 1.5 m in diameter only on Mulanje Mountain. Elsewhere, in Mozambique, Zimbabwe and South Africa, it grows as a dwarf. In addition to this unique vegetation, avifauna, small mammals, reptiles, invertebrates, amphibians and fishes exhibit a high endemism on Mulanje Mountain. Moreover, the most widely known feature about Mulanje in the whole Southern African subregion, is the rich traditional medicinal resources of the mountains, and high expertise of the local traditional medical practitioners.

This area is the second most densely populated district of Malawi with eight percent of the 80,00,000 people representing two main tribes: the Lomwe and Mang'anja. Many of the villages of the district are located between tea estates and this mountain.

A very interesting point raised by the authors of this report indicates that these people have ever applied traditional knowledge in conserving their natural resources in all their activities. But it also shows that this degree of knowledge regarding the conservation of the mountain resources varies according to the village's proximity to the Mulanje Mountain.

It appears in fact, that numerous stories, sometimes horrifying ones, connected to ancestral spirits inhabiting the mountain forests serve to keep people away, from disturbing the natural forests stands which provide shelter to the abundant and rich flora and fauna. Certain zones are even declared prohibited areas. The penalty for ignoring this edict is that either one disappears completely or may return deaf and dumb after a month or two. Nowadays most of these stories are maintained for purposes of still keeping people away. And some cases of such punishment still happen to be reported. In spite of this, many of these beliefs are no longer taken seriously due to influences of modern religion and cultural interactions with people of other cultural backgrounds. Moreover the situation has been made worse by commercial tea estates through the introduction of huge pumping systems in the deeper parts of the rivers where water spirits used to live but who have since abandoned their "homes".

However, it appears that people are well aware of what not to do for protecting this ecosystem despite commercial encouragement by the government for exploiting timber in the mountains. Indeed, the felling of timber trees for commercial purposes has outweighed the exploitation of plant resources by the local communities, and people feel that the government does not seriously consider their own well being. In addition, arable land has become scarce due to expansion of commercial tea estates, thus forcing local populations to seek more land at high altitude in the mountains. And last, but not the least, Government has undermined the authority of the local chiefs and headmen who were the main protectors of the mountain resources.

Interviews with local people reveal that they are very much aware of the importance of the plant resources on the mountain with respect to their use by them, and also regarding the role that plants play in the conservation of water and in the water cycle. They also realise that the destruction of the mountain vegetation will have an adverse effect on their own prospects for survival.

Possible actions for saving the Mulanje Mountain plants and ecosystem include involving all the people of the area (especially local leaders) in biological resource conservation work, educating the public in the value of plants to man, exploring alternative methods for the sustainable use of the plant resources, etc. Giving to local chiefs and headmen the responsability for conservation projects and programmes should also be of a great help.

IN THE COASTAL FORESTS

The following informnation is based on the report of the NMK/WWF Coast

Forest Survey; WWF Project 3256: Kenya, Coast Forest Status, conservation and management (Robertson and Luke, 1993).

The conservation of the remaining fragments of Kenya's natural coastal vegetation, particularly the forests and woodlands, is of critical importance, as it has been estimated that more than 50 percent of Kenya's rare tree species occur in the coastal Province (Beentje 1989). In these coastal regions, the rapidly increasing human population, demanding more agricultural land, and more wood products for fuel, house construction and industry, are threatening the small areas of intact vegetation.

The Sacred *Kaya* Forests

The *kayas* are small relict patches of forests in Kwale and Kilifi Districts which once sheltered the fortified villages of the Mjikanda people. Here, *Kaya* means the forested area surrounding and including the original village clearing.

Today, one only sees fragmented relicts which still retain the sacred significance, whereas the *Kayas* were probably far more extensive when the *Kaya* culture was at its height. Unfortunately, only when the clearing was no longer hidden in the depth of the forest did local people understand that their heritage was almost lost.

The Digo tribe (belonging to the Mjikanda culture) split into many small groups and set up secondary and tertiary *Kayas*, which explains the number of small *Kaya* forests in Kwale district. Through population dynamic, quite few of secondary *Kayas* were established.

As the threat from hostile people of other tribes receded, and as the population of each *Kaya* grew, the Mjikenda groups (which include the Digo tribe) moved out of the forests and established villages in the more open countryside. In so doing they cleared areas of forest and woodland but their original *Kaya* forest patches survived and were protected by customs and used for ceremonies. The most respected and senior Elders lived on in the *Kaya* clearing, or returned there for the periods of ceremony and celebrations. The *Kaya* remained the spiritual centre of each tribe of Mjikenda, sheltering the graves of important Elders, and the traditional carved memorial posts (*Vigango*).

However, over the past few decades, increased formal education, and government policy aimed at reducing dependance on witchcraft, has led to an increasing disregard for traditional values and a decline in respect for the Elders. This has led to some Elders being unable to prevent, or even cashing in on, exploitation of these small forests (with for instance the *Vigango* being nearly all now stolen or sold for tourist curios), and associated sacred groves, and some have almost disappeared.

As a consequence, there has been concern about the disappearance of the indigenous plants used in traditional medicines, as well as awareness of environmental degradation resulting from forest clearance. Some of these patches also protected many small but important watersheds of the coastal foothills, particularly on the narrow rivers cutting through the Jurassic limestone, and ameliorating the effect of the erratic rainfall of the coast by ensuring constantly flowing streams. Destruction of the forest on coral rag for the cultivation of the fragile calcium-rich soils, as at Waa, was also observed leading to barren rocky platforms, useful only for coral block diggings and then very difficult to rehabilitate. Also, developers often burnt or slashed all the indigenous vegetation and were then faced with landscaping for tourist and residential buildings with imported materials at vast expense (for instance, extra soil in which to plant the ubiquitous bougainvillea and frangipani).

In respect to conservation, the idea of gazettement as National Monuments was put forward to give the *Kayas* protection at a national level, as it appeared to be the only suitable category in Kenyan law for such small areas of natural importance.

In November 1990, a private member's motion in Parliamant advocated that all *Kayas* in the Kwale district should be registered under the Kenya National Museums as prohibited areas. This motion was approved by the House.

IN SAVANNA ECOSYSTEM

The folllowing information is based on the UNESCO Terminal Report on the Co-operative integrated project on savanna ecosystems in Ghana (UNESCO, 1996).

As in most other African countries, in Ghana, rapid population growth and expansion of economic activities have led to deforestation and degradation of the environment. In some parts of northern Ghana, the natural population increase between 1960 and 1984 has reached some 490 per cent. This has had increasingly distressing socio-economic consequences for people who are compelled to widen their resource base to the detriment of environmental conservation.

Although environmental degradation is widespread in Ghana, small pockets of residual closed canopy forests remain near human settlements and are scattered throughout the country. Many of these forest pockets are, in fact, "sacred" or "fetish" groves which have survived environmental degradation because of religious belief. The groves cover small areas, sometimes less than five hectares each, but they constitute areas where the

Guinea savanna climax and proclimax vegetation may be approximated in an otherwise extensively degraded environment. Fortunately, the rehabilitation of degraded areas ranges high in the policy of the Government of Ghana and the national Environmental Protection Agency (EPA) takes a positive stand in using sacred groves as indicator sites for potential natural vegetation and the rehabilitation of degraded areas.

The role played by the sacred groves in the socio-economic and cultural lives of many rural folks in Ghana has been possible because of the collective efforts of people to protect them. It is estimated that over 80 per cent of the sacred groves in Ghana serve as watershed or catchment areas that protect drinking sources and provide readily available essential herbs of medicinal value.

Almost all existing sacred groves in Ghana have been, and continue to be protected by taboos, traditional beliefs and some local customs. Because of the reverence people attach to the sacred groves, people's general perception of the groves have resisted encroachment or unwarranted exploitation of resources within the groves.

However, the beliefs and reverence hitherto so absolute and binding on the people seem to be eroding in the minds of the youth in particular, and rural folks in general, due to trial breaking of taboos with no immediate consequence or retributions. The belief in «modernization» seems to be eroding traditional values and virtues which consequently are demotivating the local people's drive to consolidate their long-standing beliefs which have environmental conservation values.

In Ghana, as in many other African countries, «modern» concepts of environmental protection (e.g., designation of national parks) have experienced discouraging levels of success. In some cases, local rural people were not fully integrated into the planning or implementation of these largely western concepts of protection; nor were relevant solutions to socio-economic or poverty driven factors provided that could have catered for environmental conservation at the same time.

The UNESCO-Ghanaian "Integrated Project on Savanna Ecosystems in Ghana (CIPSEG)" addressed the problem of environmental degradation by basing intervention measures (rehabilitation) on sacred groves and using an integrated and people-participatory approach. Three sacred groves and their adjoining village communities in northern Ghana were at the heart of the project. Multi-disciplinary research teams (biologists, geographers, soil scientists, agronomists, foresters, anthropologists, education experts, etc.) carried out studies with a view to assessing whether sacred groves can indeed play an important role in environmental conservation.

Environmental education on the importance of conservation and involving local people in all project stages was considered essential for the success of the project. One project result for educating young people was a project video film entitled "Going back in time". The central focus of the film is the Dagbon Sacred Grove (in the western Dagomba district) and environmental conservation. The documentary is about the village Katariga which in the past years has suffered from drought. It is through this grave situation that the senior citizens of the village remember the better past and they are now telling the younger generation how good life was in the past: very good harvests without fertilizers, consistent water supply from the wells throughout the year and reliable fish supply from the village pond all year round. This, in no small measure, raised the social status of the village in the eyes of its neighbours.

Today, people of the Katariga village spend about 80 per cent of their time in gathering fuelwood and water. The grove, which was the symbol of their unity and the source of their strength and protection, is now being tampered with impunity. The trees in the grove have all been harvested and the gods have been stripped naked. And very soon the flora and fauna around Katariga may face a similar fate. Even medicinal plants for the treatment of common colds and malaria are gradually disappearing, and the people have to trudge to Tamale (the nearest town) for expensive medical care. At the peak of this calamity, the women finally speak out and with the support of their husbands challenge the chief and his elders to do something about their situation. They set up a committee for the Protection of the Environment and with the assistance and support of UNESCO-CIPSEG they succeed.

At the field level, the CIPSEG Project promoted well adapted techniques for savanna rehabilitation: village men and women were trained in micro-catchment techniques; seminars were held on controlling bush fires and the establishment of fire belts around the sacred groves; fodder banks were set up to provide feed during the lean seasons in order to reduce the pressure of livestock on the sacred groves; and agro-forestry techniques coupled with cash-crop production (e.g., cashew nuts) were introduced to provide an economic income to the local people, especially women.

The taboos and obligations, i.e. the "do's" and "don'ts", vary from one sacred grove to the other, but there are also several common features. For instance creating shelter belts around the groves through communal labour is an obligation to everybody in the village community. The strict observance or adherence of the rules associated with the sacred groves is considered important and cannot be compromised.

FROM WORSHIPPED TREES TO REFORESTATION

As in neighbouring Ghana (see above) and Benin, sacred groves in Togo may vary from a few hundred square metres to more than ten hectares. Often one can find new and ancient sacred groves according to their specific origins. As described below, sacred places may originate from an individual tree, and over time, can grow into a discrete patch of forest including other protected plant species. However, from the species diversity point of view, this is a very selective process.

The distribution and dynamics of the primary forest tree species, *Antiaris africana* Engl. (Moraceae) outside of its original habitat has been studied by Kokou (1996) in anthropic places such as villages and cultivated fields in the savanna of southern Togo (West Africa). In Togo, this species originates both from the south-western rainforests and the dry forests of the central and northern parts of the country. Its seed dispersal is mainly assured by birds (such as the bulbul *Pycnonotus barbatus*) and it can germinate far away from its natural habitat, around houses and in the fields. Its wood is appreciated for furniture, cases and other products. The bark is used for cloths, the young leaves as feed for cattle, and different parts of the tree have medical effects. However, it is mainly due to traditional beliefs that the tree has been preserved in open places. When a seedling of *A. africana* is established, people consider it as a gift of the God or of an ancester. It is therefore cherished and looked after. The tree is often used in "Voodoo" ceremonies, as it embodies many deities. When isolated, white or red cloths are tied around the trunk, and small ritual objects (statues or pots) are laid on the ground for gifts. Liturgical plants such as *Newbouldia laevis* or *Draceana arborea* are planted around the trees; together with the spontaneous surrounding plants, which are then protected, they constitute small patches of forests: sacred groves.

On the other hand, other sacred groves originate from old primary forests and as such exhibit a large tree species diversity. Kokou (in prep.) describes four types of protected forests which, by importance of sacredness, are i) "Voodoo forests" protecting all divinities, ii) the forest of the ancestors, iii) forests to protect villages from fire (mostly planted and sometimes inhabited by a divinity) and iv) forests for community activities like hunting or collecting. In a census of 53 small sacred groves totalising 17.2 ha, the author has found 650 taxa among which 9.4% were completely unknown in the flora of Togo. Aubreville (1937) also reports that in Burkina Faso (northern of Togo in West Africa), the biodiversity of sacred groves originating from primary forests cannot be seen elsewhere in that part of Africa. Yet, when the size of these primary forest patches is not sufficently large, they are invaded by non primary forest tree species, thus altering the original biodiversity structure.

Among the common invading savanna tree species censused by these authors and by Kokou (in prep.), are *Adansonia digitata*, *Parkia* sps.,

Cassia siamea, *Delonix regia*, and *Hura crepitans*. It is interesting to find abundantly the "Neem tree" (*Azadirachta indica*) which, unlike in India, does not seem to be considered as a sacred tree in West Africa. Another species, *Draceana arborea* is commonly introduced in the "ancestor" sacred groves, with one tree of that species being planted for God's favours. Another species which confers a sacred character, *Milicia excelsa*, is known to be favoured in "ancestor" sacred forests which so become "Voodoo" forests. According to Kokou (in prep.), the main ecological consequence of the introduction of these exotic species is linked with their growth which is more rapid than primary forest species. Thus they rapidly invade and alter the natural forest regeneration and consequently the species structure.

The conservation of the four types of forests mentioned above does not seem to raise any problem according to Kokou (in prep.). Even the non-sacred type of forests (in the worshipping sense) such as those to stop savanna fires, or those kept for daily-use purposes (hunting and collecting tree materials) are rigorously controlled by the chief of the village. It is only in some forests of the coastal part, even "Voodoo" ones, that some people are reported to violate the sacred rules because of the drastic shortage of trees in the region, even though infringements are reported to be severely punished. Otherwise the situation in Togo is interesting in the sense that the four types of "sacred" forests mentioned above seem to be well respected by local populations and this seems to be the case even in periods of social and political unrest, because they are governed by the local communities themselves. In contrast, administratively protected forests such as national parks are reported not to be respected at all.

SACRED TREES GENERATING SACRED LANDSCAPES

Ramakrishnan (1996) has developed in an Indian context, the concept of sacred species and of sacred landscapes demonstrating that the first, naturally leads to the second. As this author mentions, a species of tree may be sacred for religious reasons, but may also be sacred because of socio-cultural traditions and socio-economic factors. In the following examples from Cameroon, different situations will be addressed.

According to Bernard (1996) and Seignobos et al. (1987), in the plain inhabited by the Musey group, in Northern Cameroon, close to the Tchad border (about 10° North and 15° East), one can find cultivated parks whose vegetation includes some spontaneous species of trees which are protected because of symbolic value. For example *Prosopis africana*, (Mimosaceae) which is characteristic of the Musey ethnic environment, as the rest of the parks, inhabited by other ethnic groups in this Northern area of Cameroon is mainly made of *Acacia albida* (synonym of *Faidherbia albida*, Mimosaceae; Letouzey 1982). In contrast to their neighbours, whose main economic activity is cattle breeding, the Musey tribe are agriculturalists and hunters, and used to be fighters. Traditionally, when a man dies, his grave must reflect what kind of a soldier and hunter he has been. To do so the grave is surrounded by planted posts made out of *Prosopis africana* (reputed hardwood, not rotting and parasite resistant). Thus, the accounting for courage will be demonstrated as follows: two posts for one gazelle killed, six for a panther, eight for a lion, twenty for a man, etc. Some graves may therefore exhibit more than one hundred posts according to Seignobos et al. (1987). The grave is also planted with a conspicuous marking tree which is, most of the time, a *Ficus platyphylla* (Moraceae)), but can also be a *F. dekdekena, F. gnaphalocarpa, F. thonninghii* and more rarely some other *Ficus spp.* or a *Ceiba pentandra*.

Most of these fig species represent divinities and are related to various beliefs such as fecundity and prosperity. *Ficus platyphylla* may even be considered as a person, be called "grand father" when old, and deserve a true funeral ceremony when it dies. This species also marks the boundaries of teritories or of a property. Therefore these various symbols forbid anyone to cut these trees down. Some of the other fig species also indicate power and stand adjacent to the house of chiefs or of some important persons.

Thus, the landscape consequently comprises these different sacred tree species among which the main ones are *P. africana, Terminalia macroptera* (Combretaceae) which happens to be a substitute of the former one in case of scarcity, and *F. platyphylla*, together with the other fig species mentioned above.

However, around the village of Hollom, Bernard (1996) shows that young people are evolving away from such tradition, because of education, but also because there are no more tribal wars, and less and less game to kill to fulfil the requirements of the custom. Moreover, it has become financially very attractive to cultivate cotton for which trees are cut. Therefore, *P. africana* which is not only a good fuelwood, but also serves for housing and construction is increasingly utilized. Accordingly, in this area, the tree does no longer enjoys a sacred statue despite the fact that *P. africana* is known to improve the soil quality for cotton cultivation (Bernard, 1996). Together with *Terminalia macroptera, P. africana*, is disappearing in this park landscape.

As regards the fig species associated with burial grounds, Seignobos (pers. com.) considers that they are not endangered as graves are exempt from cultivated lands, although Bernard (1996) shows clearly that in her studied area, the areas surrounding some graves are under pressure. However, although this case demonstrates that long- established measures for tree protection may be eroding according to the economic context, it seems, from what can be observed elsewhere in neighbouring areas that sacred and social symbols are still serving to preserve fig trees, at least.

In addition, in this part of Cameroon, some villages belonging to the Musey or the Toupouri groups use these parks for food-stock purposes, and particularly those parks composed mainly of fig species (*F. dekdekena*, *F. platyphylla*, *F. ingens*, *F. glumosa*, *F. sycomorus*-which also has a high symbolic value-and *F. thonningii*) at quite a high density (Dury, 1990). Indeed, this originates from not too distant periods when due to wars and climatic vicissitudes, there would have been periods of starvation without the continuous fig resources that leaves, green and ripe fruit provide, with a high feeding value both for humans and cattle. This is due to the reproduction system of figs which, in the tropics, necessitates a continuous inter-tree fig production (Bronstein et al. 1990). In fact, if wars belong to the past (although figs have been abundantly eaten in the southern parts of Chad during the recent conflicts in this country), climatic catastrophes still occur, but their consequences on food resources are nowadays partly mitigated by the donations of rice and cereals from northern countries. Yet, despite this kind of new economic system, and despite the fact that nothing can grow underneath the huge fig canopy, people have not forgotten the usefulness of fig plantations and maintain also respect for them for this reason. This situation shows that sacredness has preceded usefulness, but that both are nowadays contributing to the preservation of the fig tree landscapes.

A second situation concerns trees having an agronomic value added to the symbolic one, thus creating a particular landscape. Dury (1990) describes the the Monts Mandara (about 14° East and between 10° and 11° North) with steep granitic slopes and an important population (for such an area) of 100 to 200 people per square kilometre. The agronomic system is very intensive and cultivation occurs on flat terraces to prevent soil erosion. Due to lack of space, trees are rigorously controlled, and only useful ones are left. Eighty percent of the trees here belong to the genus *Ficus*, with *F. dicranostyla* and *F. abutilifolia* the main ones growing on the rocky walls that maintain the terraces on the steep slopes. Owing to their strong and vigorous root system, these fig trees themselves maintain the rocks of the walls. Moreover, *F. abutilifolia* roots are known to penetrate within the slits existing in the rocks and cultivators favour their growth, and even encourage seeds to germinate in these holes in order to let the roots split the rocks. This facilitates the accumulation of organic matter to increase the space for plantation of additional cereals. These fig trees also provide food resources all year long, and especially during the periods of unfavourable production for crop. The protection of *F. abutilifolia* trees is enhanced by the fact that they are believed to be planted by the spirits that subsequently inhabit them. As a consequence, it is forbidden to sit underneath this tree, and whoever would cut one down would become deaf. If a young child dies, the leaves of this tree are used as a shroud to bury the corpse.

Therefore, this mountain landscape and its ecosystem are protected both for agronomic purposes and worship reasons. An interesting point noted by Seignobos (pers. com.) is as this fig species loses its traditional economic value it gains in symbolic importance.

To summarize, in these two parts of northern Cameroon, practically each of the fig species has a symbolic value apart from the traditional economic importance. Whether positive or negative, these symbolic values are often very strong, which partly explains the continued survival and presence of these species.

DISCUSSION

This synthetic analysis of some African sacred groves raises several main observations. a) There is not a unique notion of sacredness, but several levels (or degrees) of perception and sensibility towards it, which play an important role in exploring the possibilities for conservation. For instance the belief that places are inhabited by spirits is less resistant to modern education than the regular practice of "Voodoo" in a sacred forest. Yet, ancient taboos will be a better protection for trees if they are associated with some economic importance, even a traditional value as in the case of fig trees in Cameroon. Also there are two main levels of sacred vegetation, the level of a single tree and the level of a forest patch. b) Not all sacred groves originate from natural vegetation, as is the case when sacred groves are created in Togo with *Antiaris afrina*. c) "Sacred" trees as in northern Cameroon, are distributed in such a way that they constitute parks or landscapes rather than homogenous forest patches. They may be assimilated within sacred landscapes even if different from the religious values observed in sacred landscapes of India by Ramakrishnan (1996). d) Local management by communities of sacred places, but even of places for daily hunting and plant collecting without particular worshipping seem to be more efficient in some countries or areas than administrative control. e) The new modern economic system erodes not only the religious beliefs that protect sacred places, but also respect of their ecological management by local communities. f) This last point is the most prejudiciable, as most of the natural sacred groves serve as watersheds which protect drinking sources and medicinal plants.

Comparison of the different analysed situations provides interesting perspectives on the diversity of possible cases. In the Mulanje Mountain of Malawi, it seems that ancestors have long been aware of the preponderance of protecting their ecosystem for the well-being and the survival of their community. Ancestors relied on threats from the spirits inhabiting the place to keep away people capable of disturbing the site. However, with modern education and money attraction, the power of taboos has been diminishing, if not disappearing, thus putting this area and its community in danger. Indeed, tea plantations, timber industry or water pumping can easily outweigh traditional cultural arguments and the threats of spirits.

In Kenya, sacred forest patches have first been divided and partly cleared under population pressure. However, as long as they remained a spiritual centre, they were preserved within the modern economy. And there again, it is only when traditional spiritual beliefs have been pushed aside through modernism that their exploitation took place. It is interesting to compare this situation with Malawi and Togo. In this part of Kenya, as in the study area of Malawi, things evolved the same way: first, the destruction of sacred groves which had disastrous consequences on the ecosystems, especially for water catchments, drinking sources and medicinal plants, then (but only then) triggering alarm from both local communities and the administration.

If we compare the effects of population dynamics with what is known in Togo, we see that sacred forest patches become subdivided with the increase of villages in the coastal part of Kenya. In Togo, in savanna areas new sacred places are created nearby, starting from a single worshipped tree, *Antiaris africana*, which then generates larger woody bushes in association with other trees. According to Kokou (pers. com.), these woody patches usually situated at the edge of villages, also serve a practical purpose. In savanna areas, they may constitute a barrier against fires that threaten houses, and they may also improve the environment, thus favouring the growth of medicinal plants as well as plants necessary for "Voodoo" ceremonies.

The population dynamic and the effect of modern economy in northern Cameroon do not affect all tree species, even if they are all an obstacle to economic development. *Prosopis africana* and *Terminalia macroptera* which are spontaneous species in the plains forming a natural sacred landscapes are progressively disappearing and being transformed into more extensive culture. In contrast, in the same context, the planted fig trees that originally were key resource species for lean periods, and which are still associated with very strong beliefs (presence of divinities, fecundity, prosperity, longevity, social power, etc.), are still preserved.

In considering conservation prospects from the different examples mentioned above, it seems that only wisdom at the different administrative levels, together with local people's involvement could become the substitute of sacred beliefs for ecosystem conservation when necessary. Yet, experience from the CIPSEG project in Ghana has shown that the rehabilitation of degraded areas is still possible if efforts are based on cultural beliefs and if sacred groves are used as indicators for natural potential vegetation. As mentioned in the conclusions of this project, one should seek to combine people's traditional concepts of sacred groves with modern and legal instruments to enhance the conservation of the environment. Moreover, areas for economic activities should also be designated in the vicinity of sacred groves based on sound scientific studies. The biosphere reserve concept with its elaborate land use and land management plans could be of help in this process.

To conclude, in considering natural reforestation in the vicinity of sacred primary forests, even large ones, one should be aware that many of the seedlings originating from primary forest tree species have most difficulties to colonize open places for physiological reasons. Indeed, it is more probable that some of the gaps occurring in these primary patches would be successfully colonized by savanna, thus affecting the primary forest biodiversity over the long term. To avoid this, it would seem necessary to help natural reforestation through a controlled programme of intervention.

ACKNOWLEDGEMENTS

We are most thankful to Christian Seignobos, Kouami Kokou and Christelle Bernard for the useful conversations and suggestions. We are also endebted to Robert Höft for providing relevent documents, and to Thomas Schaaf from UNESCO Paris, for his corrections and constructive comments. And, as always, Malcolm Hadley has been most helpful, especially in providing both useful documents and numerous corrections in the text. This work was supported by CNRS-ISEM (UMR 5554), Université de Montpellier II, N° 98036, France.

REFERENCES

Aubréville, A. 1937. Les Forêts du Dahomey et du Togo. *Bulletin du Comité d'études historiques, 19;* 1-113.

Beentje, H.J. 1989. Baseline knowledge of wild plants and priorities for action. In: H.J. Beentje and B. Khayota (Eds.) *Proceedings of the Workshop on Plant Genetic Resources,* Utafiti, National Museum of Kenya Nairobi, Kenya, pp. 49-52.

Bernard, C. 1996. *Etude d'un parc à Prosopis Africana au Nord Cameroun (cas du village de Holom, en pays Musey.* CIRAD-Forêt, Montpellier, France.

Dury, S., 1991. *Approche Ethnobotanique des Ficus au Nord Cameroun.* UNESCO/MAB, Paris.

Juhé-Beaulaton, D. and Roussel, B. 1992. Les forêts sacrées de l'Afrique de l'Ouest, In: *Forêts.* AGEP ed., France, pp. 250- 253.

Kokou, K. 1996. *Antiaris africana* Engl. (Moraceae) dans le paysage des savanes guinéennes au sud-Togo. In: *Le Flamboyant,* Bulletin de Liaison des Membres du Réseau Arbres Tropicaux, N0 39.

Letouzey, R. 1982. *Manuel de Botanique Forestière, Afrique tropicale.* Centre Technique Forestier tropical, Nogent sur Marne, France (3 volumes).

Majiga, C.M.B., Kalindexafe, M., Semu, L., Mfune, J.K.E. and Kamundi, D.A.I. 1995. *Interim report on the research study on the utilization of traditional ecological Knowledge of plant resources and conservation of the Mulanje Mountain.* Malawi National Commission for UNESCO.

Ramakrishnan, P.S. 1996. Conserving the sacred : from species to landscapes. *Nature and Resources, 32* : 11–19.

Robertson, S.A. and Luke, W.R.Q. 1993. *The report of the NMK / WWF Coast Forest Survey; WWF Project 3256 : Kenya, Coast Forest Status, conservation and management.* Seignobos, C., Tourneux, H., Hentic, A. and Planchenault, D. 1987. Le poney du Logone et les derniers peuples cavaliers. Coll. Etudes et Synthèses de l'IEMVT, 23, Paris, France.

UNESCO, 1996. *Terminal report on Co-operative Integrated Project on Savanna Ecosystemsin Ghana (Serial N° FMR/SC/ECO/96/209-FIT),* UNESCO, Paris.

10

Sacred Groves in Ghana: Experiences from an Integrated Study Project

T. Schaaf

Division of Ecological Sciences, UNESCO, Paris

INTRODUCTION

In Ghana, as in most other African countries, rapid population growth and expansion of economic activities have lead to deforestation and degradation of the environment. In some parts of northern Ghana, the natural population increase between 1960 and 1984 has reached some 490 per cent. This has had increasingly distressing socio-economic consequences for people who are compelled to widen their resource base to the detriment of environmental conservation. In many parts of the country, the natural vegetation has been seriously affected by bush fires, agricultural cultivation, overgrazing, fire wood cutting and even urbanisation and village sprawl.

Although environmental degradation is widespread in northern Ghana which is a dry sub-humid savannah of the Guinea type, small pockets of residual closed canopy forests remain near human settlements. Many of these forest pockets are, in fact, "sacred" or "fetish" groves which have survived environmental degradation because of religious beliefs. The groves cover small areas and may vary from 0.5 to 20 hectares each. Village communities have actively protected their sacred environments and thus, deliberately though unconsciously, have contributed to environmental conservation at large.

It is estimated that over 80 per cent of the sacred groves in Ghana serve as watershed or water catchment areas that protect drinking sources and provide readily available essential herbs of medicinal value. Almost all existing sacred groves in Ghana have been, and continue to be protected by taboos, traditional beliefs and some local customs. Because of the reverence people attach to the sacred groves, people's general perception of the groves have resisted encroachment or unwarranted exploitation of resources within the groves.

With this background situation, UNESCO carried out a project in northern Ghana, entitled "Co-operative Integrated Project on Savannah Ecosystems in Ghana (CIPSEG)" which lasted from October 1992 to March 1996. The project focused on three different sacred groves which were selected as study and intervention sites: Malshegu sacred grove, Tolon sacred grove and Yiworgu sacred grove.

INTERGRATED SACRED GROVE STUDY PROJECT

One of the main aims of the project was to assess whether the sacred groves could be indicator sites for the potential natural vegetation of the savannah area. Could they give an idea of how the savannah of the Guinea type looked like before human pressure on the savannah grew too strong? In this vein, the project's goals were to develop a scientific knowledge base on the relict fetish groves ecosystems and to study the sacred groves in terms of their plant and animal species composition.

This knowledge base was then geared towards achieving the second main aim of the project: to restore the adjacent and degraded savannah areas by using native plant species from the sacred groves' gene-pools. The project, therefore, had both a scientific orientation (study of the sacred groves' genetic resources), and a development orientation (rehabilitation of degraded environments).

In order to address these two main objectives, several scientific teams were needed which would work in a harmonized, interdisciplinary approach. As the project was embedded in the overall integrated research philosophy of the UNESCO Programme on "Man and the Biosphere (MAB)" and the counterpart institution was, in fact, the MAB National Committee of Ghana (the Secretariat is within the Ghana Environmental Protection Agency), it was no problem in setting up different study teams.

Project Partners

The Botany Department of the University of Ghana carried out an in-depth plant inventory of the three sacred groves which had been selected for the project. The Geography Department of the same university looked into the overall land use systems of the three districts in which the sacred groves are located, with a view to elaborating environmentally sound development and management plans.

The University of Science and Technology in Kumasi with its Institute for Renewable Natural Resources (IRNR) and its Forestry Research Institute of Ghana (FORIG) were concerned with assessing the area's edaphic, climatic and socio-economic conditions which were needed as baseline factors for overall development interventions. Scientists of these two universities were actively engaged in carrying out modern ecological and applied research and several of their graduate students were using the project's facilities for preparing their Master's degrees.

The newly established University of Development Studies in Tamale with its researchers in the wildlife department undertook a detailed study which were divided into 3 modules: (a) vertebrate wildlife aspects, (b) invertebrate wildlife aspects, and (c) the socio-cultural and economic aspects of wildlife, both within and outside the sacred groves.

Apart from purely ecological research, the project also attached great importance to the socio-cultural dimensions of the sacred groves. The Centre for National Culture in Tamale undertook in-depth studies of the traditional beliefs which had led to the protection of the sacred groves; the same Centre also analysed the sacred groves' functions for ceremonial purposes performed by the fetish priests. Moreover, studies focused on traditional resource use by village communities such as tree planting, ownership of tree and forestry products and marketing hereof. These studies were particularly important for the restoration of degraded savannah environments in order to meet the specific needs of village communities without violating cultural values.

Gender issues with regard to resource use were addressed by the Northern Region Rural Integrated Programme (NORRIP), since the different roles of women and men in restoration and development activities needed to be fully understood for any kind of intervention activities.

Environmental education on the importance of conservation and involving local people in all project stages was considered essential for the success of the project. The project's main counterpart institution, the Environmental Protection Agency (EPA) of Ghana, carried out multi-layered education programmes: for example, seminars were organised on the control and prevention of bush-fires which could seriously affect the sacred groves, and on the establishment of shelter belts around the groves; women were trained in tree planting, in particular using the micro-catchment technique; men were trained on the usage of fodder banks to provide feed to livestock during the lean seasons and to reduce pressure on the sacred groves; environmental awareness seminars and seminars on sustainable land use planning and implementation were geared towards the general public.

Finally, mass media were utilized to diffuse project activities to the entire Ghana nation: the Ghana Broadcasting Corporation (national television station) made several features on the project with the aim to inform on environmental conservation practices using sacred groves. A "docu-drama" was produced for video presentation using villagers of the sacred groves as actors: the story is that neglect of a village's sacred grove can lead to calamities as medicinal plants and herbs are no longer available which have been previously collected in the sacred grove, and the gods are angry. Finally, women with the support of their husbands, challenge the chief and the elders to do something about the situation. The chief then entreats his subjects to protect the grove and respect the abode of the god, for that is where the medicine comes from, restore soil fertility and create shelter belts around the sacred grove. The people-christians, muslims, traditionalists, women, men and children-all agree to work together to make the future of the village community better again. They set up a Committee for the Protection of the Environment and with the support of the CIPSEG Project they succeed.

It would lead to far here, to enumerate all the scientific results which the project yielded. However, a few results are quite interesting to mention here.

Observations and Experiences

One of the working hypotheses had been that the biodiversity within the sacred grove would be much higher than in the adjoining non-sacred areas. However, this was only half true. In terms of animal species diversity, birds, reptiles and mammals were more abundant within the sacred groves than outside the sacred groves. This was not a surprising result as the sacred groves also function as wildlife sanctuaries in which hunting is outright prohibited and trespassing of this customary law is penalized by the custodian of the sacred grove. An antelope, for example, can be hunted outside the grove, but as it enters the sacred grove, hunting has to stop.

However, as regards plant species, a higher species diversity was found at the edges of the sacred groves than within the sacred groves. We assume that the edges of the sacred groves function like ecotones where two different environmental settings meet: an ecosystem with a closed canopy cover (sacred grove), and a human-impacted ecosystem where bush fallow or agriculture occurs. Hence, the differing light conditions at the edges of the sacred groves gives rise to a more heterogenous plant diversity than within the groves. It may also be assumed that the sacred groves in this savannah environment are dry forests in their climax or sub-climax stage which are less species rich than groves with secondary undergrowth.

Research on the cultural aspects and significance of the sacred groves also provided fascinating results: Through interviews with the village elders and extrapolation of historic events, we assume that some of the sacred groves are at least some 300 years old (lack of written history in Africa makes a time assessment over longer periods difficult). They originated either as the abodes of a god or several gods. The three selected groves were the respective abodes of a python god, a leopard god and a monkey god; they can command plenty or lean harvests. Other sacred groves in the study area served and still serve as burial grounds of ancestors and have become taboo over time. In some cases, it was possible to trace back the specific point in time and the occasion when a sacred grove became taboo; during tribal warfare, a chief sought shelter in a grove where a god made him invisible from the raiding enemies and he was saved. Since that time, the grove was venerated as sacred.

The power of a chief is intrinsically linked with his function as supreme custodian of a sacred grove. No matter whether the chief is a practising muslim or christian, his power over the community derives from his role as protector of the sacred grove. Should he relinquish this function, his power as chief would be forfeited.

Some interesting taboos were brought to the fore. For example, it was considered a taboo for a young man to plant a tree in particular during the day: if the shade of the tree fell on the young man, the man would be doomed to die. This view was held high and may explain why many afforestation projects or projects on the rehabilitation of degraded lands failed in the savannah areas of Ghana. However, there is a way to go about the taboo in planting trees in the early morning hours or at sunset. The Centre for National Culture explained that this taboo did, indeed, make sense in an ecological sense. If a tree is planted at dawn or at dusk, the higher moisture content in the soil would also ensure a higher survival rate for the planted tree.

The taboos and obligations, i.e., the "do's" and "don'ts", vary from one sacred grove to the other, but there are also several common features. For instance creating shelter belts around the groves through communal labour is an obligation to everybody in the village community. The strict observance or adherence of the rules associated with the sacred groves is considered important and cannot be compromised.

Perhaps one of the most interesting lessons learned from the project was that it was not enough to convince village communities on the importance of planting trees around the sacred groves to enhance the protection of the groves and to combat environmental degradation. The grove has first and foremost a spiritual significance as the abode of a god. The god could dwell in a single tree or rock and still exert his/her power. This is a view held by many young people who would wish to extend their agricultural lands even if this extension would feed into the sacred grove. As long as the sacred tree or rock still exists, no harm is done to the religious integrity of the sacred grove.

The project, therefore, re-oriented its activities with regard to the rehabilitation of degraded lands around the sacred groves. We abandoned the idea of solely using plant genetic resources of the sacred groves for restoration activities, but focused on agro-forestry methods which also permitted cash-crop production (e.g., cashew nuts, mango etc.). These cash-crops provided an economic income to local people, especially women and young men, and permitted the restoration of a vegetation cover in particularly degraded areas. The establishment of woodlots and fodder banks were additional means to create a "buffer zone" around the sacred groves which in turn reduce the pressure on the sacred site itself. Two different situations noted are : (a) a large pressure on the sacred grove prevails to use the area for economic purposes and (b) a buffer zone has been created around the sacred grove. This buffer zone provides income benefits for people (agro-forestry, woodlots, fodder banks, cash crops etc.) so that the pressure on the sacred grove is reduced.

Experience from the CIPSEG Project has shown that the rehabilitation of degraded areas is possible if based on cultural beliefs which tie in well with the religious and spiritual views shared by a specific community. We also think that sacred groves can be used as indicators for potential natural vegetation. It is essential, however, that not only the groves but also the wider spatial area around the sacred sites are considered and embedded in integrated development schemes. Income-generating activities beyond the confinements of the sacred groves have to be developed for local people as the modern socio-economic contexts and constraints necessitate viable resource bases. Ideally, one should seek to combine people's traditional concepts of sacred groves with modern and legal instruments to enhance the conservation of the environment.

REFERENCES

Schaaf, Th.: Heilige Wälder für den Natur- und Umweltschutz. In: *UNESCO heute*, 41st year, No. II, summer 1994, pp. 181-183. 1994.

Schaaf, Th.: "Sacred Groves—Environmental Conservation Based on Traditional Beliefs". In: *Culture and Agriculture Orientation Texts.* (World Decade for Cultural Development), UNESCO Paris, 1995, pp. 43-45.

Schaaf, Th.: "Umweltschutz durch Kultur". In: *UNESCO heute,* 44th year, No. III. 1997, pp. 73-75.

UNESCO: "Co-operative Integrated Project on Savanna Ecosystems in Ghana (CIPSEG)". In: *Final Report on Symposium on Science and Technology in Africa,* Nairobi, Kenya, 14-15 February 1994, pp. 95-99.

UNESCO: Terminal Report on the "Co-operative Integrated Project on Savanna Ecosystems in Ghana" (Serial No. FMR/SC/ECO/96/209-FIT, Paris 1996).

11

A Note on the Sacred Groves in Afghanistan

Mohamed Zaman

Academy of Sciences of Afghanistan, Chemistry and Biology Department, Kabul, Afghanistan

Afghanistan, like any other country, has a tradition of conservation and sustainable utilisation of natural resources including forests and trees. The history of the country which dates back to over 5000 years has seen the importance given by different religious groups for the forests and trees and in total for the management of all natural resources in a sustainable way. For example, the Avesta religion which is considered to be the oldest religion of the country has stressed more on water management. Followers of the Avesta religion living on both sides of Hindukush used to worship water and sought assistance from it, in difficulties. Later Vidy religion, as one can see in its recitals, had given prominence to agriculture and animal husbandry. In the subsequent religions such as Zoraster and Buda planting of multipurpose plants and native wild plants on yearly basis near the religious places were considered as the sacred practice.

After the appearance of Islam, the creation and conservation of sacred groves became a part of historical and geographical tradition of the people. In Afghanistan, at present as a whole there exist more than 150 sacred groves. The major sacred groves are located in Baghlan and Badakhshan and are managed by a religion sect Ismaelia. Other famous sacred groves are Baba Wali in Kandhar, Abdullah Ansari in Herat, Mehterlam Baba in Lagman, Khwaja Seh Baba in Sarobi, Jabali Saraj area, Khaja Mohammad Parsa and Acashi Wali in Balk, Aknond Baba in Kunar, Mia Haji sahab in Shinwar, Lais Baba in Koot, Satanpur in Surkhrood, Zawa in Khogyani, Ghazni and Bamyan provinces, Khaja Safa and Baba Khan Mohammad (the white Garden in Shohada) of Kabul province, and Shah Qiswar and Gerda Selai in Paktia Province. The area occupied by these sacred groves is about 733 Km^2.

The most important trees and shrubs of these groves are *Cereis arifithii, Suaeda, Stocksia, Zizyphus, Zygophyllum, Dodonea viscosa, Conovolvulus spinasus, Astragalus gossypinus, Withanea coagulans, Acanthophyllum, Chamaerops ritchians, Acantholimon* and *Zyzyphora*.

Majority of the sacred groves are under the control of the rural communities. It was observed that in many sacred groves, the trees are being cut and used as firewood. This has resulted in the degradation of such sacred groves. It may be pointed out here that as a result of 18 years of civil war the state of Afghanistan has not been able to organize a national programme for the conservation of forests and sacred groves. However, in recent years it has been attempted to introduce the tradition of sacred groves the country. Intensive publicity through newspapers, magazine, radio and television has been organised to enlighten the people to take part in protection and conservation of these sacred groves. This process is found effective in north provinces of the country such as Ralkh, Samangan, Jawzjan and Faryab. It deserves to mention that lately the ministry of agriculture has taken steps to draft a law concerning the conservation of sacred groves. This law will be approved officially very soon and bear legal impact on the protection of the environment. There has been an increasing interest of tourism towards sacred groves. According to an official publication by the Afghan Central Census Authority in 1978, there has been 30-35 thousand tourists visiting the sacred groves and pilgrims in one year period. Therefore, conservation and sustainable management of sacred groves also help in promoting eco-tourism or eco-pilgrimage in the Country.

12

Religious and Cultural Perspective of Biodiversity Conservation in India: A Review

K.G. Saxena, K.S. Rao** and R.K. Maikhuri****

*School of Environmental Sciences, Jawaharlal Nehru University, New Delhi 110067, India

**G.B. Pant Institute of Himalayan Environment and Development Kosi-Katarmal, Almora 263643, India,

***G.B. Pant Institute of Himalayan Environment and Development Garhwal Unit, Srinagar (Garhwal), India

INTRODUCTION

Benefits of State mediated biodiversity protection measures such as establishment of reserve forests, National Parks and Wildlife Sanctuaries tend to be lowest at the local level and highest at the national and global/ regional level, while the costs of protection are highest at the local level and lowest for the national and global community. The mismatched local/ regional costs and benefits of such conservation programmes/policies are the root causes of people-conservation conflicts (Wells, 1992). Increasing threats to biodiversity loss demand new conservation approaches enabling fair share of the wider values of conservation to the local communities and positive local attitudes towards national and global conservation goals. Local community centered conservation approaches are also an ethical need when one realises that the high biodiversity potential of present National Parks and Sanctuaries derives from conservation attitudes of local communities. Nature worship has been a key force in shaping the human attitudes towards conservation and sustainable utilization of natural resources. Traditional nature worship practices in different parts of India, emerging trends and scope of these practices in promoting the national/ regional goals of conservation are discussed in this article.

NATURE WORSHIP

People's choices for the use/non-use values of natural resources are inextricably linked to belief systems governing values and the linkage between humans and the natural world. The mythological scriptures and historical accounts have described sacred forests where protection to spiritual guides was provided by the rulers of the land. Nature worship embodies an appreciation primarily for ethical, intrinsic, existence, non-use and indirect values of biodiversity and secondarily for direct or consumptive values.

The Common Perception

A variety of natural objects are regarded sacred by the Hindu community. These include rivers (e.g., Ganga, Yamuna, Saryu, Sutlej, Beas, Narmada, Kaveri, Godaveri, Mahanadi), mountain peaks (e.g., Nanda Devi, Binsar, Syahi Devi, Hariyali Devi, Trishul, Kailash in Uttar Pradesh (UP) Hills, Kailash-Kinnar in Himachal Pradesh, Amarnath in Jammu and Kashmir, Mansa Devi hill in Assam and Tirumala hills in Andhra Pradesh), lakes (Naini and Sahastratal in UP, Renuka and Chandratal in Himachal Pradesh), ocean/coastal areas (Puri in Orissa, Dwarka in Gujrat and Rameshwaram in Tamilnadu), plant species (*Ficus relegiosa, Ficus bengalensis, Elaeocarpus* spp. and *Ocimum sanctum, Azadirachta indica, Mangifera indica, Tamarindus indica, Michalia champaca*) and animals (lion, cow, ox, peacock, elephant). These are sacred in that they symbolize different gods, goddesses and superhumans. The Himalaya is considered to be the home of Lord Shankara and Ocean of Lord Vishnu. Lord Krishna has been portrayed as pastoral god roaming in forests and playing flute around *Anthocephalus kadamba*. Sun, air, water, fire, earth and all other planets/major satellites are symbolised as gods/goddesses. Sacredness sometimes has a habitat, use/non-use and biological context. *Ficus relegiosa* growing within living area is considered to be an indicator/predictor of poverty. All biomass use values of this species are not sanctioned in the Hindu religion. *Ocimum sanctum* is worshipped but one could also use it for medicinal purpose. Mortality as well as excessive regeneration of *O. sanctum* within house is believed to be a bad omen. *Ficus religiosa* entwined around *Azadirachta indica* are considered more sacred than independent individuals of the two species. Thus we get a complex religious-cultural articulation of conservation and environment in Indian mythology.

Regional Variations/Specificities

Apart from the above widely recognized values, there are a number of sub-regional variations evolved under varied ecological, social, cultural, economic and political conditions.

Indo-gangetic Plains

The Indo-gangetic plains, the oldest centre of civilization, has the highest population density and highest potential of crop production. There are three festivals on which one could observe a religious value for species richness: Harchhath (birth day of Balram, the elder brother of Lord Krishna celebrated in the month of August), Makarsankranti (celebrated on January 13-15) and Govardhan Puja (a day after Dipawali a festival falling in the month of October/November) when all possible efforts at family level are made to offer as many food grains/vegetables/fruits as possible. In every worship, the tradition is to offer flowers of as many species as possible. Coconut, arecanut, turmeric and clove, though are not grown in the region, are considered sacred and offered on a number of occasions. Banana is worshipped on Thursday and *Ficus relegiosa* on saturday. *Butea monsoperma, Ficus bengalensis* and sugarcane are worshipped on specific dates of the year. *Madhuca indica* is worshipped but largely by the socio-economically deprived sections of the society. Weeds *Calotropis procera, Argemone mexicana, Datura alba, Cynodon dactylon, Saccharum* spp., *Phragmites* sps., *Cannabis sativa* and fruit plants viz. *Mangifera indica, Ziziphus* sps., *Aegle marmelos* are valued for their products considered to be essential on many religious occasions. Tropical bamboos (*Dendrocalamus hamiltonii, Bambusa* spp.) are believed to house the bad spirits. Green bamboo culms are used only for carrying dead body. For all other purposes, the culms are kept for some time before use. Sacred groves representing relatively natural undisturbed vegetation in this ecological zone, as at present, are rare (Anonymous, 1995). Vrindavan, the forests where lord Krishna is believed to roam are practically non-existent.

Indian Desert

In the desert and surrounding arid/semiarid region characterized by low population density, poor food crop potential and pastoral livelihood, the practice of worshipping individual plant or species is not as common as in Indo-gangetic plains. A common feature of this region, particularly in the areas dominated by Bishnois, is that each village has a sacred grove, locally called 'Oran' which could be as large as 100 ha or more. For Bishnois, all animal life is sacred and felling of a green tree is not sanctioned in the code of conduct. Non-Bishnoi communities do equally believe protection provided by the deities and departed souls in sacred groves. Indigenous multipurpose tree *Prosopis cineraria* is worshipped on the occasion of Dussehra celebrated in the month of October. People believe that the goddess of prosperity, Lakshami, sits on this tree on this particular night.

Himalaya

Huge variation in the concept of sacredness is found within the Himalaya. One finds sacred groves ranging in size from 1 ha or less to a few square km in the north-east tribal areas, more so in the state of Meghalaya (Khiewtem and Ramakrishnan, 1989). In the north-west Himalaya, sacred groves are more common in district Kullu of Himachal Pradesh but the size is much smaller than the groves found in the north-east. In some parts of this district, each village has its own sacred grove together with a few larger groves worshipped by a larger community (Singh et al., 1996). There are many sulfur/hot springs (e.g., Vashistha in Himachal Pradesh and Badrinath and Sahastradhara in Uttar Pradesh hills) considered to be sacred. Customarily, people take bath in these streams before performing rituals in nearby temples. In the central Himalaya, all confluences and mountain peaks are considered sacred and human disturbances around these places are restricted. *Cedrus deodara* is considered to be the tree of God and is planted around temples. Flowers of *Saussurea obvallata*, an alpine species, is most valued for offerings. People trek miles and miles to get this flower. On the festival of a local festival U-Khyang celebrated in district Kinnaur of Himachal Pradesh, villagers trek in groups to collect flowers from all around to offer to the god. On a festival *Hariyala* celebrated in Garhwal, Kumaon and Himachal Pradesh, five or seven or more food crops are germinated in small clay pots, kept in a dark place till they etiolate and then offered to the God by each household independently. A huge landscape with snow covered peaks, glacial lakes, river system and varied vegetation types is considered sacred by the Budhists in Sikkim (Ramakrishnan 1996).

Indian Peninsula

Sacred groves are common in the forested landscapes of Andhra Pradesh (Anonymous 1996), Kerala (Ramachandran and Mohnan, 1991), Karnataka, Maharashtra (Gadgil and Vartak, 1976; Daniels et al., 1993; Roy Burman, 1995), Madhya Pradesh and Orissa. Prayers for snake gods are common in these groves. On the festival of Ganesh Chaturdashi seeds of as many weeds as possible are collected by each family and this could be viewed as a mechanism of restricting weed populations.

TRADITIONAL PERCEPTIONS ON THE FUNCTIONS OF THE SACRED

Broadly speaking, treatment of natural object as a sacred entity means a belief that man is subordinate to the nature. Though direct, indirect, existence and non-use values of species and ecosystems are seldom mutually exclusive, specific functional values are attached to some objects.

Azadirachta indica is valued specifically for its air purification and medicinal values. It is believed that gods residing in this tree are able to prevent as well as cure diseases. Some of these traditional values now have scientific support too. Tamarind leaves are chewed to have a melodious voice but the souls residing in this tree are considered dangerous. The folklore say that one cannot survive long if he sleeps below the tamarind canopy. Souls residing in Champa and *Ficus bengalensis* are believed to be generous and people pray these species for satisfaction of their needs. *Ficus relegiosa* is considered to be the tree of enlightenment. It is believed to quench the thirst of departed souls. Fierce souls are believed to reside in *Ficus relegiosa* and hence its biomass use is uncommon. The individuals who plant this tree are regarded to be superhumans. Since this species could grow even in wall crevices, people consciously clean the living space to avoid the possibilities of its regeneration. Regeneration of *Ficus relegiosa* could imply improper maintenance of the living space. *Ficus bengalensis* is valued for its giant size, long life and generosity. It symbolizes generosity as it provides shelter and food to a variety of organisms including man. While *Ocimum sanctum* is worshipped for all purposes and every day, other species are generally worshipped for specific purposes and during specified periods. Most of the sacred species grow in energy rich open environments, have wide geographical range of distribution and abundant natural regeneration.

Religious importance to weeds and poisonous species like *Datura stromanium* and *Calotropis procera* implied conservation of plants whose consumptive uses were not known to the humans. This practice could be viewed as a traditional way of enforcing conservation for the sake of posterity.

In general, system scale sacred entities (sacred groves/forests, water systems and landscape) are worshipped by a larger community whereas small scale sacred entities (sacred individuals or products of sacred species) are worshipped independently by individuals or families. Majority of sacred groves are essentially sacred for people in immediate vicinity. There are a few that are sacred for the regional or global society. Sacred forest of Hariyali Devi is worshipped by the entire Garhwali society. Nanda Devi peak is worshipped both by Garhwal and Kumaon society. The entire Himalaya is considered to be sacred by the global Hindu community. Since the historical times the head priests in the most sacred areas viz. Kedarnath and Badrinath in the Uttar Pradesh hills and Pashupatinath in Nepal are from the south (particularly Namboodri village in Kerala), though the land has always been ruled by the native kings. All along the trek routes from foot hills to the shrines in the Uttar Pradesh Himalaya, feudal lords had donated some lands, called Gunth lands, to the religious authority. These Gunth lands were meant to meet the food requirements of the devotees from different parts of the country and of the priests living in and around the shrines. These cultural practices could enforce cohesion in a large heterogeneous Indian society together with least human interference in some ecosystems. In some remote high hills in the mountains settled by the Buddhist or animist tribal societies, land rights traditionally rest with the religious supreme enforcing a feeling in common man that the entire landscape is sacred. Some areas are excluded from use by the society and are termed as Gompas. Religious supreme allots land considering the requirement of a given family, as assessed over a period of three years and considering equity and sustainable resource utilization in traditional system. While a detailed account of traditional land rights and use system in Tibet is available, Indian high altitude territory remains largely unexplored (Goldstein and Beal, 1990). Thus social and religious functions of sacred areas were linked with their ecological functions.

Concern for water and soil management seems to be integrated with the religious, biodiversity and social motives behind sacred area concept. In semi-arid/arid Rajasthan ponds are integral components of sacred groves. So is the case in many arid/semiarid regions of the Indian peninsula. Sacred groves in the hills, in general, are located on and around hill tops and thus ameliorate the erosive effect of water protecting the agricultural slopes and dwellings downslope. In some areas, sacred forests are the recharge sources of springs, the source of potable water. Ensuring availability of water, the fundamental basis of life, seems to be an important factor forcing protection of the groves from human disturbances.

Though direct material benefits from biological capital in sacred areas have been known to the local communities, existence and no-use values of biodiversity prevailed over the use values in traditional system. While in few cases such as Orans of the desert region or sacred forests in some parts of the Himalaya biomass uses were not permissible at all, limited collection of medicinal plant products and deadwood were religiously sanctioned as the last resort for survival in most cases.

EROSION OF TRADITIONAL VALUES

Sacred areas representing climax forests at present are only few and restricted to Meghalaya, Kerala, Tamil Nadu, Karnataka and Maharashtra. With passage of time, use values were emphasized more and more over the non-use values, reaching to extremes where the original concept of sacredness of natural objects was altogether replaced by the concept of sacredness in artificial objects like temples and the idols or icons therein. Sacredness of ecosystem functions/processes in terms of air purification has been replaced by artificial functions like sacred pyre. The concept of sacred areas changed to the concept of sacred individuals and species in the alluvial plains possibly because it was relatively easy to convert natural ecosystems to agriculture. The Indo-gangetic population which represents the mainstream continued to regard ecosystems elsewhere such as Himalaya or coastal areas sacred. Perhaps, increased demands of food, improved security, intermixing of small scale societies and high population densities led to evolution of the reductionist approach to sacredness. With gradual evolution of state controlled social welfare measures, religious value of sacred natural objects as a source of spiritual power and means of social service belief in sacredness of natural ecosystems was further eroded. With improved accessibility and urbanization, sacred areas turned into tourist places to serve economic interests. *Gangotri, Badrinath, Kedarnath* and *Tirumala* hills are illustrative of this situation.

Ecosystem and biodiversity protection had been on the agenda of the state for a considerably long period of time but the instruments adopted to promote conservation did not match with the local perceptions and values. National policies such as bringing 33% of country's area under forest cover and 5% area as protected area and setting national targets of economic growth did not consider the traditional land systems. The new land tenure and land rights policies did not recognize the age old values of local efforts of conservation. There are a few new private initiatives of establishment of sacred groves but these seem to be driven more by the motive of land grabbing rather than by any socio-cultural or religious motive. Large scale religious transformation to christianity is said to be a major factor causing degradation in many parts of the country but this factor alone cannot explain the erosion of traditional conservation practices.

LOCAL COMMUNITY CENTERED CONSERVATION STRATEGIES

Consumptive and nonconsumptive uses of natural resources have been looked as interdependent concerns by the traditional societies. There are limits to utilization of natural resources in a subsistence economy. Such limits are not a concern in market economy until the threats of exhaustion of the capital or negative impacts of excessive resource uses are felt. The new land tenure/ land rights system considering conservation and production as alternative land use choices at a macro-scale (nation or biogeographic zone) does not match the traditional allocation systems. Conservation targets such as bringing 5% of geographical area of the country under protected area would not have any attraction to the local communities unless they are convinced about weaknesses of their indigenous resource use and conservation practices. Though there is increasing pressure for enhancement of people's participation in forest regeneration/conservation and protected area management, emphasis is more on raising the authority of local communities than on scientific upgradation of traditional and transitional resource use practices (Poffenberger et al., 1995). There is a need of scientific evaluation of traditional conservation practices as all traditional practices may not be the best options in the present day world. Subsistence economies certainly cannot survive in the new circumstances. Monetary economy is increasingly becoming an attraction to the traditional people too.

Convention on Biological Diversity essentially treats biodiversity as a means of economic development rather than as a service to humanity. Traditional knowledge and conservation practices have served as the source of clews for many pharmaceutical innovations but local communities have not been able to benefit from such indigenous potential (Farooquee and Saxena, 1996; Rao and Saxena, 1996; Dobriyal et al., 1997). Cultural causes alone cannot rejuvenate conservation ethos in the present situations unless conservation is promoted as a means of sustainable economic development. We would contend that the people resisted exploitation of traditionally conserved species and ecosystems (e.g., sacrifice of life by Bishnois at Khejadli in Rajasthan in historical times, mass movement against British forest policy in Kumaon, Chipko (hugging the trees) movement in Garhwal, many other events of opposition to establishment of National Parks where it is mandatory to displace the indigenous population) more for asserting their claims on the resource base they had conserved, for a larger share of monetary benefits from new economic opportunities from the conserved capital and threats to their livelihood likely from exploitation by the aliens unknown to the local communities rather than due exclusively to religious or cultural force.

ACKNOWLEDGEMENTS

We are thankful to the Director, G.B. Pant Institute of Himalayan Environment and Development for administrative support. The views in this article are of the authors and not of the organizations they are affiliated with.

REFERENCES

Anonymous, 1995. *Sacred Groves of Kurukshetra Haryana-Relevance and Perspective*. Regional Centre, National Afforestation and Ecodevelopment Board, New Delhi.

Anonymous, 1996. *Sacred and Protected Groves of Andhra Pradesh.* WWF, Andhra Pradesh. State Office, Hyderabad.

Daniels, R.J.R., Chandran, M.D.S. and Gadgil, M. 1993. A strategy for conserving the biodiversity of the Uttara Kannada district in south India. *Environmental Conservation, 20:* 131-138.

Dobriyal, R.M., Singh, G.S., Rao, K.S. and Saxena, K.G. 1997. Medicinal plant resources in Chhakinal watershed. *Journal of Herbs. Spices and Medicinal Plants, 5:* 15-27.

Farooquee, N.A. and Saxena, K.G. 1996. Conservation and utilization of medicinal plants in the high hills of central Himalaya. *Environmental Conservation, 23:* 75-80.

Gadgil, M. and Vartak, V.D. 1976. The sacred groves of Western Ghats. *Economic Botany, 30:* 152-160.

Goldstein, M.C. and Beal, C.M. 1990. *Nomads of Western Tibet—The Survival of a Way of Life.* Odyssey Productions Limited, Hongkong.

Khiewtem, R.S. and Ramakrishnan, P.S. 1989. Socio-cultural studies of the sacred groves at Cherrapunji and adjoining areas in the north-eastern India. *Man in India, 69:* 64-71.

Poffenberger, M., Josayma, C., Walpole, P. and Lawrence, K. 1995. *Transitions in Forest Management: Shifting Community Forestry from Project to Process.* Asia Forest Network, Berkeley.

Ramachandran, K.K. and Mohnan, C.N. 1991. *Studies on the Sacred Groves of Kerala with particular Reference to Conservation of Rare, Endangered and Threatened Plants of Western Ghats, India.* Project Report, Centre for Earth System Studies, Kerala.

Ramakrishnan, P.S. 1996. Conserving the sacred: from species to landscape, *Nature and Resources, 32:* 11-19.

Rao, K.S. and Saxena, K.G. 1996. Minor forest products, management-problems and prospects in remote high altitude villages of central Himalaya. *International Journal of Sustainable Development and World Ecology, 3:* 60-70.

Roy Burman, J.J. 1995. The dynamics of sacred groves. *Journal of Human Ecology, 6:* 245-254.

Singh, G.S., Saxena, K.G., Rao, K.S., and Ram, S.C. 1996. Traditional knowledge and threats of its extinction in Chhakinal watershed in the north-western Himalaya. *Man in India, 76:* 1-17.

Wells. M. 1992. Biodiversity Conservation, Affluence and Poverty: Mismatched costs and benefits and efforts to remedy them. *Ambio 21:* 237-243.

13

Species Composition of Sacred Groves, Their Diversity and Conservation in Bangladesh

A.K.M.N. Islam, Md. A. Islam and A.E. Hoque

Ecology Laboratory, Department of Botany
University of Dhaka, Bangladesh

INTRODUCTION

Bangladesh occupies a unique geographic location. Almost all the mighty rivers such as Ganga and Brahmaputra, drain through Bangladesh and the country functions as a funnel. It is virtually the drainage outlet for a vast river basin complex of the Ganges, the Brahmaputra and the Meghna and their tributaries. During the monsoon season, the country is flooded and looks like a temporary marsh. The damages to crops, houses and humans are immense.

There are different religious groups such as Muslims which constitute 86.6% of the total population of the country followed by Hindus (12.1%), Budhists, Christians and others. About 20 tribes inhabit in various parts of the country. The total tribal population is 8,97,828; of which about 50% are residents in three Hill Tract districts.

In the Country the natural forests are represented by three main types e.g., the hill forests in the east; the Sundarbans mangrove forest in the South West and plainland forests in the central and northern region. In addition, there is also a category of unclassed state Forests in the Chittagong Hill Tracts (0.73 million ha) and have been placed under the control of District Councils. This category has long been subjected to shifting cultivation. The hill forests cover an extent of 0.67 million ha. The dominant tree is *Dipterocarpus* sp. followed by *Sweitinia floribunda, Lagerstroemia speciosa, Hopea odorata* etc. Plainland forests cover 0.12 million ha. This is also known as inland sal forest, *Shorea robusta* is the dominant species associated with *Adina cordifolia. Dillenia pentaphylla, Terminalia bohera, Mallolotus philippinensis.* The Sundarbans mangrove forests (0.57 million ha) are located in the south western region. All these forests are declining at a faster rate. In this context, revival of traditional systems of forest and vegetation conservation is important. Management of scared groves and sacred plants is one such traditional system existing in the country.

SACRED GROVES AND SACRED PLANTS

The idea and concept of protection of patches of woods/forest/herbaceous plant community are ancient. Such groves persist in Indo-Pak-Sub-continent, particularly in locations where Hindus and Muslim are residing. In the shrines of Muslim saints, some plants and animals are considered as sacred. *Plumeria acutifolia* in the Biozid Bostami's mazar (Chittagong) is strongly considered as sacred. The visitors after their worship wrap a piece of cloth on the branches of *P. acutifolia* with the belief that their desires will be fulfilled.

The size and area of the sacred grove may vary with a range from a few trees to a large patch of forest. Some times, even a single tree and its surrounding is considered as sacred. *Ficus religiosa* is one such tree species well known as a sacred in the country. It is of interest to note that the Sacred Groves or sacred trees function as a home for birds and mammals, hence indirectly for conservation of living organisms.

In various parts of the country, a number of sacred groves/trees were noted. In few of such places removal of any plant material is strongly forbidden. In the Sundarban mangrove forest, approximately 1 km^2 vegetation on the extreme southern tip called Dubla Island, is a sacred grove. Once in a year (7th November) thousands of Hindus (low caste) come to the Dubla Island and donate sweets on a tidal land, pray and then leave. The vegetation in 1 km^2 is highly sacred to the people. The dominant tree is *Excoecaria agallocha* followed by *Heritiera fomes. Bruguiera sexangula, Ceriops decandra* and on the sandy beach some members of Gramineae are also present. There may be many such groves in the country which should be explored and should be protected in their natuaral form.

Some societies give strong emphasis to a large number of natural and wild plant species for a number of reasons; for example food and medicine, religious beliefs, selected trees for meditation. For example *Ocimum sanctum* (Tulshi) is a sacred plant and common in all Hindu houses, not only as Goddess but also as medicinal plants particularly for cough. *Sesbania grandiflora* is cultivated in Hindu house, for the following reasons:

i) The plant flowers all through out the year; flowers are used during worship (Puja) and the festival of Hindus is spread over the twelve months. Flowers are also used as vegetables.
ii) The plant is dome shaped and its fruits are used as vegetables.
iii) Leaves are used as fodder; the plant is fast growing and provides good fuel wood. The plant increases soil fertility through nitrogen fixation and grows well in fallowland, marginal land.

Azadirachta indica, Glycosmis arborea and *Streblus asper* are highly valued in the village community. The branches of all these species are used as tooth brush (Meswak). Village people have the belief that cleaning of teeth with the leaves and branches of *Azadirachta indica* means even in the old age the teeth will persist. *Sarcolobus globosus* (Bowali lota) the tribal community is highly valued as antifertility drug. Fruits are sold in the market and is usually eaten by female members of the family to avoid pregnancy. *Ficus racemosa* is considered as sacred in the North Bengal areas; Fruits are sold in the market and are eaten as vegetables to control diabetics. In the Hindu temples *Garcinia xanthochymus* (Tamal; medium sized tree) is very common. The twigs and woods are used in fire (Jagga) by the priest (Takhor) in the beginning of worship (Puja). *Heteropanax fragrans* a rare plant of the deciduous sal forest is considered as sacred for its medicinal value by the tribal (Garo) people.

CONCLUSION

Rodgers (1994) has stressed the importance of identification of and sociologically recognised plants as they are often ecologically important as well. Use of such species for rehabilitation programme will give desired results as one can expect a stronger participation of the local community for such attempts.. The easiest way in this direction is the thorough inventory and documentation of sacred groves and sacred plants, and revival and establishment of new sacred groves in the region.

REFERENCE

Rudgers, W.A. 1994. The Sacred Groves of Meghalaya. *Man in India, 74:* 339-348.

14

Sacred Sites in Bangladesh: Country Report

A.B.M.E. Hussain

Department of Botany, Jahangirnagar University, Savar
Dhaka-1342, Bangladesh

Bangladesh, although geographically a small country, possesses as a rich cultural diversity and a cornucopia of several ethnic minorities. In the country many sacred groves or sites are being protected and safeguarded through time immemorial, either by the acts of worshipping, adoring, and religious/ spiritual beliefs, or by offerings/rituals. At present these sacred sites/ landscapes, either small or large in size are distributed in a scattered manner throughout the whole country. While some of these sites are associated with one or few trees (e.g., Dhakeswari Temple in Dhaka). The hillock with so many graves of known and unknown followers of Hajrat Shah Jalal encompasses a good number of various tree species, notably *Azadirachta indica, Albizia lebbeck, Melia sempervirens* and *Phoenix sylvestris*) in Ramkot Vihara, an ancient Buddhist temple and cultural centre, known as 'Ramkot Vihara and Forest home' a dense vegetation with a water body is existing. Mahasthangarh is yet another sacred place where the rich vegetation is still existing which needs to be studied for the species richness and diversity. Chandranath temple and Bholanath temple at the top of the hills near Chittagong town studied possess many tree species and also some rare or threatened plant species. Near the Kantanagar Temple at Dinajpur there exists a big lake called Ramsagar. Recently this lake has been declared as a protected area under wetland ecosystem by the Ministry of Forests and Environment of the Government of Bangladesh.

There are many more sacred sites associated with vegetation in the country. But unfortunately, no thorough investigation or even a preliminary survey is undertaken on their status and importance in conserving biodiversity and cultural diversity.

15

Role of Sacred Groves in Conservation and Management of Biodiversity in Sri Lanka

H. Withanage

Environmental Foundation Ltd., No.3, Campbell Terrace
Colombo 10, Sri Lanka

INTRODUCTION

Sri Lanka, an island of 65,610 Sq km situated in the Indian Ocean off the South Eastern Coast of India, famed from antiquity. It was referred to about four thousand years ago in the Hindu epic, the Ramayana, as the domain of the ten-headed demon king Rawana. This country is rich in natural resources and biological diversity with about 3000 plant species, 430 bird species, 84 species of mammals and 133 species of reptiles. Compared to any other country in Asia, Sri Lanka has more endemic species; 27% endemics among the angiosperms and 14% of mammals, (Kamini and Hemantha, 1999). Like in many other tropical countries, destruction of the Sri Lanka forests is one of the most pressing problem which has to be addressed today. At the beginning of the century about 75% of the land area was closed in dense forest cover. This dwindled to about 50% in 1945 and now stands at less than 20%. This has had a direct impact on fauna and flora and many species face the threat of extinction due to loss of their natural habitat. However, it may be pointed out here that in the country Buddhism has had a profound influence on the attitude to plant and particularly animal life. Over the centuries there have been wars, invasions, quarrels and killings of among humans but the basic fact that we must respect the right of other living beings to live is still present among the majority of Buddhists, the dominant inhabitants of the century, and has been one of the most important influences for the preservation of a diverse life forms on this island. In the present paper the role of religion in conserving biodiversity through different ways and means is discussed.

ROLE OF RELIGION ON CONSERVATION

The people of Sri Lanka are of diverse ethno-cultural composition. The Sinhalese form the dominant ethnic group. The Tamils rank next in size while the Moors, Malays and Bergers are small but significant minorities. The majority of the Island's inhabitants are Buddhist by religion. Hindus are also numerous. Christians and Muslims are few by comparison.

Although the Island's recorded history begins with the coming of the Sinhalese, they were not its first inhabitants. Early Sinhalese chronicles refer to groups such as the Yakkhsas, Deva,Naga, Rakshasas whose existence in Lanka finds corroboration in the Hindu epic, the "Ramayana". Archaeological evidence indicates that the earliest known inhabitants belonged to Stone Age cultures. Of these, Neolithic man has been identified with the greatest degree of certainty. Stone Age man probably engaged in gathering, hunting and fishing and may have practised primitive non-sedentary forms of agriculture. Living reminders of the Stone Age are the Veddahs (forest dwellers), now on the verge of extinction and complete assimilation by more modern groups. The inhabitants of pre-history appear to have been derived from Australoid Negrito and Mediterranean stock. These were duly absorbed by immigrant groups from northern and southern India (Wijesekara, 1965).

The Sinhalese comprise the largest ethno-cultural group in Sri Lanka. They speak Sinhala, which is derived from the Indo-Aryan languages, Pali and Sanskrit. They claim Aryan antecedents and popularly trace their origins to the Indian Prince Vijaya whose story is enshrined in the Buddhist chronicle, the Mahavamsa, which is the story of the early Sinhalese. This account tells of Sinhabahu and Sinhasivali, the offspring of the princess and a lion in northern India (Bernard Swan, 1987).

In the 4th century BC the island was colonised by Prince Vijaya and 700 of his followers who were said to have been banished by Vijaya's father, the king of Sinhapura. A princess of the Yaksha tribe betrayed her people for Vijaya, only to be discarded by him in turn as he sent to India for brides for himself and his followers. Thus according to the ancient chronicle the Mahawamsa, the Sinhala race was founded.

In very early times the Veddahs occupied much of the forested area of Lanka, particularly the Wet Zone as the Dry Zone was the more populated and agriculturally developed area in the ancient Kingdoms of Sri Lanka. Sabaragamuwa (Saparagamuwa), which includes a large part of the

Wet Zone Montane forests, means "land of the wild people". With the clearing of the Wet Zone and its settlement by the Sinhalese and others, the Veddahs were driven to the Dry Zone which had by then gone back to the jungle. During the rebellion in the Uva Province against the British in 1818, it is said that the Veddahs too fled the province and went to the eastern part which was then heavily forested. This area (Bintenne) then became their homeland and it is here that various people came to study and write about them, for example Dr. C.G. Seligman in 1911 and later, Dr. R.L. Spittle and Gamini Punchihewa with whom we spoke.

Before the approach of cultivation have destroyed their habitat, they were all forest dwellers. As with forest dwellers around the world, although they use the products of the forest, they are the chief protectors of the forest as well, for that is their home and their chief source of food obtained by hunting and gathering. What they do not require as food they leave alone, never killing for the sake of killing though very determined when they must appease their hunger (Kamini and Hemantha, 1991).

In the 3rd century B.C. the Arahath Mahinda, son of the Emperor Dharmasoka came to Sri Lanka bearing with him the teachings of Gautama Buddha who lived and preached in India in the 6th C BC. The story of Mahinda's meeting with the King Devanampiyatissa who was on a hunting expedition is probably the best known legend in Sri Lanka. The words of Mahinda to King Tissa illustrates best the Buddhist attitude towards life, particularly animals: "Oh great king, the birds of the air and the beasts have an equal right to live and move about in any part of this land as thou. The land belongs to the people and all other beings and thou art only the guardian of it."

Mahinda was sent to Sri Lanka to preach the doctrines of Buddhism, (one of the first recorded missionaries) and his mission was a total success. The King and all his court were converted to Buddhism and the religion was established here and is still the religion of 70% of the people of the country.

Soon after, Mahinda sent for his sister the Princess Sangamitta who brought with her a branch of the Bo tree (*Ficus religosa*) under which Buddha mediated as he reached enlightenment (Nirvana). Ever since, the Bo tree has been venerated in this country, every temple having one around which lamps are lit and flowers offered. No Buddhist will willingly cut down a Bo tree or even lop a branch if he/she can help it. Sangamitta Theri also established an order of female monks.

Many kings in the ancient times have decreed that no animal shall be slaughtered within certain limits of the city etc. King Nissankamalla 12th C. AD and King Aggabodhi C. AD. are two such kings. The Emperor Dharmasoka after his conversion to Buddhism decreed that certain animals should not be killed. He had many decrees carved on stone pillars. For example "not a single living creature shall be slaughtered or sacrificed. Even husks (of coconut) with living things therein should not be burnt"

Buddhism considers humans as the highest life form among animals. But humans can also be born as animals according to their Karma and *vice versa*. The Buddhist attitude should be one of Maitriya (loving kindness) towards all beings {animals (human and non human)} and this attitude is best expressed in the teaching known as the Karaneeya Metta Sutta (the preaching of the Buddha are recorded as "Suttas").

Where the following line is often repeated "Sabbe satta Bhavantu Sukhitatta "May all beings be happy and free from fear." All beings are embraced in this Maitriya or loving kindness. "The long, the very large, the medium sized and the very small-those visible and those invisible-those far away and those near us, the born and the yet unborn- may all beings be well and happy and free from fear". Again "Even as a mother would cherish her only child with her life, let him/her cultivate a boundless love towards all beings".

The Buddhist view of the unity of animals and man is also borne out by various incidents in the Jataka (re-birth) stories, where the Buddha aspirant Bodhisatva passes through several births on his journey to becoming a Buddha. He is said to have been born as Naga Raja (King of Cobras) as an eagle, as leader of a monkey clan, etc. The stories depict these animals (and particularly the Buddha aspirant) as being as wise or wiser than the humans (Kamini and Hemantha, 1991).

Among the five "precepts" that the Buddhists repeat (similar to the commandments except here one undertakes to do certain things) the first one is "I will not destroy life". The precept is taken very seriously by the majority of Buddhists. So much so that the thought of killing, even to put an animal out of its suffering is reprehensible to many Buddhists. Also it is firmly believed that if one kills any creature, however small, it is an "akusala", which may be translated as a sin- an unmeritorious act and therefore not good for your "Karma" which you will take with you to your next birth. Most sufferings or happiness is thought to be due to your actions in the past birth or even in this life.

This precept has led to the establishment of a national consciousness against killing animals of whatever form, unless necessary. Although in this modern age pesticides are used rather freely for crops and in the home, in the past one tried to lure them away with Mantras (chantings). However, animals like the Russell's viper will be killed if it happened across as it is treacherous and thought to attack without warning or reason. But one will organise search parties to go out and kill every viper in an area because one person was attacked.

Certain animals are never killed unless there is danger of attack. Indeed one may say that this is true of most animals. A policy of live and let live operates. A cobra is almost never killed. The killing of a cobra is a de-meritorious act. Legend has it that as Buddha sat meditating before he reached the state of enlightenment, a thunder storm broke out and it was also very cold. The king of cobras -Muchalinda- came and wrapped himself round the Buddha, protecting him from the cold and also spread its hood over his head to shield him from the elements. Buddhists are always grateful to the cobra for that kindness to their teacher.

There is no doubt that the respect for all forms of life and the feeling that all creatures have the right to live and that life should not be taken away stems from the religious teachings of early childhood.

With regard to plant life there are many stories which reflect the attitude of the Buddha to plants. One of the best known is his comment on the forest, quoted in the "Forester" journal of the Department of Forest Conservation: "The forest is a peculiar organism of unlimited benevolence that makes no demands for its sustenance and extends generously the products of its life activity. It affords protection to all beings, offering shade even to the axeman who destroys it".

Another story is that Lord Buddha requested a young novitiate monk to go out into the forest and bring him a plant or animal that he considered totally useless. The young novice spent several days searching, and returned saying to his teacher that he had failed in the task as he could not find a single plant that was not of some use. Buddha said to him that he had not failed, but learned the lesson which he wished to teach that there is no plant or animal that is not beneficial to any other plant or animal.

There are many quotations from one of the books of scriptures, notably the Tripitaka. There is one sutta which is called the Ruk Ropana Sutta-meaning the sutta regarding the planting of Trees. In this it says that a person approached Buddha and asked him "who would you say is a man, who day and night, does good (ping) meritorious work ?" Buddha answers "He who would grow and create forests and flower gardens .... who plants and gives forests ..... such a person, day and night, does good."

The attitude of being grateful towards trees, particularly to the species *Ficus religiosa* (Bo tree) for services rendered, i.e. providing shade and cool while he meditated under it, was demonstrated by Buddha. After having attained enlightenment (the state of Nirvana) Buddha paid homage to the tree for seven days.

Ficus religiosa, the first plant which was brought to Lanka by the princess Sangamitta, is now widespread and highly venerated in Lanka.

Lamps are lit round many Bo trees and it is difficult to find a Buddhist who would cut a *F. religiosa* even if it was in the most awkward place. Other *Ficus* sp. notably the Banyan, also appreciate a share of veneration, but mainly because certain gods (Devas or demons) are thought to reside in them.

Buddhism also teaches that one must be free of desire (Thanha)- the desire to possess. It is to teach this concept that giving (Dana) is held in such high esteem as being the most desirable of virtues to be practiced by Buddhists. With this is also brought the concept of consuming less. "Buddhist monks are expected to live frugally. It is forbidden in the order to eat after mid noon. They should wear the simplest clothes. In the days of Buddha, the monks gathered clothing which they sewed together and dyed with natural dyes from the forest trees such as *Artocarpus heterophyllous*. They wear only sandals or more often go barefoot. An umbrella and a fan made of palm fronds should be their only possessions. Therefore, a virtuous Buddhist monk is one of the smallest consumers in the world. Unfortunately over the years the enthusiasm of kings and other groups of the laity for Buddhism has led to some monks being given all that Buddha forbade them regarding comfort, food and worldly desires.

In the Vinaya Pitaka (the book regarding discipline of monks) it is stated that a fully ordained bikkhu (monk) should not deliberately kill, even an ant or an unborn animal. He should not drink or use water that has any small animals in it. Cutting trees or having them cut is also forbidden to bikkhus. In the Themiya Jataka it is said that he who has slept under a tree and then cuts branches from it commits an act of betrayal. Rev. Pannasekera also says that the belief that certain divine beings (devas) are associated with trees is given in the teaching called the Vanasanyukta Nikaya (teachings regarding the forest.) This was necessary as the monks spent much time in the forest in meditation. The teaching states that Bikkhus who choose a forest for meditation should choose one in which there is a diversity of trees (Kamini and Hemantha, 1991).

SACRED GROVES AND TEMPLE FORESTS IN SRI LANKA

The forest was considered as a place of peace and healing. When monks got restive and difficult it is said that Buddha himself led them to a forest in the Himalayas. The beauty and calm of the forest set their minds at peace.

Buddhist religion plays a great part in traditional and cultural practices of the country as it has been present here for over 2000 years. Intermixed into these are also old beliefs in Gods and Demons which were also present in the time of Gautama Buddha, and which were not totally discarded by him. In the stories of that time, certain celestial beings called 'devatawa' were frequently referred to, and even some of the suttas (discourses) are in response to questions asked by these persons/gods. Gods are generally good, beautiful, kind and many of them have selected trees which they inhabit. However, they are known to get very angry if the tree which is their home is harmed. These trees then should not be touched, unless absolutely essential. If even a small portion is propitiating gesture shall be made to the inhabitant. Any tall majestic tree could be said to be inhabited by, or be under the authority of, a 'devatawa' or a demon (*yakka*). Certain trees are more likely to be affected than others, e.g. Na (*Mesua ferrera*), Nuga (*Ficus* spp.), Rukkhathana *(Alstonia scholaris*) Bo (*Ficus religiosa*) Sapu (*Cananga odorata*) and Ahala (*Cassia fistula*).

There are many stories related about Bikkhus and others who cut these trees at the time of the Buddha and consequently suffered a terrible fate as a result of the wrath of the gods who lived in the trees. According to the Buddhists there are six types of deities associated with the trees. They are Pushparama-associated with hermitages and the trees around them; Andavana-associated with the forest; Naleru- associated with certain trees; Aralu, Nelli-Associated with medicines; Thal, Pol-associated with palm trees and non-flowering plants.

Before the arrival of Arahath Mahinda in 3rd century B.C the concept of sacred groves was not established in Sri Lanka. However there is evidence that the ancient tribes believed and worshipped trees, forests, stones, wind, rain, fire, moon, sun and stars as well as gods, devils, the souls of their ancestors and many other objects and powers. But after the arrival of Arahath Mahinda this was slowly changed. According to Mahawamsa, in the 3rd century B.C. Mahamewna Uyana, which is in the city limits of Anuradhapura (which was built by Minister to the King Pandukabaya name Anurudha) was established and handed over to the Monks. Later the most venerated Bo tree was planted in this Mahamewna Uyana. This is the first sacred grove established in Sri Lanka. Today most of the chethiyas and other ruins can be found within these limits.

Mihinthale, Samanola peak (Adam's peak), Dimbulagala, Sigiriya, Vadihiti Kanda, Sithul Pawva, and many other places where we find monasteries and worship places were established as sacred groves. At present there are thousands of temple forests (sacred groves) but unfortunately they are dilapidating very rapidly due to the rapid increase of population and the consequent demand for land and resources, etc. At present these sacred groves are under threat. Some of these sacred groves are already protected under the Fauna and Flora Protection Act as Strict Nature Reserves, National Parks, Sanctuaries, etc. (Forestry Planning Unit, 1991).

DOMBA GAS KANDA FOREST RESERVE (BODINAGALA)—KALUTARA

Dombagas kanda forest reserve is located in the wet zone of the south-western part of Sri Lanka. The area is well known to the public because of the monastery situated within the reserve. A detailed study was conducted by the Young Zoologist Association of Sri Lanka from 1990–1991 and the report is available. Vegetation of the forest reserve comes under the category tropical rain forest.

The Monastery has a history which dates back many centuries. There are many legends attached to it. The earliest known human remains in Sri Lanka which date back 34,000 years (Balangoda Man *Homo sapiens balangodensis*) were discovered recently in the Pahiyangala cave located about 4 miles south-east of the reserve. Invertebrates, 75 butterfly species were recorded. Icthyofauna 36 spp., 9 endemic amphibia 16 spp., 6 are endemic., reptiles out of 8 endemic species of Agamids 02 were recorded. Out of 18 endemic snakes 03 were recorded Avifauna 151 being recorded of which 9 are endemic.Mammals 15 spps. There are 300 plant species in the reserve.

KALUGALA PROPOSED RESERVE

Located in Western province, Kalutara District, Matugama forest range, it can be reached from Baduraliya, second largest forest reserve with an estimated area of 3449 ha of which 239 ha ridge top forests. Only 55 ha are close canopy forests. 2471 ha are heavily exploited and 322 ha are scrublands without trees. 40 ha are planted forests and 320 ha comprises of cultivated areas. An Aranya (hermitage) with about 15 monks occupy 100 acres, popular place of pilgrimage and a meditation center. Streams draining the eastern slopes feed the Pelaeatte ganga and few streams from the western slopes join Maguru ganga.

Floristically a very unique forest with *Dipterocarpus insignis, Cotyleobium scabrisculkum, Vatica affinis, Schumacheria angustifolia, Denrobium macarthiae, Raphidophora pertusa* and *Mapania immera*.

Here 30 fish species of which 12 endemic, 9 Amphibians spp., of which 2 endemic, 34 Reptiles of which 19 endemic, 55 Birds sps. with 12 endemics and 37 Mammals with 4 endemics are recorded in the region.

Yagirala Forest Reserve

Located in western province, total area of the Forest Reserve is estimated as 2658 ha. Mean elevation of the forest reserve is 143 meters from air photos. In the forest department management plans the extent of ridge top forests is 2043 ha. There is no good closed canopy forests. Heavily exploited forest areas constitute 1653 ha and scrubland 127 ha. An Aranniya (hermitage) is situated within the forest and the road leading to the Hermitage has many interesting species of endemic and indigenous species.

Viharakelle Forest Reserve

It is located in southern province Matara district with 1,400 acres of forest. A total 175 species were recorded along the trail from one temple to the other. These species represent 161 genera and 57 families; 70 species of plants are endemic.

Some sacred groves located within forests are protected under the forest ordinance. However, there are many other groves which are not protected under the Fauna and Flora Protection Act or under the Forest Ordinance.

Legends and Beliefs about Sacred Groves

There are a number of legends and beliefs regarding sacred groves. Many people witnessed lights at night which move from one tree to another, one forest to another or one sacred place to another sacred place, and especially on "Kemmura days". Kemmura days are Tuesdays and Fridays according to the local astrology. People believe that these lights are the sign of gods and other deities.

In some forest areas once you enter and harm it you cannot find the way out. According to the belief, if you walk over "Man Mula Vel"(a creeper) you cannot find the way out.

In some sacred groves people believe that there are treasures which were buried by the ancient Kings and other Aristocrats. According to legend, that king of aristocrats came with an honest worker who dug pit or cave in a rock base and put the jewellery and gems and other treasures there. Then asked the worker to protect it and kill him and bury the body next to the treasure. After that he became a "Bahirawa" and protected the treasure. If someone tried to take the treasure he would face many problems, such as elephants or other wildlife attacks or other unexpected powers. In many sacred groves these kinds of places are known and people who believe these legends never go to such places. (But today many people try to remove them.)

Also people believe that deities live in the large trees of some forests and if they cut those trees or clear the area they will have to face many problems. If people want to cut the forest such as for chena cultivation they worship the deities before cutting.

However, although the old generations believed these stories and protected these sacred groves, present generations spoiled by the market economy do not believe and do not protect them. Therefore most of the sacred groves are under threat today.

Among the thousands of sacred groves some are popular while others are not. But since the forest cover and the diversity are reducing very rapidly, it is important to protect all these areas. Therefore past knowledge and familiarity with legends are very important in establishing these cultural values.

SACRED HILLS

Ritigala

In the plains of the north and east isolated hills are found. Two of the most historically and ecologically interesting are Sigiriya, the rock fortress and abode of one the 5th century kings which contains some of the best preserved and lovely frescoes in the country, and Ritigala hill {Ritigala kanda (572 m)}, ecologically very interesting as it shows floristic and faunal characters of both wet and dry zones. Legend has it that when King Rama (of the Ramayana) and his brother Lakshmana battled King Ravana of Lanka, Hanuman the monkey warrior was sent to fetch a herb growing in the Himalayas in order to treat Lakshmana who had been wounded in battle. When Hanuman got to the Himalayas he did not know how to identify the herb, so he scooped up a whole chunk of the Himalayan mountain and placed it on the top of Ritigala kanda. While bringing this- he travelled by air- a piece is said to have fallen off and dropped on to the southwest of Lanka and is today known as Rumassala which is said to be rich in medicinal plants. The top of Ritigala is different from the dry zone vegetation of the lower slopes, hence this legend (Kamini and Hemantha, 1991).

Ritigala Kanda (2514 feet) referred to in legends of the Ramayana, has also been preserved because of the tales of Yakshas or demons inhabiting it. Dr. R.L. Brohier in his 'Seeing Ceylon" relates the tale of the peasant who got lost in the jungles of Ritigala and was warned off the place by a Yaksha who first gave him food and drink, then told him to never set eyes on the area again. He was also told that he could go home, 'but nothing must be taken out of the forest.' Legends such as these have preserved the unique vegetation of this peak.

From its complete isolation and abrupt rise on all sides, it presents a more imposing appearance than would be expected from its actual height (2514 feet). Yet strangely, its summit is frequently bathed in mist-more likely during the south-west monsoon when the country surrounding it is parched and dry. The story is so old that it is associated with Hanuman traditions of the Ramayana epic, which was, written more than a thousand years before the Christian era.

Ritigala is the highest ground intervening between the central mass of the Ceylon mountain system and the very similar hills of Southern India. Hence the legend that Hanuman jumped to India from Ritigala to take the joyful message to Rama that he had discovered where Sita was being held captive in Lanka by the king Ravana.

The legend apparently sparks from the fact that the cap of Ritigala really does present a characteristic little oasis of vegetation distinct from the dry zone forest covered slopes lower down. Even in recent, to honour this legend, holy sanyasis braved the arduous climb to search the summit of Ritigala for a herb they called "sansevi" which possessed the powers to confer long life.

More recently, but still in the distant past, Ritigala which called Arishtha—the Arittha Pabbata (dreaded rock) of the Pali histories-was one of the principal low-country territorial abodes of the "yakkas", as the aboriginal veddas of Ceylon were called. The Mahavamsa (Geiger, ch. X. p. 72) presents a graphic description of a great battle which took place between Pandukabhaya and his uncles around 307 B.C., in which the Yakkas of Ritigala rendered much assistance to the young warrior, who became victorious. They received in turn much favour when he later became king. On Pandukabhaya's death, the chieftains of this clan of aborigines lost their influence and appear to have been gradually driven away from their habitations by the increasing Indo-Aryan population (Brohier, 1965). It so happened, that when Buddhism was firmly established in Lanka, Ritigala, which was by then devoid of habitation, was selected as a suitable spot for building viharas. The Mahavamsa mentions two: the "Lanka Vihara" (177 B.C.) and the "Arittha Vihara" (50 B.C.).

A thousand years later Sena (A.D. 831) added many sacred buildings which adorned the mountainside, and bestowed them all, together with grants of land, on "the humble Pansukulika order of the priesthood" -by which is meant "an order of bhikkus who had taken a vow that the robes they wore would be made of rags from refuse-heaps or from cemeteries, and pieced together". The Mahavamsa also tells that the king gave to this ecclesiastical establishment "royal privileges and honours, and a great number of keepers for the garden, and servants, and artificers".

This rich history illustrates that Ritigala is a spot steeped in history, and one of a few of the older historical names still being used in Ceylon. It is therefore of great importance from a philological angle.

There seems little doubt that this institution which was earlier used as a stronghold by contending aboriginal clans, and later as a place of refuge for fugitive princes and religious devotees, was laid waste during the Chola invasions of the 11th century. Ever since then the forest, which grew on slopes of this hill-range, have hidden from view much which bears testimony to the truth of history and legend. Until recently (if not even so today) there was a belief prevalent amongst the scanty village population in the neighbourhood that Ritigala was haunted by the spirits of the Yakka tribes who originally inhabited it.

The story was passed around, that one day a villager living nearby was benighted on the fringe of jungle at the base of the hill-range. Seated half-dozing under a tree in the darkness, he was startled by the sounds of the barking of dogs, the cries of children, and the usual bustle of a busy village. He thought it strange, knowing that there was no village nearby, and was bewildered when a little later a Yaka in the form of a man carrying a lighted chulu in one hand, and a bath-mula in the other, approached and laid before him a large quantity of rice and curry, with plantains and oranges. The benighted villager was enjoined to eat his fill, but sternly warned to depart in a southerly direction before the day dawned, and not to take away any of the food left uneaten lest some evil befall him. The villager was petrified by the apparition; however, being hungry, as soon as his fear subsided he ate as much of the food laid before him as he could. Thereafter, before the first glimmer of daybreak, he got up and proceeded in the direction he was bidden. It did not take long before he found himself back in his village.

On being reprimanded regarding his overnight absence, he told his fellow villagers with extraordinary clearness all that had happened to him. Of his stream of words all they remembered was the prohibition that 'nothing should be taken out'. Few villagers therefore ever roamed the Ritigala forests in search of honey, or game, or wild-fruit, or even to gather brush-wood, for fear of encountering the wrath of the "Yakka spirits". As a consequence, the entire hill-range remained for many centuries more or less untrodden by man.

The first person in recent times to break the taboo and invade the peace of this singular low-country hill-top, was a survey officer, James Mantella, who was stationed on trigonometrical duties at Ritigala from the 11th of May to the 10th of July 1872.

Apparently James Mantell, besides conferring the sobriquet Kodi-bendapu-kanda (the hill which was flagged), on the summit, had possibly much to say of the salubrity of a plateau 400 feet below Ritigala's wind-smitten zone where he resided for two months. It was perhaps he who kindled the idea that it would make an ideal close-at-hand retreat for officials stationed in the unhealthy climate of Nuwarakalaviya.

Another surveyor, J.B.M. Ridout, who went up in the course of his work, left a record of his visit in the Journal of the Royal Asiatic Society 1892. A.P. Green an entomologist, a botanist, and the Archaeological Commissioner H.C.P. Bell were, as far as is known, the only other persons to climb Ritigala in the last century. The botanist, who made the trip in 1887, the famous Henry Trimen, the author of the rare Handbook of the Flora of Ceylon, published in 1893. He says, in a paper read before the R.A.S. Ceylon Branch, that he availed himself of some leisure to make the ascent knowing that the remarkable vegetation on the summit had never yet been seen by botanically trained eyes. He confirmed that the flora on the cap was characteristic of the hills in the neighbourhood of Kandy-stunted trees draped with pendent mosses and different from those of the low-country hill-tops. Bell, on the other hand, explored the wooded slopes of Ritigala-kanda range, from "end to end" in August 1893. He discovered more than 32 caves and historically valuable lithic records which he describes in details in the Sessions Paper XXXVIII of 1904 (Brohier, 1965).

In entanglements of undergrowth surrounding the main ruins there were lichen-scarred and bleached pillars, exquisitely fashioned steps, colonnaded passage-ways with stone balus-trades—all of which were collectively described by our guide as maligawas. A graceful peace pervaded this sanctuary. A tangible sadness hung in the air and invaded the senses, as one in a deep melancholic contemplation pondering the extreme age in stone and mortar which lay about under one's very feet. Truly, the heavy forest which grows on the lower slopes of Ritigala had been kinder to these ruins than man has proved himself to be.

And of the numerous gal-geval (rock caves which were once lived in)—the very spirit of the place which grips one's affection, I can write much. One of them, called the Na-maluwa which nestles under a battling rock, still exhales a spiritual dignity. From it there is a magnificent grove of very old Nagas (the graceful iron-wood tree) stretching a considerable distance to another cave. The grove of ancient trees too bears testimony to the antiquity of both cave and the images found in them.

Because of its interesting features, ritigala was proclaimed as a strict Natural Reserve by Gazette notification No 8809 of 7 November 1941 and administered accordingly by the Department of Wildlife. Total area of this reserve is 1,528 ha. In the 417 taxa have been recorded (Jayasuriya, 1991) of these, 51 are vascular plants including 28 ferns and fern allies and one gymnosperm. The 337 flowering plants belong to 81 families and 260 genera: Euphorbiaceae 29 spp., Orchidaceae 29 spp., Poaceae 23 spp., Rubiaceae 19 spp., Moraceae 14 spp., Acanthaceae 13 spp., and Asteraceae 10 spp. Endemism is very high relative to the rest of the dry zone; evidently the presence of many endemic wet zone species contribute to the diversity of its flora. Of the angiosperms, 54 spp., (16%) are endemic to Sri Lanka and 3 of these are confined to ritigala. About 28% of the total recorded are known to be used in Aurvedic medicine.

The article also suggest that "In view of geographical location, great habitat and species diversity, high endemism, unique conglomeration of species, role of biological refugium, and the presence of a large number of medicinal plant resources, it is strongly recommended that this reserve should be elevated to a National Heritage Wilderness area.(Jayasuriya, 1984)

Samanola Kanda (Adam'S Peak)

God Saman is believed by Buddhist and Hindu alike to be the tutelary divinity of the Peak Wilderness. God Saman is the only local god among the 3.3 billion gods which originate in Hindu legends. Both the Peak and the mountain bear his name-Samanthakuta and Samanola.

The sacred Rhododendrons which grow on the higher slopes are dedicated to him. In his honour the butterfly takes the name of "Samanalaya". This accounts for the story told whenever masses of butterflies are seen travelling in one direction-that they are all making for the Peak, where they will dash themselves against the precipitous sides and die (Brohier, 1965).

The height of Adam's Peak is 7,360 feet above the sea level. The summit is of elliptic form and is surrounded by a parapet above five feet high. Immediately within the enclosure, which is called the maluwa, a level space of irregular breadth runs all the way round. The centre is occupied by a mass of gneiss about nine feet high at the highest point, on top of which one can see the outline of a footprint.

In traditional belief the Peak had obtained the name Samanthakuta before the second Buddha appeared around 2099 B.C. The name has since been preserved with little variations.

Buddha is said to have arrived at Kelaniya about 577 B.C. and to have stopped on to the Peak. The Mahavamsa records the event as follows: "When the Teacher, compassionate to the world, had preached the doctrine there (at Kelaniya). He rose, and left the traces of his footstep plain to sight on Samanthakuta. Afterward he spent the day as it pleased him, on the side of the mountain with the brotherhood. He said forth to Digavapi." This tradition has become an article of faith.

A curious story is told relating how the sacred mark on the summit of the Peak was discovered. King Walagambahu, who ascended the throne a century before the Christian era, was driven into exile by the Malabar invaders.

For many years he wandered about, a fugitive, in the mountain wilderness, living on herbs and fruits. One day, while in a cave on the slopes of the Samanala mountains, he saw a deer in the distance. Perhaps, for want of something to do, he ventured to approach the animal. But strangely, the deed kept slackening or increasing its pace. or stopping altogether just exactly as the king did, in his effort to approach it.

Eventually they reached the top of the mountain. On the very summit of the Peak the animal, as if by miracle, vanished. Walagambahu hurried to the spot and there discovered the mark on the rock. It was then revealed to the King that in this strange manner the guardian divinity had made known to him the presence of the sacred relic.

It is a popular belief that the true impression was left on a large precious stone (menik-gala) produced for the purpose by Saman and that this relic lies buried beneath the large rock on the summit. The exposed hollow is believed to be an artificial print cut on the order of Kirti Sri Nissanka, a Sinhalese monarch who undertook a pilgrimage to the sacred mountain.

The Hindu tradition is based on a belief that Siva in one of his manifestations retired to this mountain for the purpose of certain devotional austerities and that, to mark the event, he proved authenticity by leaving the impress of his foot.

This memento of the presence of Siva on the spot came to be called by the Hindu Sivaties-Sivan-Oli-Padam (the sacred footprint of Siva). The mountain accordingly was endowed with the name Swargham (the Ascent to Heaven). The Vishnuite directed his devotion to Saman-Worshipped in India under the name Lakshmana.

It should be noted that these Hindu beliefs are not universally accepted for they are not contained in the Puranas. They probably date from the Cholian invasions, A.D. 1205, when a good portion of the mountain regions was occupied by the Hindu sojourners.

According to other traditions the peak is the pinnacle on which Adam alighted when he was cast out of Paradise, and where he remained standing on one foot until years of penitence and suffering had expiated his offence-thus forming the footprint. And so about the tenth century the Peak came to be called Baba Adam-malai, or literally "Father, Adam's Mountain", and became a place of pilgrimage to the followers of the Prophet.

Many more of these old legends and beliefs associated with this solitary eminence remain to be discovered. They cluster around, shrouding it in romance as do the morning mists.

Some of these stories are so old that they have gone beyond recall; new ones take their places. So will it ever be through the successive ages that the ceaseless pilgrimage continues.

ECOLOGICAL ROLE OF SACRED FORESTS

Rice was and still is the staple food of the people and is the main type of agriculture practised to this day. The rice fields are fed by streams which have their origin in the forests. The importance of the forest as a source of water was always recognised. Main forests and village forests which shelter springs and streams were never fully cleared, though forest products were collected and used.

Sri Lankan agriculture was totally based on local species and varieties up to the 1960s until the green revolution was started. Therefore ancestors recognised the value of the forest as a germplasm bank. Sri Lanka had more than 3000 rice varieties and more than 300 banana varieties. Most of other vegetables, fruits, yams etc. have come from the forests. Agriculture also needs sticks, leaf fertiliser, organic weed and pest control plants. Therefore, people respected the forest as a germplasm bank.

Almost all the medicinal plants which are used in Aurvedic medicine come from the forest as well. Therefore the forest as a germplasm bank was widely accepted. People kept the village forests intact. Until 3o years ago all the villages had village forest above the village tank. Paddy fields were situated in the lower lands of the village tank and houses were built on the high land. The village forest was used for chena cultivation. But in old chena cultivation farmers never removed the bigger trees. After one season they moved to another plot of land and let the forest regenerate for 4–5 years before returning to the same land. Bigger trees were viewed as the mother trees, and their seeds regenerated the forest.

But current demand for land to cultivate chena as well as replanting of the same land without letting the soil refertile has been very destructive and the village forest no longer exists today.

SOCIAL BELIEFS AND COMMUNITY MANAGEMENT

In many rural areas, particularly in the dry zone, people still respect forests, trees, gods, other deities etc. It certainly helps protect the sacred forests. However, the people who come from outside are very destructive.

Media also plays a major role in conservation in these sacred groves. They publish interesting stories about these sacred groves time to time. They certainly open the eyes of authorities and the people.

In many areas there are no programmes to protect these groves. Recently A group called RITICO (Ritigala Community Organisation) a community based organisation, started a community based project in Ritigala Hills to protect the valuable Ritigala forest. Because of the medicinal value of the plants available in Ritigala, it is under threat from both local and national level collectors. The idea of the project is to support the surrounding community and manage the forest by reducing the pressure. The project is supported by the Asia Foundation which is administered under USAID.

ECO-TOURISM TO SACRED GROVES

The local tourism industry which exists today for many of these sacred groves are mainly for religious purposes. Buddhist pilgrim groups visit these groves to prepare food for the monks in hermitages. Although they cut trees for fire wood they never cut trees unnecessarily.

However, the tourists who visit these areas should maintain discipline and they should not disturb the monks. The old people maintain these traditions and they educate the younger generations. But most of these groups are coming from rural areas. Regulated tourism should not be a problem to some of these groves.

Unfortunately most tourists both local and international do not respect monks, trees, wild animals so if we lose control when the groves opened for tourism, we just might lose the groves altogether. Because of the above reasons, Monks who inhabit those hermitages do not allow people to go into the forest.

Eco-tourism has been identified by the government, Ministry of Forest and Environment, Ministry of tourism and many other regulated authorities and the necessary policies are already in place.

According to the new National Forestry policy, objective 1.1 "To conserve forests for posterity, with particular regard to biodiversity, soils, water, and historical, cultural, religious and aesthetic values. Policy pare 2..2 states that "The traditional rights, cultural values, and religious beliefs of people living within or adjacent to forest areas will be recognised and respected". Also policy 6.3 para states that "Nature based tourism will be promoted to the extent that it does not damage the ecosystems and as it provides benefits to the local population". Therefore the policies in place are in full support of eco-tourism amidst to several limitations in implementing them.

CONCLUSION

In many sacred groves (temple forests) there are monasteries, Aranayas (Hermitages) and other worshipping places. Monks who inhabit these places protect these sacred groves voluntarily. However most of the sacred groves are under great threat today and authorities or community organisations have failed to protect them. Some drawbacks are the lack of data, demarcation lack of commitment, no man power, high population density, increase of the demand for timber and other forest products, etc.

A number of sacred groves throughout the island need to be studied properly. Under the review of Forest Management plan for environmental conservation some of these were studied but data are not freely available. They include some of the most popular sacred groves such as Salgala forest reserve-Kegalla, Kegalla Sanctuary-Kegalle, Buddangala Sanctuary-Ampara, Kudumbi gala Sanctuary-Ampara, Nimalawa Sanctuary-Hambantota (next to Yala National Park) Madhunagala Sanctuary-Hambantota, Situl Pawva-Kataragama, Vadahity Kanda-Kataragama., Mihintale Sanctuary-Mihintale etc.

A detailed research on sacred groves is very important since the data are not available and since they are degrading and disappearing rapidly. It is unfortunate to mention that most of the traditional values have been erode within the last 50 or so years and the resources are seen now purely for their economic value. Traditional beliefs are now confined to rural areas as well as older generations. But it is very important to uphold traditions and beliefs in order to protect our resources and to establish a sustainable society.

REFERENCES

Bernard and Swan, B. 1987. *Sri Lankan Mosaic, Environment, Man Continuity and Change.* Marga Institute, Sri Lanka.

Brohier, R.L. 1965. *Seeing Ceylon.* Lake House Investments Limited, Colombo.

Forestry Planning Unit 1991. *Review of Forest Management Plans for Environmental Conservation,* August 1991, Forestry Planning Unit, Colombo.

Jayasuriya, A.H.M., 1984. *The Sri Lanka Forester.* Volume XVI Nos. 3 & 4 (New Series). Jan-Dec. 1984. Forest Department, Colombo.

Jayasuriya, A.H.M., 1991. *Review of the Flora and Phytosociology of Ritigala Strict Natural Reserve and Some Suggestions for its Conservation and Management.* The Sri Lanka Forester Volume XX Numbers 1&2 January- December 1991 Sri Lanka Forest Department.

Kamini M.V. and Hemantha W. 1991. *Biodiversity and Biodependence,* Environmental Foundation Ltd., Sri Lanka.

Wijesekera, N.D. 1965. *The People of Ceylon.* M.D. Gunasena & Co. Colombo.

16

A Note on the Sacred Plants of Maldives

A.H. Hussein

Ministry of Planning, Human Resources and Environment
Male (Republic of Maldives)

The Republic of Maldives (latitudes 7°6′30″ and 0°41′48″ S and longitudes 72°31′30″ E and 73°44′54″ E) consists of a chain of 26 coral atolls, are located in the Indian Ocean, 500 km Southwest of the Southern tip of India. There are 1190 islands in the Republic, rising from an aseismic submarine ridge (Chagos-Laccadive Ridge) 300-400 m deep with several major channels around 1000 m deep separating the atolls. These islands form a double chain of atolls, tapering from north to south to single atolls. The capital island, Male, is located in the centre of the country. The islands vary in size from 0.5 sq.kms, to around 2 sq. kms and 198 of these islands are inhabited and 992 are uninhabited.

Mangrove, marshes and wetlands, coral reefs and seagrass beds are the major ecosystem types in Maldives and they are harbouring a rich flora and fauna. A large number of plants available from these ecosystem types are traditionally being used for medical purposes. For the Maldivians such medicinal plants of traditionally importance are also sacred. However, unfortunately no attempt has been made so far to prepare a data base of these plants. Many of these plants are becoming rare or even extinct due to the changing landuse pattern in the country. At the same time, one can also note a drastic erosion of traditional knowledge and practice about use and management of these plants. Therefore, an attempt should be made to document the socio-cultural, socio-economical and ecological features of all sacred plants or medicinal plants of the country.

17

Sacred Groves in Mongolia: Country Report

U. Gongorin

Land Management Department, Ulaanbaatar city, Mongolia

Mongolia is located in the center of the continent of Asia (41°35'-52°09'N latitude and 87°44'-119°56'E longitude). It covers an area of 1,566,500 km^2, making it the 18th largest country in the World. Landscape of this country is diverse covered with high mountains, boundless steppes, vast valleys and the Gobi desert. Mongolia has one of the lowest population densities in the world. Although Mongolia's population (2.3 million people) is small compared to any of it's Asian neighbours, the population growth rate of 1.8% per year is one of the highest in east Asia. Urban population growth has been accompanied by a rapid growth in natural resource consumption. Main economic sector is animal husbandry. 15% of population are rural nomads who are the symbols of livelihood patterns to conserve the nature. Mongolians have local beliefs and practices on environmental protection because their lifestyle makes them to be in close relationship with nature and the environment. Particularly the continued ecological health of Mongolia's grasslands is vital for both livestock and wildlife. In most of the villages in the country sacred groves are common. These sacred groves are protected due to local beliefs and practices. Mongolian people maintain careful local practices and teaching traditional conservation measures to young people. However, due to the major social and economic transformation that began in the early 1990s people of Mongolia have started facing financial hardships. In 1992 and 1993, for example, the gross national product declined by 13% each year. These difficulties bring increased pressure on natural resources. People began to forget the sacred groves and use its resources for income generation. There are no governmental management systems to protect and help local people to conserve such sacred groves.

Apart from sacred groves in the villages, there are a few sacred places, some of which have been declared officially as sacred sites and protected by the Government. Three such sacred places are described below:

1. Khan Khentii Strictly Protected Area

This is the birth place of Chingis Khaan located in the Tov and Selenge provinces covering about 1.2 million hectares with stands of taiga, high mountain forest vegetation and mountain steppe. This landscape is the catchment area of the major rivers like Kherlen, Onon, Tuul Minj and their tributaries. Khan Khentii is one of the last large wildernesses of Mongolia, almost uninhabited and whose mountains are worshipped by Mongolians.

The reserve contains 10% Mongolia's forests and defines the southern edge of Siberia's taiga. Therapeutic hot springs, long used for medicinal purposes, lie along the Onon River and elsewhere in the protected area. More than 1150 species of plants characteristic of both taiga and steppe have been unidentified to date. The protected area is home to more than fifty species of mammals, including endangered musk deer and moose, brown bear, wolf, fox, lynx, badger, wolverine, sable, weasel, roe and elk. The 253 species of birds unidentified to date include whopper swans, spoonbills, great white egrets, and numerous raptors. Twenty-eight species of fish inhabit local lakes and rivers.

2. Bogdkan Mountain Strictly Protected Area

Bogdkhan Mountain Strictly Protected Area is Mongolia's oldest nature preserve. Located on the southern edge of Ulaanbaatar, Mongolia's largest city, the protected area encompasses the beautiful Bogd Uul mountains. The total area of the site is 41, 600 ha covered by mountain forests and steppe.

Although officially protected in 1778, Bogdkhan was first recognized as a sacred mountain, where logging and hunting were prohibited, in the twelth or thirteenth century. The highest point in the protected area, Tsetseegun mountain (2268 meters), is one of four holy peaks that surround Ulaanbaatar.

Species here are characteristic of the taiga, mountain forest, and steppe zones, including over 500 species of vascular plants and animals including, 47 mammals, 116 birds, 4 restless, and 2 amphibians. Five mammal species-including musk deer, roe deer, sable, and mountain hare-are endangered or threatened-as are 20 bird species, and 16 species of plants.

Some 70 families, including reserve rangers, live in or adjacent to the protected area. Also, a large number of Mongolian and foreign tourists visit the Bogdkhan each year. The area also includes hiking trails close to downtown Ulaanbaatar, an astronomical observatory, overnight accommodations at the Nukht Eco-tourism Center and tourist facilities at Manzhir Hiid monastery.

3. Otgon Tenger Strictly Protected Area

The peak Otgon Tenger Uul is 4,021 metres above sea level, covered by eternal snow. It is one of the worshipped mountains of Mongolia. The peak and its environs have been protected since 1992, to conserve the high mountain ecosystem. The Otgon Tenger of the Mountain Range but also its spurs, and lake of Badar Khundaga. The Otgon Tenger has many rare plants including *Saussurea involucrata, Adonis Mongolia,* and Juniper (Juniperus).

Strategies for Conservation of Sacred Groves/Landscapes in Mongolia

The sacred groves and sacred sites in Mongolia have a role to play in the conservation of biodiversity and cultural diversity. At the same time many of these sites are having the potential to develop eco-tourism in the country. However, there is no statistical data and research materials on sacred groves of the country to develop proper strategies for managing and conserving them sustainably. The local system of management of sacred groves and sacred sites are also facing the economic and social difficulties to carry out their functions effectively. In this context, a comprehensive research on sacred groves and sacred place on the national scale is needed. Similarly, involvement of the governmental and non-governmental agencies for educating people about the role of sacred landscapes, developing and implementing plans for of sustainable management of sacred groves/sites is required. It would be appropriate to establish a research center within the Moangolian Green Party in cooperation with UNESCO to initiate new projects to indertake reseach on problems related to sacred groves in Mongolia.

18

Sacred Groves of Kerala—A Synthesis on the State of-art-of Knowledge

P. Pushpangadan, M. Rajendraprasad and P.N. Krishnan

Tropical Botanical Garden and Research Institute, Palode,
Thiruvananthapuram-695 562, Kerala, India

INTRODUCTION

Sacred groves are known under different names in different parts of India country as 'Dev' in Madhya Pradesh, 'Deorais' or 'Deovani' in Maharashtra, 'Sarnas' in Bihar, 'Orans' in Rajasthan, 'Sidharavana' or ` Devarkadu' or 'Pavithravana' in Karnataka, 'Sarpakavu' and 'Kavu' in Tamilnadu and Kerala. Sacred groves range in size from a few trees to dense virgin forests of hundreds of hectares. Ecologically, the sacred groves of India can be considered as a self-generating and self-sustaining ecosystem. These ecosystems with their complex array of interaction, influence the flora and fauna of the region as well as microclimate of that locality. Religious taboos associated with sacred groves went a long way in preserving many of them in their pristine form and condition even today. Existing knowledge on the sacred groves of Kerala state of India has been reviewed in this article.

DISTRIBUTION OF SACRED GROVES IN KERALA

Sacred groves and associated ponds constituted a unique network of ecological system that intervined with the life and the culture of the people of Kerala. 'Ayyappan Kavu' the sacred groves dedicated to Lord Ayyappa, used to be the most common in Kerala in the past. 'Thottamppattu', a devotional song invoking Lord Ayyappa names over 108 important 'Ayyappan Kavus' and also mentions that such Kavus are numerous in the length and breadth of Kerala. Many sacred groves are associated with the temples and ancestral home (Tharavad) temples of Namboodiri, Nair and Ezhava families.

Literature pertaining to the distribution of sacred groves of Kerala is scanty. One of the first scholars to document sacred groves was the first Inspector General of Forests D. Brandis who wrote about occurrence of sacred groves in 1897 (Rao, 1996). The first authentic report on the sacred groves is the Census report of Travancore of 1891 in which Ward and Conner (1927) reported 15,000 sacred groves in Travancore. Historical records, legends and the folk songs, particularly certain devotional songs like 'Thottampattu' sung in praise of Lord Ayyappa throw light on sacred groves of ancient Kerala. `Thottampattu' (believed to have been composed between 500-600 AD) names 108 major 'Ayyappan Kavu's and mention about numerous other 'Ayyappan Kavus' distributed all over Kerala. Only few 'Ayyappan Kavu' survive today. Kochi and Malabar regions have also numerous sacred groves. But no authentic inventory of sacred groves of Kerala was ever done. Ramachandran and Mohanan (1990) listed 239 important sacred groves. Induchoodan and Balasubramanyan (1991) made a survey on endemic plants of sacred groves.

Rajendraprasad (1995) estimated 2000 reasonably well preserved sacred groves in Kerala. Balasubramanyan and Induchoodan (1996) estimated 761 important sacred groves in Kerala with floristic wealth of over 722 species belonging to 217 families and 474 genera. They have also reported that among the 722 species recorded 153 are endemic to peninsular India. The biodiversity potential of sacred groves was found to be very good when compared to well protected evergreen formation of South India. For example 90 km^2 Silent Valley encompasses 960 species of angiosperms compared to 722 species in 1.4 km^2 of the area occupied by the sacred groves.

VEGETATION STRUCTURE AND DYNAMICS

Ecological investigation of the sacred groves of Kerala was carried out by workers like Induchoodan (1988), Menon and Sasidharan (1994), Rajendraprasad (1995) and Rajendraprasad et al., (1994, 1996a and 1996b). Induchoodan (1988) carried out a detailed ecological investigation of 'Iringole Kavu', one of the biggest existing sacred groves in Kerala. Menon and Sasidharan (1994) evaluated the optimum productivity of the sacred grove systems.

The general structure of sacred groves of Kerala is extremely complex. It is in fact the most elaborate of all plant communities in structure and richest in species. The impression of sombreness and monotony of the sacred groves of Kerala is mainly due to the overwhelming predominance of woody plants and to the uniformity of foliage, and the absence of marked seasonal canopy dynamics. The abundance of phanerophytes indicates the non-seasonal continual favourable climate. The maximum similarity between the flora is noticed in south-north direction rather the west-east lowland-highland direction. Only occassionally an aggressive species may find an opportunity for relative rapid expansion. Species like *Vateria indica* and *Hopea ponga* invade disturbed sacred groves.

Physiognomy

The sacred groves of Kerala comprise, (a) trees which grow distinctly in three tiers, (b) shurbs and herbs, (c) climbers and stragglers, (d) epiphytes, and (e) parasites (Rajendraprasad, 1995). Tree lines are stratified canopies. The height of different strata of trees varies from place to place. Thus height of stratum 'A' is about 30 to 40 m normally, but may exceed 50 to 60 m sometimes (common emergent tree species are *Hopea ponga, H. parviflora, Artocarpus hirsutus, Alstonia scholaris, Ailanthus tryphysa, Anacolosa densiflora* etc.). Similarly the height of the 'B' stratum is 20 to 30 meters (common members are *Hydnocarpus pentandra, Strychnos nux-vomica, Holigarna arnottiana, Carallia brachiata, Mimusops elengi* etc.) 'C' stratum consists of plants having the average height to 10 to 15 meters (common members are *Samadera indica, Polyalthia korintii, Lannea coromondelica, Aporosa lindleyana, Calophyllum calaba, Meiogye ramarowii* etc.). Stratum 'A' usually has a more or less discontinuous canopy though there is a considerable variation in this respect due to the growth and spreading of climbers. The 'B' stratum may be normally continuous or more or less discontinuous. The 'C' stratum is always more or less continuous and may sometimes appear as the densest layer. The tree crowns tend to be umbrella shaped in stratum'A' and conical in stratum 'C'.

The herbaceous layer consists of a variety of annuals and perennials, the remaining include woody and herbaceous perennials. Their appearance is also according to their light demand. The luxuriant herbaceous vegetation is found in opening of canopy where photosynthetically active radiation is adequate. In the interior, only shade tolerant species and seedlings of upper stratum members are abundant. Climbers and stragglers are one of the most conspicuous features of the sacred groves of Kerala. They compete actively with the trees for space and light and considerably affect the structure and function of the whole system. The most common members are *Gnetumula, Anamirta cocculus, Cissus pallida, Tetracera akara, Strychnos minor, Connarus monocarpus* etc. Epiphytes constitute a common group of the sacred groves of Kerala. The epiphytic vegetation include mosses, lichens, pteridophytes, orchids and other flowering plants. The tree top epiphytes include mainly Orchidaceae, the curious stragglers (mainly *Ficus* and *Fagrea* spp.) which begin their life as epiphytes and often develop afterwards into independent trees rooted in the ground. Important members are from genera like *Hoya* spp., *Pothos* spp., *Bulbophyllum* spp., *Vanda* spp., *Taeniophyllum* spp., *Drynaria* spp., *Bolbotis* spp. etc. constitute the dominant epiphytic members of the sacred groves of Kerala. The epiphytic semiparasitic members are common in sacred groves which normally include members from the families Loranthaceae, Moraceae, and Loganiaceae. *Semiparasitic Santalum album* is widely seen in the groves.

In the upper canopy members, the trunk, as a rule, is straight and slender and does not branch till near the top. In other strata mixed type of bole structure is noticed. The base in some cases is provided with butteresses and stilt roots. Caulliflory is rare. The bark is generally thin and smooth and rarely has deep fissures. A majority of the mature trees, shrubs and saplings have leathery dark green leaves with entire or nearly entire margins. Large and strikingly coloured flowers are uncommon. Most of the trees and shrubs have inconspicuous often greenish or whitish flowers. Solitary flowers are very rare.

Biological Spectrum

The biological spectrum of sacred groves of Kerala closely resembles normal spectrum of tropical forest except in the case of therophytes. A high percentage of therophytes indicates the human impacts. Sacred groves comprised 45% phanerophytes, 8% chameophytes and 6% geophytes comparable to values generalised by Raunkiar (1934). The other lifeforms are 3% epiphytes and 15% lianas. Floristic wealth of the 168 sacred groves of Kerala comprises of well over 318 species of higher plants belonging to 247 genera and 86 families. The largest family of flowering plant is Rubiaceae. The other abundant and widely distributed dominant families are Fabaceae, Asteraceae, Orchidaceae, Laminaceae etc. When we consider the status of distribution, about 1.3% of the genera are very common in occurrence, 5.2% common in occurrence, 9.3% often present, 19.6% seldom present and 64.6% are rare. *Alstonia, Tabernaemontana, Vateria, Holigarna, Abrus, Sizygium, Ixora* are some of the most comon genera.

The most common annuals are from families like Rubiaceae, Asteraceae, Poaceae, Cyperaceae, Acanthaceae, Amaranthaceae, Malvaceae, Dioscoreaceae, Araceae and *Zingiberaceae*. The shallow rooted annuals appear at the beginning of the rainy season. These sprout after the summer rain and complete their life cycle during the rainy season, because the soil moisture percentage is almost below the wilting coefficient during the summer. Perennial undershrubs such as species of *Dioscorea, Amorphophallus, Anaphyllum, Gloriosa, Curcuma, Zingiber, Costus* etc. show disappearance of foliage with the onset of drought. The sacred groves on sandy soil close to coastal areas have psamophytes like *Ipomoea, Boerhaavia* etc. which possibly control the soil erosion and raindrop impact. Another characteristic feature of these sacred groves of sandy area is the occurrence of *Amaranthus spinosus, Ziziphus oenoplia, Z. rugosa, Calotropis gigantea, Bryophyllum pinnatum* and *Opuntia* spp.

The sacred groves found in the mid and high land area are mostly with sandy loam soil which show presence of *Sida* spp., *Naregamia alata, Cassia* spp., *Kunstleria keralensis, Geophila repens, Begonia malabarica* etc. Many seasonal Pteridophytes like *Selaginella* spp., *Pteris* spp., *Drynaria* spp., *Bolbitis* spp. are also found in these sacred groves. Besides this, a number of cosmopolitan annual species like *Chromolaena, Hyptis, Ageratum, Mikania, Aerva, Cassia, Passiflora, Tridax, Vernonia, Achyranthus, Clerodendron* etc. are common. Fungal fruiting bodies may be found during certain seasons or through the year on soil or diseased trees.

Sacred groves in the high lands of Kerala have typical Western Ghats forest type vegetation. Dominant flowering species include *Hopea parviflora, Antiaris toxicaria* and *Anacolosa densiflora*. On the other hand the sacred groves of midland region are dominated by *Artocarpus hirsutus, Hopea ponga, Hydnocarpus pentandra, Holigarna arnottiana, Vateria indica* etc. The coastal Kerala is characterised by *Calophyllum inophyllum, C. calaba, Carallia brachiata, Artocarpus hirsutus, Samadera indica, Holigarna arnottiana* etc.

The pantropical families are megathern families that are exclusively found in sacred groves of Kerala. These are Anonaceae, Bombacaceae, Combretaceae, Connaraceae, Cyperaceae, Dichapetalaceae, Dilleniaceae, Clusiaceae, Myristicaceae, Pandanaceae, Rhizhophoraceae and Zingiberaceae. There are some families very common in sacred groves which are tropical with sub-tropical and temperate outlines. They include Ebenaceae, Icacinaceae, Lauraceae, Fabaceae, Caesalpiniaceae, Mimosaceae, Oleaceae, Passifloraceae, Piperaceae, Rubiaceae, Rutaceae, Sapindaceae, Urticaceae, Verbenaceae and Vitaceae. The cosmopolitan families in sacred groves are Moraceae, Rubiaceae, Euphorbiaceae, Loganiaceae, Rutaceae, Bixaceae, Lauraceae and Fabaceae.

Among the intertropical elements like mangrove family Rhizophoraceae one genus viz; *Carallia* is somewhat widespread. Another element is Dipterocarpaceae of which two genera *Hopea* and *Vateria* show wide distribution in sacred groves. Pandanaceae with one genus is a paleotropical family found mostly in sacred groves located in swampy coastal areas.

Some of the noteworthy pantropical families and their representative genera which are seen in sacred groves are: Anacardiaceae, (*Mangifera*), Anonaceae *(Anona* spp., *Artabotrys* spp., *Uvaria* spp.), Apocynaceae *(Alstonia* spp.,*Rauvolfia* spp.,*Tabernaemontana),* Bombacaceae (*Bombax* spp.), Combretaceae (*Combrutum* spp. and *Terminalia* spp.), Dilleniaceae (*Tetracera* spp.). Intertropical family Ebenaceae is represented by *Diospyrus*. Pantropical Clussiaceae by *Calophyllum* and Lauraceae by *Cinnamomum* and *Litsea*.

Species Diversity Pattern

The climax evergreen forest of the Western Ghats showed Shannon Wiener indices (H') between 3.6 to 4.3 and Simpson's indices (D) between 0.86 to 0.90. In sacred groves, the H' value varies from 2.8 to 3.6. Within the same sacred grove itself the indices show variations, the variation is generally positively related to the disturbance. In partially disturbed once the D value varies from 0.68 to 0.92 and H' value varies from 2.6 to 4.2. But in undisturbed groves D value varies from 0.84 to 0.86 and H' value varies from 3.3 to 3.6 (Rajendraprasad, 1995).

Similarity Index between Flora of Different Sacred Groves

Rajendraprasad (1995) carried out extensive investigation on floristic pattern and composition of 168 sacred groves representing different agroclimatic zones of Kerala and made a detailed analysis of similarity between flora of five selected agroclimatically divergent sacred groves. Similarity index between different sacred groves from different conditions showed a minimum value of 32% and maximum of 68% (Table 1). The floristic difference between the highland and coastal groves was maximum. The flora between highland and midland showed maximum similarity while the flora of midland to coast showed moderate similarity. Besides this, southward-northward dissimilarity is less marked compared to coastal- high land dissimilarity. Such trends indicate longitudinal continuity of the flora of Kerala.

Soil Fungi

Soil fungi of the sacred grove play an important role in the soil fertility. The greatest number of moulds are found in the surface layer, where the organic matter is ample and aeration is adequate. Rajendraprasad (1995) reported 365999 fungal individuals per gram dry soil in coastal sacred groves, 2447600 per gram soil in mid altitude and 654000 individuals per gram soil in high altitude sacred groves. The sacred groves in Southern Kerala have an average of 360999 individuals per gram soil against an

Table 1. A comprehensive account of similarity index of different sacred groves, divergent in agroclimatic conditions, of Kerala.

Sl. No.	Names of sacred groves (community)	Agroclimatic regions	Similarity index (%)
1	Vandanam-Iringole	Coastal—Midland of Kerala	51.79
2	Vandanam-Kolani	Coastal—Highland of Kerala	32.00
3	Vandanam-Pachaloor	South-Central Kerala (coastal)	42.60
4	Vandanam-Puthiyaparabil	South-North Kerala	36.99
5	Iringole-Kolani	Mid-Highland Kerala	68.06
6	Iringole-Pachaloor	Mid Central—South Coastal Kerala	51.19
7	Iringole-Puthiyaparabil	Mid Central—North Coastal Kerala	39.59
8	Kolani-Pachaloor	High-South Kerala	47.06
9	Kolani-Puthiyaparambil	High-North Kerala	36.56
10	Pachaloor-Puthiyaparambil	South-North Kerala	41.12

average of 296333 fungal individuals per gram soil in North Kerala groves. The common fungal species identified in the sacred groves are *Abisidia corymbifera, Aspergillus ornatus, A. fumigatus, A. niger, A. japonicus, A. nidulans, A. restricus, A. flavus, Chaetomium datum, Circinella simplx, C. lunata , Curvularia lunata, Humicola fusco-atra, Poecilmyces varioti, Penicillium nigricans, P. luteum, P. lividum, P. achraceum, P. chrysogenum, P. citrinum, P. purpurogenum, P. lilacinum, P. urticae, Phoma medicaginis* and *Tarula herbarium.*

ECOLOGICAL FUNCTIONS OF SACRED GROVES

Soils of the sacred groves vary depending on the geology, geomorphology and vegetation. Sacred grove soils are mostly reddish brown in colour, old, deeply weathered, leached and acidic. They are generally sandy or sandy loam in texture. The sacred grove soils show high porosity and low bulk density compared to the soils of nearby areas (Rajendraprasad, 1995). The channels created by higher organisms and thick litter cover together enhance the water retention, root system development, gaseous exchange and heat conductance. The soil moisture retention characteristics of sacred groves are higher when compared to the adjacent area (Rajendraprasad, 1995). Moreover the system is relatively closed in the case of nutrients and water. The sacred groves act as micro-watershed in local areas. They are always associated with a freshwater ecosystem meeting the water needs of the local communities. They also reduce temperature on a small scale. Without sacred groves and associated ponds, the rain water would be wasted as runoff, eroding the surface soil and flowing rapidly in the near by areas. The sacred groves as a watershed are of great ecological importance.

Sacred groves play a dynamic role in balancing the ecosystem including the agroecosystem of the region. Sacred groves is the abode for various organisms whose food chain is connected through a prey-predator interaction. The birds and bats find their natural nesting place in the sacred groves. They in addition to their scavenger role check the insect and pest population. The bird droppings rich in phosphorus replenish the phosphorous deficient soil of the region. Snakes and mangoose find their home in sacred groves. The snake controls the rodent population, which if left unchecked will destroy the crops of the locality. The snake population is kept under check by the mangoose. Insect flora, particularly the bees make their hives in sacred groves and facilitate the cross pollination of many plant species of the locality.

RITUALS, WORSHIPS AND CELEBRATIONS

There are many myths, legends and faith associated with the sacred groves of Kerala. All sacred groves of Kerala are dedicated to Gods or Goddesses or to certain ancestral or natural spirits. The deities in the sacred groves are at times represented by some trees like *Alstonia scholaris, Adenanthera pavonina, Hydnocarpus pentandra, Commiphora caudatum, Caryota urens, Holarrhena antidysenterica, Strychnos nux-vomica, Ficus tinctoris, Mimusops elengi* etc. A stone slab installed at the base of the tree is the altar on which the offerings including the animal sacrifices are made. These trees are also considered to be the abode of ancestral or natural spirits and demons. The sacred groves owned collectively by the villagers are mostly dedicated to Lord Ayyappa and called as 'Ayyappan Kavu' or 'Sastham kavu' and to Goddess Bhagavathi called 'Bagavathi Kavu' or 'Amman Kavu'. One interesting feature about 'Ayyappankavu' is the freedom to enter this sacred grove to offer worship irrespective of any castes or creeds. Sacred groves owned by the tribal communities are dedicated to 'Vanadevatha'; the Goddess of the forest, or to natural spirits or demons or ancestral spirits. The fishermen caste 'Dheevara' or 'Arayan' also maintain sacred groves in the coastal areas of Kerala. These groves are called 'Cheerma' or 'Cheerumba' and the patron deity is 'Cheerma'. Cheerma is the Goddess of smallpox and other epidemic diseases. The sacred groves owned by the families are dedicated mostly to Snake Gods (Naga) or Goddess or both. Sacred groves of the tribals inhabiting near and around the forest areas are known as 'Madam Kavu' or 'Yekshikavu'.

The sacred groves of North Kerala are mostly associated with Goddess whereas the sacred groves of South Kerala are associated mostly with snake worship. Many sacred groves associated with Siva temples also have serpent Gods. The various patron Gods/Goddess or Spirits associated with the sacred groves can be grouped as below:

Dedicated to Snake Gods: 'Nagam', Nagaraja, Nagakanya, 'Sarpam', 'Nagayakshi', 'Karinaga Yakshi', Karinaga, Nagini.

Dedicated to Goddess: 'Amma', 'Ayalakshi', 'Ayiravalli', Bhadrakali, Bhavani, Bhagavati, Bhuvaneswari, 'Chandi', 'Chamaundi', Devi, 'Durga', Mahishasura Mardini, Mariamma Mookambika, Rakteswari, Vana Durga, Vanadevatha.

Dedicated to Gods: Ayyappan, Sastha, Parai-Daivam, Malai Daivam.

Dedicated to Spirits: Arukola, Marutha, Madan, Yakshi, Gandharvan, Yogeeswaran, Muthappan.

Many sacred groves have more than one diety, the patron diety and two or more assistant deities.

The local people observe a strict code of conduct in protecting the sanctity of sacred groves. Human infrastructures are normally not allowed inside the sacred groves except to perform rituals and offer prayers and offerings to propitiate the deities. No materials, plant or animal origin, are permitted to be taken out of the sacred groves except on certain exceptional cases or occasions and that too only after consulting the local priest. No one is allowed to cut or remove any plants or kill animals. Even the fallen twigs, branches of trees or leaves are not removed. Violation of the rules, disturb or dispel the sanctity of the sacred grove and its immediate surroundings were considered to be unpardonable sins that will invite the wrath of the patron diety or spirits by bringing epidemic disease, famine, natural calamities or sufferings to the people. Menstruating women keep away from the sacred grove area till their polluting period is over. This faith and belief guided non-interference in such islands of biodiversity is an excellent example of micro level *in situ* conservation practices evolved by the Indian rural folk.

The rituals and rites performed in the sacred groves vary with the region, caste and patron diety of the sacred grove. 'Nurum Palum' is an important offering made to the Snake Gods in 'Sarpakkavus'. 'Nurum Palum' is offering of rice powder, turmeric powder, cow's milk, tender coconut water, 'Kadali' banana and Ghee. This ritual is performed on 'Ayilyam' star of the local almanac. A ritualistic devotional dance called 'Pambumthullal' is also performed by girls once in every 10 or 12 years or as and when required as per the predictions of the priest to propitiate the Snake Gods. Elaborate preparations are made to organize the 'Pambumthullal'. The surroundings of the serpent deities are cleaned and decorated beautifully with tender coconut leaves, banana stem/pith, leaves of mango, jack fruit, peepal trees etc. The girls and the main priest who are to perform the serpent dance 'Sarpam thullal' observe 41 days of strict discipline before the ceremony. The ground in front of the serpent deities are cleaned and thinly plastered with a paste of fresh cowdung and mud in the morning. By the evening the plaster will be dry. Some invocational 'Pooja' is made and offerings of fruits and rice preparation are made. Brass lamps or stone lamps are lit and designs of various sketches of Gods and images of serpent Gods are drawn on the plastered floor with powders of 5 different colours-rice powder for white, paddy husk burned charcoal powder for black, dry green leaf powder for green, red coloured sand for red and turmeric powder for yellow. When the design is completed it is called the 'Sarpakkalam'. Late in the night at about 10 p.m. the girls take their seat at one side of the 'Sarpakkalam' and the priest takes his seat on the opposite side. Husband and wife of a Pulluvar caste sit in a corner with their traditional musical accompaniment called 'Pulluvakudam' and start their devotional song 'Pulluvan Pattu' to invoke the serpent Gods. The girls with unlocked hairs hold tender inflorescence of Arecanut and close their eyes and concentrate on the 'Pulluvanpattu'. Within half an hour the girls begin to shiver and then shake their body which is considered as the sign of being possessed by serpent Goddess 'Nagani' and the girls are now known as 'Naganis'. The music emanating from the playing of the strings of 'Pulluvakudam' provides the background. The drum beating with a peculiar rhythm is also played and the girls start dancing in a peculiar manner. They move like snake through the columns of the design and reach in front of the male priest who also begins to show the sign of being possessed by the 'Nagaraja'-the serpent God. He also begins to shake his body violently and dance with the speeding rhythm of the drum beat. The girls one after other get exhausted of their dance and fall down. The 'Nagaraja' then ask for milk. After drinking the milk he calms down. At this point of time Nagaraja is ready for blessing the devotees and answer questions and clear doubts pertaining to matters like misfortunes, diseases or difficult problems of the society or individual members. 'Nagaraja' tells the devotees that the diseases or misfortunes faced by the devotees are due to violation of the sanctity of the sacred groves or killings of the snakes or dispelling or disturbing other elements of the sacred grove. He asks them to do certain penance, reserve the sanctity and ensure protection of the plants and animals inhabiting the sacred groves. The whole ceremony last till late midnight or till day break. But this elaborate ceremony has now become rare and some of the villagers feel that the misfortunes and deterioration in the social and family life are due to the disappearance of these rituals and ceremonies.

Fig. 1. "Serpent God"—the common dwelling deity of sacred groves of Kerala.

Fig. 2. Abundant and luxuriant growth of different life forms in the background of a Devi temple in a sacred grove of Kerala.

Fig. 3. "Muchillode Bhagavathi"—A common 'theyyam', typical example of artistic heritage of India-performed in sacred groves of Kerala.

Fig. 4. A sacred grove of Kerala (Cheemeni in Kasargod district). Recently protected by the Government and Dewaswam Board by the involvement of local people, shows a dense multilayered growth and variety of trees.

Similar rituals and ceremonies are also performed in sacred groves associated with Goddesses or Lord Ayyappan or the Spirits or Demons. The 'Kalam' design drawn in such places are mostly the images of Goddesses or Ayyappan or Spirits or Demons. The 'Kalam' thus made are called by the name of the patron diety of the sacred grove whose image is drawn in the 'Kalam'.

Animal sacrifice is also a part of the ritual in sacred groves associated with Goddess, ancestral Spirits or Demons. After being possessed by the patron diety the priest in such places ask for blood for which fowls are often given. The fowls are usually sacrificed at the altars. After this the priests bless the devotees and answer the questions of devotees and provide solutions to their problems. Some times goats are sacrificed at the altar of the sacred grove.

Most of the sacred groves associated with Goddess in North Kerala perform another ritual dance called 'Theyyam' or 'Theyyattam'. 'Theyyattam' literally means the dance of the God. 'Theyyam' is a distortion of the word 'Daivam' in Malayalam or Tamil word for God. Theyyam is also known as 'Thira' or 'Thirayattam'. Only the male members of some particular caste like Vannan, Malayan, Cheravan, Chingathan, Velan, Mannuthan, Anjuthan, Koppalan, Pulayan, Pampathar and Paravan alone participate in this devotional dance. More than 100 types of 'Theyyam' representing various Goddess, Spirits and Demons are known. The performers of 'Theyyam' are supposed to be possessed by the deities they represent. They move in measured steps and rhythmic dances from time to time they get posed till the end of the performance. Most interesting feature of 'Theyyattam' is the use of resplendent costumes and gorgeous colours and magnificent facial make up and a towering head-gear called 'Mudi' of the performers. This makes 'Theyyam' as one of the most spectacular pageant that stands out almost unique among the ritual dances of Kerala, perhaps of India. 'Theyyattam' is an exceptionally vigorous dance in which the dancer moves foreward and backward every now and then. The dancing reaches heights of frenzy and the movements become quicker and quicker with the resounding music produced by the rhythmic beating of a number of traditional drum, cymbals etc. With the dance occupying the central position with the surging devotees and other spectators move along in a procession around the village and finally come back to the sacred grove. When the dancer takes seat on a heavy wooden stool call "Peedham" and proclaim his appreciation of the ceremonies. The devotees thereafter prostrate or bow before him and make offerings and pray for blessings or to fulfil their wishes or solve their life's problems. 'Theyyam' while blessing the devotees asks them to preserve the sanctity of the sacred groves and preserve them in pristine manner.

CONCLUSION

Sacred groves in essence epitomize an all embracing concept and practice of the ancient Indian way of *in situ* conservation of biological and genetic diversity. They are patches of virgin forests preserved and protected by the local people on religious grounds. They are thus *'sanctum sanctorum'* of rare, endangered and endemic plant species, many of which have disappeared from the region outside the groves. Sacred groves used to be a common feature of every village and hamlets in ancient Kerala. In addition to the numerous family owned sacred groves there were at least one sacred grove in every hamlet that was owned commonly by the local communities. In the past the sacred groves played a dynamic role in integrating the society at family and community level. It functioned as an integral part in the social and cultural life of the local people. It had nurtured and fostered through centuries the unsophisticated imagination and rhythmic impulses and spiritual aspirations of sensitive people whose supreme expression evolved in the form of numerous varieties of indigenous folklore, visual and performing arts. The annual rituals and celebrations of the Gods and Goddesses associated with sacred groves provided a venue for them to promote these folk arts.

A holistic understanding on the current status, structure, function and dynamics of sacred grove ecosystems is an essential prerequisite for assessing their ecological role, productive potentials and conservation values. In fact even complete inventory and distributional pattern of sacred groves of Kerala which stand as the living testimony of our cultural heritage is yet to be made. There is an urgent need for launching a highly coordinated action-oriented multidisciplinary programme on sacred groves of Kerala.

REFERENCES

Balasubramanyam, K. and Induchoodan, N.C. 1996. Plant diversity in sacred groves of Kerala. *Evergreen* 36: 3-4.

Induchoodan, N.C. 1988. *Ecological Studies of the Sacred Groves.* M.Sc. Thesis, Kerala Agricultural University, Trissur.

Induchoodan, N.C. and Balasubramanyan, K. 1991. Sacred groves-saviour of endemics. *Proceedings of the Symposium on Rare, Endangered and Endemic plants of Western Ghats.* Kerala Forest Dept. Wildlife Wing, Thiruvananthapuram. pp. 348-453.

Menon,V.S. and Sasidharan, A. 1994. *Studies on the Photosynthetic Performance of Selected Trees and Woody Vines in the Sacred Groves of Kerala-An attempt to Evaluate the Optimum Productivity of these Ecosystems.* Final Report submitted by Tropical Botanic Garden and Research Institute to the State Committee on Science, Technology and Environment, Kerala State, Thiruvananthapuram.

Rajendraprasad, M. 1995. *The Floristic, Structural and Functional Analysis of Sacred Groves of Kerala.* Ph.D.Thesis, University of Kerala, Thiruvananthapuram.

Rajendraprasad, M., Krishnan, P. N. and Pushpangadan, P. 1996a. Floristic variations in the sacred groves of Kerala: A case study in five agro-climatically divergent sacred groves. *Paper presented in the National Seminar on Sacred Groves, 21st April 1996*, Hyderabad.

Rajendraprasad, M., Krishnan, P.N. and Pushpangadan, P. 1996b. Floristic wealth and diversity in the sacred groves of Kerala. *National Seminar on Conservation of Endangered Species and Ecosystems.* Dec. 5-7, 1996, Banaras Hindu University, Varanasi.

Rajendraprasad, M., Menon, V.S., Krishnan, P.N. and Pushpangadan, P. 1994. Sacred groves: A nutrient and water resource for Kerala. *Proceedings of the International Congress on Kerala Studies, 27-29 August 1994.* Vol. IV. pp. 33-34.

Ramachandran, K.K. and Mohanan, C.N. 1990. *Studies on the Sacred Groves of Kerala.* Final Project Report submitted by Centre for Earth Science Studies to Ministry of Environment and Forests, Govt. of India.

Rao, R.K., 1996. *Sacred and Protected Groves of Andhra Pradesh.* WWF-India, A.P. State Office, Hyderabad.

Ward and Conner, 1827. *Memmoirs of the Survey of Travancore and Cochin States.* Cited from Census Report of Travancore 1891.

19

Sacred Groves of the Western Ghats of India

M.D.S. Chandran, M. Gadgil** and J.D. Hughes****

*Department of Botany, Dr. Baliga College of Arts and Science, Kumta 581343, Karnataka, India

**Centre for Ecological Sciences, Indian Institute of Science, Bangalore 560012, India

***University of Denver, Denver, Colorado 80208, USA

INTRODUCTION

The genesis of sacred groves in the Western Ghats may go back to hunting-gathering societies which attributed sacred values to patches of forests within their territories as they did to several other topographic or landscape features like mountain peaks, rocks, caves, springs and rivers. The practice of setting aside patches of forests as sacred groves would have strengthened with the spread of agriculture, when slashing and burning of forests began on a massive scale. There is no simple explanation for this. The reasons could be religious and cultural compulsions as well as subsistence and ecological needs. In the Western Ghats, sacred groves are very characteristic of the agricultural landscape. Despite the rising popularity of worship in temples (several of them constructed in place of groves), the groves, although diminished in area, still persist as an integral part of the eco-cultures of most pre-Brahminic agricultural societies. In the worship associated with the groves, many traces of hunting traditions are still found.

Communal hunting is conducted at least once a year to appease the deities of the groves among many farming communities like the Halakkivokkals and Namadharis of Uttara Kannada, and the Kodavas of Coorg. In the Kasaragod district of northern Kerala, special sub-committees of hunters are formed in connection with the annual festivities of the *kavus* (groves). The animals hunted are sacrificed to the deities. The practice of sacrificing domestic animals like fowls and goats may be carried out on special community occasions, or at other times in fulfillment of the vows taken by individual devotees. Where the deities have been Sanskritized the hunting and animal sacrifices have been discontinued. However, in many Kerala temples, *pulliveta* (a symbolic hunt) is conducted, and gourds instead of live animals are sacrificed to the deities. The association of deities of the groves with weapons like tridents and arrows or spears is reminiscent of the hunter-gatherer tradition. The devotees of the famous Aiyappa temple of Sabarimala in Kerala may carry a wooden arrow with them; just as devotees carry tridents to the shrines of Murukan (identified with Subrhamanya, son of Shiva), a god of many Tamil Nadu hilltops. The Tamil people of Yalappanam in Sri Lanka identify most of their deities with tridents (Sivathambi, 1991). Although early agriculturalists in the Western Ghats caused deforestation, they also preserved sacred forests in honor of village gods.

Sacred groves, called *kans* and *devarakadus* in Karnataka, and *kavus* in Kerala, are associated mainly with agricultural communities. In the preagricultural Western Ghats, clad in primeval forests, sacred groves *per se* would not be ecologically differentiated. But the beginnings of agriculture here, over three millennia ago, would have seen the shrinkage of forests and biodiversity, setting in of soil erosion, fertility loss, drying up of the watershed and changes in the microclimate. Therefore the practice of protecting sacred groves increased in significance with the arrival of agriculture. The groves, in addition to their role as the abodes of gods, would have protected a range of landscape elements with their characteristic biodiversity. The larger groves would also have functioned as resource patches where non-timber forest produce could be harvested in a regulated fashion.

At present, sacred groves are found in a wide range of situations, from the estuaries of the West Coast to the montane heights of the Western Ghats at over 2,000 m elevation. The ecosystems covered by the groves range from mangroves and fresh water swamps to different forest types.

Mangroves of the West Coast have steadily declined due to heavy population pressures, reclamation for agriculture and, of late, shrimp farming. Yet, occasionally, one may see a sacred grove of mangroves still retained. It may just be a clump of trees. A grove of over half a hectare is still found in the Aghanashini estuary of Kumta. Unnikrishnan (1995) reported a unique mangrove sacred forest of about 7 ha in northern Kerala.

Sacred groves of evergreen-semievergreen forest types are commonest in southwest India. Groves of deciduous trees are found along the drier eastern slopes of the Western Ghats. In the Pune district of Maharashtra, Gadgil and Vartak (1976) studied 23 groves situated in varied locations from the floor of the river valley through slopes at different heights to the top of the plateau. The groves cover vegetation types from stunted forest on the exposed hill crest at 800-1500 m to tall luxuriant growth in the ravines. In another study covering 233 sacred groves of Maharashtra, Gadgil and Vartak (1981) observed a range of vegetation from semievergreen to dry deciduous type in rainfall regimes from 5,000 mm to 500 mm, protected as sacred groves. Burman (1992) considers sacred groves of the Maharashtra Western Ghats to be much more numerous than Gadgil and Vartak (1976, 1981) observed in their various studies. He found that all Gond and Mahadeo Koli tribals and Kunbis have sacred groves almost in every village, their size ranging from a clump of trees to 60 ha, the median being 1.5 ha.

Myristica swamp is an endangered fresh water swamp ecosystem found in a few locations in the southern Western Ghats. These swamps may give rise to perennial streams. Evergreen forest trees with various root anomalies in the form of stilts and breathing roots, and many other plants, are associated with the swamps. Due to conversion into arecanut gardens and rice fields, most such swamps have vanished. Some survive, enjoying some degree of protection because they are associated with deities. Katlekan, in Uttara Kannada, the abode of a Bhuta (a spirit), has several rare trees endemic to the Western Ghats, such as a species of wild nutmeg, *Myristica fatua* var. *magnifica,* and *Gymnacranthera canarica*. Aravanchalkavu, Andalurkavu and Theyyotukavu, sacred groves of northern Kerala, are associated with *Myristica* swamps (Chandran and Gadgil, 1993a, b,c; Unnikrishnan, 1995).

SACRED GROVES FOR CONSERVATION OF BIODIVERSITY

Set amidst a mosaic of landscape elements like shifting cultivation fields and fallows, permanent agricultural areas, savanna and secondary forests, the sacred forests protected several valuable food plants including mango *(Mangifera indica)*, jackfruit *(Artocarpus integrifolia)*, *Garcinia* spp., *Caryota urens*, an important starch and toddy producing palm, spices like pepper *(Piper nigrum)* and cinnamon *(Cinnamomum* spp.*)*, and scores of medicinal plants. Despite rising human pressures, sacred groves shelter many elements of the biota which may have vanished elsewhere in the local landscape.

Gadgil and Vartak (1975, 1976) found a grove in the Kolaba district of Maharashtra harboring a solitary specimen of the liana *Entada phaseoloides*. People came from an area of about 40 km radius to collect its bark to treat cattle bitten by snakes. Another Maharashtra grove had two magnificent trees of *Canarium strictum,* otherwise present only in Uttara Kannada 200 km to the south. One grove in the Pune district and another in the Yeotmal district were found to support ancient teak forests which

Fig. 1. *Dipterocarpus indicus* endemic to the Western Ghats of India, has its northern limit in some of the sacred forests of Uttara Kannada (Karnataka).

Fig. 2. *Myristica fatua* var. *magnifica*, a highly threatened trees of the Western Ghats of India in a threatened ecosystem—The *Myristica* Swamp. This fresh water swamp is part of a sacred forest in Uttara Kannada, Karnataka.

had vanished from elsewhere in those districts. These teak specimens represent genetic variants of considerable importance in improving the stock of this important timber tree.

Dipterocarpus indicus is an evergreen timber tree endemic to the southern Western Ghats. Palynological study by Caratini et al. (1991) shows that over 3,500 years ago it was also a common tree of Uttara Kannada in the central Western Ghats. Today it is mainly confined to two *kans* in the district, which form the northern limit for this tree. In the entire 10,200 km^2 area of Uttara Kannada, with closed forests covering over 60%, the natural population of another dipterocarp, *Vateria indica,* occurs only in an isolated one hectare grove in Mattigar village in Siddapur taluk, which is the northern limit for this Western Ghat endemic. Yelakundlikan, a sacred grove of 4 ha in Shimoga district, in the neighborhood of Uttara Kannada, is another remarkable sanctuary for this tree.

As noted above, Katlekan in southern Uttara Kannada is perhaps the only sacred grove which shelters *Myristica fatua* var. *magnifica*, an endangered endemic tree of the Western Ghats. Exclusive to the *Myristica* swamps, it also exhibits its rare presence in the Travancore forests of the southern Western Ghats. Katlekan has many other endemics: a fragile shade palm (*Pinanga dicksonii*), *Semicarpus auriculata, Dipterocarpus indicus* and *Gymnacranthera canarica*. Katlekan and its neighboring evergreen forests form the northern limit in the Western Ghats for the lion-tailed macaque (*Macaca silenus*.), an endangered primate (Chandran and Gadgil, 1993a).

Kunstleria keralensis, a climbing legume, reported from a sacred grove in southern Kerala, is a new genus record for India and a new species altogether (Mohanan and Nair,1981). *Blepharistemma membranifolia, Buchanania lanceolata* and *Syzygium travancoricum* are rare species found only in the sacred groves of Kerala (Nair and Mohanan, 1981). Mohanan also discovered a rare species of cinnamon, *Cinnamomum quilonensis*, in some of the *kavus* of Alapuzha district in Kerala (Unnikrishnan, 1995). Chandran (1993) found a high level of endemism in the *kans* of Uttara Kannada. The Kallabbekan in Kumta taluk, over 50 ha in extent, despite being in the midst of arecanut-spice gardens of a populated village, is rich in endemics like wild nutmegs *(Myristica malabarica)*, cinnamon, *Garcinia gummi-gutta* and pepper. Despite the degradation they have suffered in recent times, these *kans* still are gene pools for pepper of world importance. Many of the evergreen-semievergreen *kans* of Shimoga occur in a rainfall zone of 1150-1740 mm where one may normally expect moist deciduous forests. Someren (1871) described Induvallikan as a remarkable patch of evergreen in the otherwise deciduous Belandur forest of Shimoga. The *kan* had magnificent evergreen trees like *Artocarpus* spp.,

Calophyllum tomentosum, *Cinnamomum* spp., *Elaeocarpus* spp., *Persea macrantha* and *Vateria indica*. Of the *kans* of Yellapur in Uttara Kannada, Puri et al., (1989) stated: "The kan forests are the patches of evergreen forests left in the midst of moist deciduous forests. Due to the presence of local deities in the kan forests the species are not felled."

Some of the notable wild relatives of cultivated plants found in the sacred forests are *Garcinia* spp., mango, *Artocarpus* spp. and *Piper* spp. The ground layer in the sacred groves often harbors wild turmeric (*Curcuma* spp.), wild ginger (*Zingiber* spp.)., and cardamom (*Elettaria cardomomum*). Ponds close to sacred groves often have wild rice (*Oryza* spp.).

Animal Diversity of the Sacred Groves

The landscape of the pre-colonial Western Ghats, where sacred groves were enmeshed in secondary forests, fallows and grasslands, would have been ideal for wildlife. Animal life was very rich in the region until the close of the nineteenth century. But landscape changes resulting from state takeover of forests and hunting for sport by British sportsmen and local gunmen caused the decimation of wildlife (Campbell, 1883; Chandran and Gadgil, 1993b). Decline of the groves, impoverishment of the village commons, invasion of the grassy clearings by forest growth following the ban on shifting cultivation, and raising of plantations of timber trees and other commercial crops may have contributed to the overall wildlife decrease.

It cannot be expected that isolated sacred groves would shelter any major mammalian wildlife. Nevertheless, they harbor numerous birds, butterflies and bats, apart from primates and minor mammals. A survey by Chandran and Gadgil (1993c) with the help of ornithologist Ranjit Daniels, in 25 km^2 of rural landscape on an undulating terrain at 600 m in the Siddapur taluk of Uttara Kannada showed that even the small fragments of formerly large groves form good habitats for birds. In this case study area, under traditional land use, sacred forests occupied 6% of the land. Today the 54 surviving groves occupy only 0.3% of the area. These forests are heavily impacted by human activity, reduced to almost scrub and tree savanna or converted into plantations of exotic trees. We recorded 107 species of birds in just two days of sampling efforts. This is a good number considering the total number of 416 bird species recorded by Daniels (1989) from the entire district, over 400 times larger than the case study area. Of these 107 bird species, 33 were typical forest species, which included crested goshawk (*Accipiter trivirgatus*), lesser serpent eagle (*Spilornis cheela melanotis*), bronzed drongo (*Dicrurus aeneus*), fairy bluebird (*Irena puella*), and Nilgiri flowerpecker (*Dicaeum erythrorhynchus*).

White bellied blue flycatcher, an endemic bird of the Western Ghats, was found in a 0.6 ha evergreen grove in Kalyanpur village which also has a tiny pond inside it. Eight species were migratory, including eastern swallow (*Hirundo rustica*), Indian golden oriole (*Oriolus oriolus*), and greenish leaf warbler (*Phylloscopus* sp.). Most of the forest birds were found in the sacred groves of the region; other birds occurred in the general landscape. The bird diversity was highest in the groves of Mattigar (1 ha) and Kalyanpur (0.6 ha).

A detailed account of the animal diversity in the *kavus* of northern Kerala is given by Unnikrishnan (1995). In a pond in the Melothumkavu of Kasargod district, hundreds of white tortoises are protected. The worshippers of the grove feed these tortoises. The serpent groves of Kerala are well known for various snake species, including cobra, viper, krait and python. Nine species of frogs have been reported from these *kavus*. Jafar Palot, a naturalist, sighted a rare bird, the white bellied sea eagle (*Haliaeetus leucogaster*), considered divine by the fisherfolk, in some of these *kavus*. The disappearance of tall trees from the densely populated Kerala coast has affected this bird adversely. Out of twelve nests found in the region eight were located in the large trees of the *kavus*, which also are home to fruit bats and hornbills. More than half of about 400 species of birds recorded from Kerala have been spotted in the *kavus* of Northern Kerala. These include rare forest birds like crested goshawk and three-toed forest kingfisher (*Ceyx erithacus*). Thavidisserikavu is perhaps the only sacred grove in Kerala which has the Nilgiri langur (*Presbytis johni*), a threatened species (Unnikrishnan, 1995).

Sacred Groves and Watershed Protection

Most sacred groves of the Western Ghats are associated with ponds, streams, springs or rivers. Wingate (1888a) stated that the *kans* of Uttara Kannada were of "great economic and climatic importance. They favor the existence of springs, and perennial streams, and generally indicate the proximity of valuable spice gardens, which derive from them both shade and moisture." The Government of Bombay (1923) highlighted the watershed value of the *kans:*

> "Throughout the area, both in Sirsi and Siddapur, there are few tanks and few deep wells and the people depend much on springs .... Heavy evergreen forests hold up several feet of monsoon rain....if an evergreen forest is felled in the dry season, the flow of water from any spring it feeds increases rapidly though no rain water may have fallen for some months...."

The *kavus* of Kerala, including the smaller serpent groves in the premises of houses, are associated with water bodies. The people believed that the groves were responsible for permanent water bodies in their habitation. The older generation of people believed that desecrating a *kavu* would cause drying up of the pond. A serpent *kavu* or an abode of snakes was an indispensable adjunct to well-to-do Nair and Nambudiri families of Kerala. The sage Parasurama is said to have advised his Brahmin followers of the West Coast that the places allotted to the Nagas (snakes) were to be left untouched by the spade, thus enabling the underwood and creepers to grow luxuriantly (Pillai, 1940). There is no doubt that, unlike the case in most other parts of India, the Nambudiri Brahmins of Kerala adopted the indigenous practice of sacred groves, and themselves became the keepers of the groves.

SACRED GROVES AND SUBSISTENCE

Strict taboos existed against any form of interference in the smaller sacred groves of the Western Ghats. But the larger groves played a major role in supplying a variety of non-wood produce of subsistence value to local communities. This produce included fruits of various species like mango, *Artocarpus* spp., *Garcinia* spp., spices like cinnamon, pepper, various edible seeds, medicinal plants, toddy from the palm *Caryota urens*, etc. Rattan canes (*Calamus* spp.) and reeds like *Ochlandra* were collected for basket or mat weaving. Pepper, exported in large quantity from the West Coast from the time of the Roman empire, was an important product of the *kans* of Uttara Kannada and Shimoga (Chandran and Gadgil, 1993a). The collection of non-timber forest produce from the village *kans* was not an open access affair. In the *kans* of Sorab, Brandis and Grant (1868) found, the local inhabitants used to pay taxes or *warg* to the state for the privilege of collection. Each privilege holder, or *wargadar,* operated in a specific part of the *kan* without infringing into another's area. Even today, despite its state of decline, the villagers of Kallabbe in Kumta derive a good income from their *kan,* collecting marketable nontimber produce like wild nutmeg *(Myristica malabarica), Garcinia gummi-gutta*, pickle mango, mushrooms, honey and various medicinal plants. The *kan* indeed provides employment for a number of poor families here.

The sacred groves shelter several medicinal plants of great value not only for the primary health care of the village communities, but many also important in modern pharmacopoeia. These include *Dioscorea* spp., *Piper longum, P. nigrum, Saraca asoca, Garcinia gummi-gutta, G. indica, Symplocos racemosa, Strychnos nux-vomica, Vateria indica,*

Asparagus racemosa, Nervilia sp., Zingiber zerumbet, Syzygium travancoricum and Hydnocarpus alpina. The ponds of the groves may have lotus *(Nelumbo nucifera)* and water lily *(Nymphaea stellata)*, which are reputed in native medicine.

On a landscape of slash and burn cultivation and fire-prone secondary forests the evergreen sacred groves would act as natural fire breaks, sheltering tracts of fire-sensitive species and safeguarding the watershed. Fire affects the regeneration of forests through burning of seeds, seedlings and trees and indirectly through increasing surface temperature of soil, reducing organic matter, modifying soil texture and facilitating erosion. Not fully realizing the role of *kans* in traditional land use system of the Western Ghats, D. Brandis, the first Inspector General of Forests of British India, along with another officer, L. Grant, wondered in 1868 at the two kinds of forests that Sorab taluk in Shimoga: the dry (deciduous) and the evergreen *kans*. During the dry months, fires raged through the dry forest. But, "No fires enter the evergreen forest, leaves, branches and fallen trees accumulate and gradually decay, forming ultimately a rich surface layer of vegetable mould." They wondered "why a certain locality should be covered with evergreen, and another in the immediate vicinity with dry forest?" (Brandis and Grant, 1868).

With the state consolidating its hold over forest resources and prohibition of shifting cultivation in most of the central and southern Western Ghats during the latter half of the nineteenth century, fire ceased as a notable ecological factor. The myriads of evergreen sacred forests may well assist in the natural restoration of evergreen forests in their surroundings through supply of seeds and favourable microclimatic conditions. Evergreen forests are on the return in the central Western Ghats, free from the pressures of coffee, tea and rubber plantations which are widespread in the southern parts (Chandran, 1993; 1997b).

THREATS TO THE SACRED GROVES OF THE WESTERN GHATS

From early in the nineteenth century, the British laid claim to the forests of the Western Ghats, including the sacred forests. British forestry, in its early stages, was focused on harvesting important ship-building and other marketable timbers like teak *(Tectona grandis)*, rosewood *(Dalbergia latifolia)*, poon (*Calophyllum tomentosum*), anjili *(Artocarpus hirsuta)*, etc. Teak was taken mainly from the secondary deciduous forests of the river valleys and plateaus (Gadgil and Chandran, 1989). It is to be assumed that local communities enjoyed their age-old rights and privileges in the sacred forests up to the mid-nineteenth century, when they began to lose them.

Following the Indian Forest Act of 1878, which consolidated a state monopoly over forest resources, the local people, unable to meet their biomass needs because they were excluded from former village commons, turned to the sacred forests. For instance, in the drier eastern parts of Sirsi and Siddapur taluks in Uttara Kannada, where every village had its own evergreen to semi-evergreen sacred *kans*, the state monopoly over the timber-rich secondary forests of deciduous trees made the people depend on their *kans* for routine use. Later the Government of Bombay permitted the villagers to gather dry fuelwood from them. By the 1920s, many *kans* were already full of canopy gaps that favored *Lantana*, a prolific exotic weed which sheltered pigs to the ruin of the groves (Collins, 1922). Some decades later *Eupatorium*, an even more aggressive exotic, invaded. The degraded forests and grazing lands in the vicinity of villages, which earlier had been under control of local communities, and presumably utilized in a sustainable fashion, were converted into "minor forests" by the state and thrown open to the public. The breakdown of the village community management of property resources resulted in unregulated exploitation of these common lands, liquidating the forest growth (Gadgil and Iyer, 1989). Naturally this would have increased the pressure on the village sacred forests *(kans)*. The *kans* were no longer separated from the rest of the landscape by trenches, and people and cattle freely started moving in and out of them.

Simultaneously, the arecanut cum spice gardeners, who were already assigned some *betta* or leaf manure forests by the government, started encroaching into the *kans*. In the words of R.T. Wingate (1888b):

> "Anyone acquainted with the interior of Kanara [Uttara Kannada] cannot fail to be struck with the gradual encroachment being made on the evergreen. The people ignore their *betta* assignments proper, and resort to the evergreens in the rains .... for....*soppu* [green leaves for manure].... They cut and hack the evergreen forest trees. The consequence is that the sunlight penetrates the natural shade, the pepper vines and other wild fruits die off, the springs dry up, and the evergreen eventually disappear".

In densely populated northern Kerala, cattle grazing, crisscrossing footpaths, collection of green and dry leaf manure and fuelwood affect the ecology of the *kavus*. Mining of china clay and laterite bricks is ruining other *kavus*. The widening canopy gaps favor deciduous tree species and weeds like *Eupatorium* and *Mikania*. The declining evergreenness of the *kavus* will have telling consequences upon their ecology (Unnikrishnan, 1995).

In the *devrais* of Maharashtra, Gadgil and Vartak (1976) noted that taboos relating to the groves began to weaken, especially since Indian Independence in 1947. Removal of leaf litter and dead wood became a common practice since western Maharashtra was deforested to meet the demand for charcoal from nearby urban centers.

The *devarakadus* of Coorg had fluctuating fortunes under state management. Their area had increased from 4399 ha to 6278 ha between the years 1873 and 1905 under the management of the Forest Department. Although record of encroachment into the *devarakadus* and alienation of portions of them for coffee plantations goes back to the 1880s, their major shrinkage took place between 1905 and 1985 when they were brought under the management of the Revenue Department. This decline is more noticeable after Indian Independence in 1947. Kalam (1996) noted the case of Mahalingeshwar Devarakadu in Palur village. Nearly 61 of its 68 ha area had been encroached upon. Coffee cultivation had spread into it and within its confines were 43 houses. Yet another *devarakadu* under severe encroachment since 1982 was in Bharadi village. First a planter encroached upon 11 ha for planting coffee and pepper and another 4 ha was encroached by a plantation company. This was followed by an onrush of migrant laborers. Currently 130 families live in this *devarakadu*. The settlement is said to be complete with a burial-cremation ground and a kindergarten. Recently 1.6 ha area of Chamakavu, a grove in Kasaragod district was cut to make room for a school playground. Vareekarakavu of Kannur district, which had an area of 20 ha, two decades ago, was exploited for timber by the owners, as well as encroached by the people, so that it got reduced to just about one ha.

Encroachment in the *kavus* of southern Kerala has become a rampant affair. The breakup of the old joint families into nuclear families has resulted in a lack of manpower to protect family sacred groves. Poachers and encroachers are threatening the very existence of many *kavus*. Despite being aware of the ancient wisdom that the destruction of the *kavus* will spell doom to the wells and ponds in their vicinity, the householder, to cope up with soaring land prices, is often compelled to clear off the serpent grove in the corner of his property. This is done after observing necessary rituals propitiating the serpent and by installing serpent idols in tiny enclosures in the place where the entangled grove once stood (Nayar, 1987; Unnikrishnan, 1995).

Human settlements and expansion of agriculture, apart from tree felling and gathering of other biomass, are affecting the *kans* of Shimoga too. At Sorab, the town itself is expanding into the Hiresekunikan. The remains of the *kân* today, with widely spaced trees, are infested with *Eupatorium*;

roads and footpaths crisscross the grove and housing complexes are sprouting within its precincts.

State Forestry in Sacred Groves

The impact of state forestry, especially on the larger groves, became evident from the mid-nineteenth century. Brandis and Grant (1868) found the *kans* of Sorab in the Mysore kingdom (under British suzerainty) in a state of neglect. Perhaps due to the higher taxation imposed on the local privilege holders or *wargadars,* many *kans* were deserted. Pepper vines were no longer cared for and the villagers did not maintain the cattle-proof trenches around the *kans.* Someren (1871) found several unoccupied *kans* in the Belandur area of Shimoga.

The British began to employ contractors for collecting non-wood produce from the *kans.* As (Wingate 1888a) observed:

"I am still of the opinion that the system of annually selling by auction the produce of the *Kans* is pernicious one. The contractor sends forth his subordinates... who hack about the *kans* just as they please, the pepper vines are cut down from the root, dragged from the trees and the fruits then gathered while the cinnamon trees are all but destroyed.... I was greatly struck with the general destruction among the Kumta evergreens; they were in a far finer state of preservation fifteen years ago".

Wingate (1888a) also alludes to the fact that a proper demarcation of the *kans* was not conducted by the government, implying that several sacred groves cum safety forests merged with ordinary forests and lost their identity (Chandran and Gadgil, 1993a). Since most evergreen species of the groves have perishable wood, the state was not initially interested in timber extraction from the *kans.* However, the resource shortage faced by the common people after the forest reservations resulted in tree felling within the *kans.* Rao (1919), described the "disastrous" effects which followed fellings in the *kans* of Shimoga, including decline and disappearance of the water supply.

During 1940s *Dipterocarpus indicus* from Katlekan in Uttara Kannada was supplied to the railways and a plywood company. A forest working plan of 1966 for Sirsi and Siddapur taluks in Uttara Kannada included 4,000 ha of *kans* for felling industrial timbers (Shanmukhappa, 1966). Another working plan for Sirsi included 670 ha of *kans* belonging to 10 villages for selection felling of well grown trees (Thippeswami, 1963). A *kan* in Siddapur was even clear-felled and converted into an Eucalyptus plantation (Chandran and Gadgil, 1993a). During 1976 the village *kan* of Kallabbe in Kumta taluk, rich in endemics and in fine state of preservation, was leased out to a plywood company which extracted hundreds of large trees from it. Although the villagers went to the High Court of Karnataka and won the case against the state in asserting their rights over the *kan*, the years of dispute witnessed the breakdown of traditional management of the *kan*, which is on the road to ruin.

From 1970s the state took to timber harvesting from the *devarakadus* of Coorg which were already on a state of decline. Some of the groves had expensive trees like rosewood *(Dalbergia latifolia)* and sandalwood *(Santalum album)*. The government felt it was necessary to exploit these and reimburse a share of the sale proceeds for improvement of temples in the Coorg district. To this the Divisional Forest Officer of Hunsur responded.

Devarakadus are spread over 104 villages in Virajpet taluk of Coorg district. In each village there are about 2 to 12 *Devarakadus* and they are in small bits and many of the bits are less than an acre. There are valuable tree growth in these *Devarakadus*. The *Devarakadus* are sacred forests usually assigned to some deity or temple. Therefore the village community protect the *Devarakadus* and usufructs are used for betterments of the temple and community purposes. Large scale exploitation of trees may not be advisable, as village people may agitate against... and they may lose their faith in the protection of such forests and consequently may indulge in destruction of forests. Therefore, selection felling of overmatured, dead, dying and wind fallen trees of all species only may be taken up. Softwood timber may be given on contract to some of the wood based industries and firewood out of lops and tops of the trees be sold at the spot by conducting petty auction sales (Kalam, 1996).

Thus began an organized timber exploitation by the state in the sacred forests of Coorg. The Chief Conservator of Forests (General) for Karnataka in 1975, regarding sharing a portion of the proceeds with the temples wrote, "it is not possible to make payments to the temples in the absence of any specific provisions".

SOCIO-CULTURAL CAUSES OF DECLINE OF THE SACRED GROVES

Religion had an overwhelming influence on the preservation of forest patches, in addition to other ecological and economic values attributed to them. A notable feature of Indian culture is the continuation of many prehistoric religious practices, despite the growth of dogmatic religions alongside them. Vedic Hinduism, with its text-based dogmas, appeared in the Indian sub-continent during the fourth millennium BP. Despite its proclaimed faith in gods abstracted from the elements of nature, like water and wind, sun and moon, planets and stars, Vedic Hinduism for the next 1,500 years or so was on a course of collision with the various earlier regional cults of India, which were more intimately related to local ecosystems. There was, however, no outright rejection of folk cults related to nature. By the time the great Epics like *Mahabharata* and *Ramayana* were composed, the Hindu religion, already a nature based one, had incorporated many more cults related to trees, animals, bodies of water and various other natural aspects. The Hindu religion, the most ancient of the dogmatic religions such as Jainism and Buddhism to develop in the subcontinent, went on hybridizing with the various indigenous creeds. Such a cultural transformation and changing worldview of nature among the people of the Western Ghats were among the causes for the decline of the sacred groves. (Chandran and Hughes 1997). Following temple building and consecration of the new deities the groves get cut, exploited in other ways, ignored or even subjected to other land uses. There are also many instances where groves continue to be protected even where temples have become the centers of worship.

Sacred groves, no doubt, are on the decline. But there is a redeeming factor structured into the Indian culture which promises to reawaken the sacred in nature. Indeed the efforts in this direction began millennia ago. In the *Vibuti Yoga* of the *Bhagavad Gita*, Sri Krishna raises his followers to exalted levels of spiritualism in nature by revealing to Arjuna his cosmocentric vision. The divine spirit permeates the entire universe including the elements of the local landscape. Krishna declares, "Of all the trees I am Asvatta" *(Ficus religlosa)*. In fact the Indian landscape, both urban and rural, is dotted with myriads of sacred Asvatta, whose worship could be traced back to the Indus Valley culture. *Ficus* spp. have been, of late, recognized by ecologists as a keystone species of the tropical forests, fruiting at critical times and sustaining animal life (Terborgh, 1986). Krishna also considers himself to be the Sun among the celestial bodies, Marichi among the winds, Ocean among the water bodies, Meru among the mountains, Ganga among the rivers, Himalaya among unmoving things, Airavatai among elephants, thunderbolt among weapons, Vasuki among poisonous snakes, Makara (shark?) among fishes and so on.

This is undoubtedly an effort to unify the people of diverse faiths of the Indian subcontinent into the acceptance of one supreme god, but with a difference, that is without rejecting the various indigenous cults which made them live in kinship with the elements of nature. Here we find the wisdom of Krishna who bridges the religion of the Vedic people, who primarily worshipped various cosmic powers, with that of the non-Vedic people, who mostly found divinity in the elements of the local landscape, including the plants and animals.

In fact the Hindu scriptures highlight the sacredness of trees. Most streams and rivers are sacred too for the people. Indeed the term *Tirthayatra* is understood as a pilgrimage to holy waters, which naturally are associated with watershed forests. The *Matsyapurana,* for instance attaches great importance to the planting of trees and even to the celebration of tree festivals. The same Purana states, "A son is equal to ten deep reservoirs of waters and a tree planted is equal to ten sons" (Kane, 1962).

The Hindu *Dharmashastras* exhorted people to plant gardens of sacred trees in the premises of temples. These temple gardens may be called "Star Forest," "Planet Forest," and "Zodiac Forest," and they contain sacred trees of diverse kinds, favorites of the various heavenly bodies (Chandrakanth et al., 1990). There are forests of other miscellaneous kinds. A drawback of these text-based Hindu sacred forests is that the trees recommended for planting in them come from diverse ecosystems. At the best their planting together may constitute an arboretum rather than a sacred forest. For instance the Star Forest has, among the 27 species recommended, *Acacia catechu* and *Calotropis gigantea* from the desert, *Pinus longifolia* from the snowy heights of the Himalayas, and *Piper longum* and *Mimusops elengi* from the tropical evergreen forests. Whereas numberless original sacred groves, which are often patches of natural vegetation, like the *kavus, kans, devarais* and *devarakadus* of the Western Ghats, are perishing due to changes in religious outlook, commercial exploitation, encroachments and negligence, the concept of the ritualistically reconstituted sacred forests like the Star, Planet and Zodiac forests are likely to find favor. The Forest Department of the Government of Karnataka, while permitting industrial felling and other alterations of *kans* and *devarakadus,* recently made elaborate plans for creating new temple forests in Uttara Kannada. Such a forest of trees of ritualistic importance has already been planted in the Bekal village of Sirsi taluk. This forest, or rather a garden of trees coming from diverse bioclimatic regions, is in no way an adequate substitute for many of the *kans* which even today shelter rare species and rare kinds of ecosystems, play important roles in soil and water conservation, and perform various other ecological functions. The sacred grove is therefore to be seen in a proper historical, religious, cultural and ecological context.

In the northern Indian plains, considered the heartland of Brahminic Hinduism, temple construction began in the Mauryan period and intensified during the Gupta period. The trend gradually spread to the rest of India as well. Nakeera, a poet of the Sangam literary period in the early Christian era, stated that Lord Muruka (identified as a son of Shiva), could be found in the deep wood, in a place surrounded by waters, rivers, tanks, meeting places under trees and new grown groves. He was the owner of all hilly tracts with rich groves (Ramachandran, 1990). The Brahminic influence in the Western Ghats-West Coast region began in a major way from the fourth century A.D. with the Kadamba kings of Uttara Kannada, who were instrumental in introducing Brahmins in this tract. Whereas on the one hand temples began to appear as worship centers, on the other hand some Brahmins also adopted the cult of the sacred groves, true to the cultural syncretism of the Hindus. In Kerala for instance, it is almost mandatory for the Namboodiri Brahmins to have a sacred grove in the compound of their traditional household or *illam*. Namboodiris are the priests for many famous temples situated in well preserved sacred groves as in the Iringolekavu of Perumbavoor in Ernakulam district.

But barring such exceptions, various cultural transformations along the Western Ghats-West Coast have adversely affected the sacred groves. The spread of dogmatic world religions like Christianity and Islam, which derided such practices, caused the decline of the groves in their pockets of influence through the last thousand years or more. Christian and Muslim households are, generally, not associated with sacred groves.

The conversion of sacred groves into temples is seen as a major cause for weakening links between gods and groves in Kerala by Unnikrishnan (1995). Shrines and temples are being constructed inside the groves in Coorg (Kalam, 1996). Bhuthasthanas, or small shrines devoted to the reigning spirits of the sacred groves have become common place in Dakshina Kannada. Sacred groves in the interior of hilly Uttara Kannada are relatively immune from such changes. But many village communities here too are yearning to keep pace with the all pervasive cultural changes. In the urban parts of Uttara Kannada such changes have already taken place. Thus one will not be surprised to see a mother goddess (considered an incarnation of Parvati, Shiva's consort) in the form of a three meter high termite mound, housed under a concrete roof, in a temple complex, of Kumta town. Yet another larger termite mound worshipped as Betedevaru [Hunter God] by the Halakkivokkal peasants, is found under a tiled roof in a humbler shrine of the same town.

SACRED GROVES: A FADING LEGACY AND SURVIVALS IN THE WESTERN GHATS

Where indigenous people have depended for long periods of time on local environments for sustenance they have developed a stake in conserving, and in some cases enhancing, biodiversity (Gadgil, et al., 1993). Folke and Berkes (1995) consider traditional ecological knowledge as differing from scientific knowledge in being moral, ethically based, spiritual, intuitive and holistic and having a large social context. In contrast to the traditional ecological perception of nature found in indigenous societies, modern scientific management, with its roots in the utilitarian and exploitative world view, assumes that humans have dominion over nature (Gadgil and Berkes, 1991; McNeely, 1991). Conservationists, often voicing a view that separates the natural world from the human world, called for endangered species to be protected in pristine natural habitats, protected areas sanitized of human influence. Of late, however, ecologists are beginning to form more holistic views of biology, together with the realization that humans are everywhere a vital part of the ecological landscape (Johnson, 1995).

The practice of worshipping the elements of nature, including patches of primal vegetation, dates back to antiquity. The decline or outright ruin of sacred groves in most of the world may be correlated to the transformation of ancient agricultural societies into formidable political and military powers, whose strength was derived from the non-sustainable exploitation of the natural resources of their own and neighboring territories. Mention may be made of ancient Greece and Rome, where deforestation and ecological crises, including decline of the sacred groves, has been noted (Hughes, 1994). The arrival of the major world religions such as Christianity and Islam, which denounced and decreed against paganism in general, resulted in the groves largely vanishing from the face of Europe and most other lands which came under the sway of these religions.

In the subcontinent of India Buddhism and Jainism, which came into existence about 2,500 years ago, while not negating the traditional worship associated with the sacred groves, promoted stupas or temples as worship centers with the groves gradually losing their importance. It is notable that Buddha, who himself was born in a sacred grove, Lumbinivana, often delivered his sermons from the *Chaityas* of the sub-Himalayan belt, which were originally sacred forests dedicated to sylvan deities like the *Yakshas*. Subsequently the *Chaityas* became centers of Buddhist monasteries and stupas. The pipal tree *(Ficus religiosa)* is sacred to Hindus, Buddhists, and Jains alike to this day.

Jainism developed on similar lines, incorporating the agricultural cults, while deifying the *Thirthabkaras* in the temples *(Basdis)*. The Jains of the Western Ghats region, mainly belonging to the agricultural and trading classes, even today pay obeisance to the deities of the groves such as *Bhutas, Yakshis,* and serpents. Serpent stones can still be found under the sacred trees in front of the Jain temples.

Hinduism is built up on the foundation of the Vedas, which lay much emphasis on the worship of the elemental forces of nature. Through the passage of millennia, the text-based Vedic Hinduism expanded within the Indian subcontinent, incorporating within its manifold cults associated with sacred groves and sacred waters. In the cultural syncretism of today's Hinduism, most of the deities of the groves have been accepted as incarnations of the major Hindu gods or as their kin or minions. As temples were built to house major gods like Shiva, Vishnu, Parvati, Lakshmi, Ganesha, etc., the lesser gods were given subsidiary positions within the sacred grove adjoining the temple or under the canopy of sacred trees.

Centralization of religion, which came rather late in the Western Ghats region, went hand in hand with the concentration of political power. Except for tiny bits of sacred groves which remained under the control of village communities or individual households, the state, particularly from the British period, asserted control over the sacred forests. Since the state did not consider the groves, like the *kans* of Uttara Kannada and Shimoga, as sacred, they merited treatment not much different from the rest of the forests, and were subjected to various commercial pressures or were thrown open to meet the biomass needs of the local communities.

The sacred groves of the Western Ghats can still be rescued and restored if it is realized that:

(1) The sacred groves, under various names such as *kans, kavus, devrai devarakadu*, etc., are in the last phase of their existence. Despite their waning state many of them still harbor the best natural vegetation for given regions, and provide refugia for rare species of flora and fauna. The groves protect the watersheds and act as storehouses for medicinal plants and non-timber forest produce, as well as enhancing landscape heterogeneity and safeguarding microclimates.

(2) The historical role of sacred groves in the development of the syncretic culture of modern India has been grossly neglected. On the contrary, the groves, despite rising fascination for them among scientists and environmentalists, are viewed as something associated only with primitive worship.

(3) The people in general consider regular worshipping of the major gods in large temples as a sign of their social advancement, and worship of the lesser gods of the groves as an unavoidable periodical necessity to avert their displeasure. They like the Brahmin priesthood to trace the lineages of the gods of the groves into relationship with the major gods of Hinduism.

If the Indian elites realize this situation and consider that the Hindu gods, despite their present popular conception, have their genesis from nature, nature in general and groves in particular will merit more serious thought from the angle of their conservation. With such realization it may be possible to forestall the destruction of what may accurately be termed as living museums of Hinduism, the stages in the evolution of which are enshrined in the sacred groves of the Western Ghats.

REFERENCES

Brandis, D. and Grant 1868. *Joint report No. 33, dated 11th May 1868, on the kans in the Sorab taluka*. Forest Department, Shimoga.

Burman, R.J.J. 1992. The institution of sacred grove. Journal of Indian Anthropological Society, 27: 219-238.

Campbell, J. M. 1883. *Gazetteer of the Bombay Presidency, Vol. 15: Kanara* (parts 1&2), Government Central Press, Bombay.

Caratini, M., Fontugne, M., Pascal, J. P., Tiscot, C. and Bentaleb, I. 1991. A major change at ca. 3500 years BP in the vegetation of the Western Ghats in North Kanara, Karnataka. *Current Science* 61: 669-672.

Chandrakanth, M.G., Gilless, J.K., Gowramma, V. and Nagaraja, M.G. 1990. Temple forests in India's forest development. *Agroforestry Systems* 11: 199-211.

Chandran, M.D.S. 1993. *Vegetational Changes in the Evergreen Forest Belt of the Uttara Kannada District of Karnataka State*. Ph.D. thesis, Karnataka University, Dharwad.

Chandran, M.D.S. 1997a. Transitions in Post-Indus Ecological Traditions of India. Paper presented at the Conference on Ecological History and Traditional Sciences of India, 27-29 March, Centre for Science and Environment, New Delhi.

Chandran, M.D.S. 1997b. On the ecological history of the Western Ghats. *Current Science* 73: 146-155.

Chandran, M.D.S. and Gadgil, M. 1993a. Kans- safety forests of Uttara Kannada. In M. Brandl (Ed.) *Proceeding of the IUFRO Forest History Group Meeting on Peasant Forestry' 2-5 September 1991*, No. 40, Forstliche Versuchs- und Forschungsanstalt, Freiburg, pp. 49-57.

Chandran, M.D.S. and Gadgil, M. 1993b. State forestry and the decline in food resources in the tropical forests of Uttara Kannada, Southern India. In: C.M. Hladik, A. Hladik, O.F. Linares, H., Pagezy, A., Semple, and M. Hadley (Eds.) *Tropical Forests, People and Food*, MAB Series, Vol. 13, UNESCO-Parthenon, Paris, pp. 733-744.

Chandran, M.D.S. and Gadgil, M. 1993c. *Sacred Groves and Sacred Trees of Uttara Kannada (a pilot study)*. Report submitted to the Indira Gandhi National Centre for the Arts.

Chandran, M.D.S. and Hughes, J.D. 1997. The Sacred Groves of South India: Ecology, Traditional Communities and Religious Change. *Social Compass*, forthcoming.

Collins, G.F.S., 1922. A report on the general condition of forest administration in Siddapur taluka (15 June); Forest Settlement Office, Karwar.

Daniels, R.J.R. 1989. *A Conservation Strategy for the Birds of Uttara Kannada District*. Ph.D. thesis, Centre for Ecological Sciences, Indian Institute of Science, Bangalore.

Folke, C. and Berkes, F. 1995. Mechanisms that link property rights to ecological systems. In: S. Hanna, and M. Munasinghe, (Eds.) *Property Rights and the Environment: Social and Ecological Issues*. The Beijer International Institute of Ecological Economics & The World Bank, pp. 121-137.

Gadgil, M. and Berkes, F. 1991. Traditional resource management systems. *Resource Management and Optimization* 18: 127-141.

Gadgil, M., Berkes, F. and Folke, C. 1993. Indigenous knowledge for biodiversity conservation. *Ambio* 22: 151-156.

Gadgil, M. and Chandran, M.D.S. 1989. *Environmental impact of forest based industries on the evergreen forests of Uttara Kannada district: A case study*: final report. Department of Ecology and Environment, Government of Karnataka.

Gadgil, M. and Iyer, P. 1989. On the diversification of common property resources use by the Indian society. In: F. Berkes (Ed.), *Common Property Resources : Ecology and Community Based Sustainable Development*. Belhaven Press, London, pp. 240-255.

Gadgil, M., and Vartak, V. D. 1975. Sacred groves of India : a plea for continued conservation. *Journal of Bombay Natural History Society* 72: 314-320.

Gadgil, M. and Vartak, V. D. 1976. The sacred groves of Western Ghats in India. *Economic Botany* 30: 152-160.

Gadgil, M. and Vartak, V. D. 1981. Sacred groves of Maharashtra: An inventory. In : S.K. Jain (Ed.) *Glimpses of Ethnobotany*, Oxford University Press, Bombay, pp. 279-294.

Government of Bombay, 1923. Revenue Department Resolution No. 7211, (May).

Hughes, J.D. 1994. *Pan's Travail: Environmental Problems of the Ancient Greeks and Romans.* John Hopkins University Press, Baltimore

Johnson, N. C. 1995. *Biodiversity in the Balance : Approaches to Setting Geographic Conservation Priorities.* Biodiversity Support Program, WWF, Washington.

Kalam, M.A. 1996. *Sacred groves in Kodagu district of Karnataka (South India).* Pondy Paper in Social Sciences 21, French Institute, Pondicherry.

Kane, P.V. (Ed.). 1962. *History of Dharmasastra.* Vol. 5, Part 1. Govt. Oriental Series, Class. B., No. 6. Bhandarkar Oriental Research Institute, Pune.

McNeely, J.A. 1991. Common property resource management or government ownership : Improving the conservation of biological resources. *International Relations*, 211-225.

Mohanan, C.N., and Nair, N.C. 1981. *Kunstleria* Prain—a new genus record for India and a new species in the genus. *Proceedings of the Indian Academy of Sciences B-90:* 207-210.

Nair, N.C. and Mohanan, C.N. 1981. On the rediscovery of four threatened species from the sacred groves of Kerala. *Journ. Econ. Tax. Bot.* 2: 233-235.

Nayar, P.K.B. 1987. *Religion, Mythology and Ecosystem.* Report to the Department of Environment and Forests, Government of India, New Delhi.

Pillai, T.K.V. 1940. Travncore State Manual. Vol.1. Government of Travancore, Trivandrum. Puri, G. S., Gupta, R. K. and Meher-Homji, V. M. 1989. *Forest Ecology*, Vol. 2. Oxford & IBH, New Delhi.

Ramachandran, C. 1990. *The forest, our divine source.* Seminar on Ecology and Ancient India, 28 April, Institute of Oriental Study, Thane.

Rao, M.S.N. 1919. *Working plan report of Belandur State Forest,* Ananthapur Range, Shimoga District. Forest Department, Shimoga.

Shanmukhappa, G. 1966. *Working plan for the unorganized forests of Sirsi and Siddapur.* Government of Mysore, Bangalore.

Sivathambi, K. 1991. Divine presence and/or social prominence—An inquiry into the social role of the places of worship in Yalappanam Tamil society. *Journal of Institute of Asian Studies* 9: 1-22.

Someren, G.J., Conservator of Mysore and Coorg, 1871. Letter No. PWD, Rev. For. No. 1507 to the Chief Commissioner of Mysore, Forest Department, Shimoga.

Terborgh, J. 1986. Keystone plant resources in the tropical forest. In: M.E. Soule (Ed.) *Conservation Biology: The Science of Scarcity and Diversity.* Sinauer Associates, Inc., Publishers, Sunderland, Massachusetts.

Thippeswami, S.C. 1963. Sirsi town firewood supply plan. Government of Mysore, Bangalore.

Unnikrishnan, E. 1995. *Sacred Groves of North Kerala - an EcoFolklore Study* (in Malayalam), Jeevarekha, Thrissur.

Wingate, R.T. 1888a. *Settlement proposals of 16 villages of Kumta taluk, No. 210, Forest Settlement Office*, Karwar.

Wingate, R.T. 1888b. *Extract from the Assistant Officer's letter no. 46, dated 5th March to the Survey and Settlement Commissioner.* Forest Settlement Office, Karwar.

20

Role of Sacred Groves in Biodiversity Conservation with Local People's Participation: A Case Study from Ratnagiri District, Maharashtra

A. Godbole, A. Watve, S. Prabhu and J. Sarnaik

Applied Environmental Research Foundation, Pune, Maharashtra, India

INTRODUCTION

Sacred grove is a traditional institution which permits management of biotic/ natural resources through people's participation. The sacred groves have drawn attention of the environmentalists and conservationists due to their value as repositories of trees and other plant species otherwise becoming rare in the surroundings. The real need of people's participation in conservation has also been realised recently. Due to failure of pure legal protection in guaranteeing conservation, it became necessary to search for the solutions in the traditional conservation and resource management systems based on indigenous Knowledge of the local communities.

After such realisation, serious research efforts have been started all over India. In the initial stages these studies were focused on inventories of sacred groves and their floristic composition, but very little has been thought about their present status and their possible role in resource conservation (Gadgil and Vartak,1976; Induchudan,1988; Nipunage, 1988). In late 1990s sacred groves have been interpreted from the socio-historical perspectives (Freeman 1994, Kalam 1996). These later viewpoints strongly opposed the ecological wisdom of local people and possible role of this wisdom in the conservation of sacred groves. However, it is necessary to assess the status of sacred groves from ecological, social and conservation perspective in order to develop a strategy for revival and incorporation of this traditional institution in conservation. With this background, a case study has been conducted in the Ratnagiri district, Maharashtra. The present article discusses the results of this case study.

CASE STUDY

During preparation of the preliminary inventory of sacred groves of Ratnagiri District it was observed that Sangameshwar Tehsil has more than 250 Sacred groves and many of them are still preserved in better condition. These Sacred groves are located in Ratnagiri part of Western Ghats in India. Since the sacred groves are meticulously protected 'traditional forests reserves', their floristic composition is quite different than the surrounding degraded areas, which support dry deciduous or scrub vegetation (Sharma and Kulkarni, 1980). It was observed that floristic composition of Devraies in Ratnagiri District is unique in many ways. A brief floristic survey of 11 selected sacred groves in the region has been carried out. A list of plant species encountered in these sacred groves is given in Table 1. These floristic studies clearly indicate the extent of biodiversity available and preserved through the effective functioning of traditional institution of sacred groves. It is therefore imperative to protect and conserve these groves.

Table 1 : Species recorded from sacred groves in Sangameshwar Tehsil of Ratnagiri district of Maharashtra, India.

Species	Number of sacred groves
Antiaris toxicaria	4
Dracaena turniflora	1
Anamirta cocculus	1
Aporosa lindleyana	1
Olax imbricata	2
Gymnema khandalense	1
Nervilia infundibuiliformis	1
Hydnocarpus pentandra	2
Khema attenuata	1
Dysoxyllum binectariferum	4
Meiogyne pannosa	1
Entada scandens	4
Gnetum ula	5
Syzygium caryophyllatum	1
Tetrameles nudiflora	1
Clerodendron infortunatum	1
Dimorphocalyx lawiannus	1
Syzygium phyllolaenum	2
Amorphophallus campanulatus	1
Acampe sp.	1
Aeginetia indica	2

Local People's Perception and Contemporary Understanding

The sacred groves play very important social role though they may not feature very distinctively in the day to day activities of the local people. (Roy Burman, 1995). In general the local people are unable to assign any direct benefits of sacred groves to the village or community. Existence of sacred groves and the maintenance of this traditional institution lies in the community's perspective and its role in the religious and social life of the community.

Sacred groves from the study area may be classified into three categories as per the functions attributed to them by the local people.

a. Sacred Groves Maintained for Village Deities in a Formal Way

This kind of groves normally have proper temple (old or renovated) within the grove. Important village festivals like Navratri, Holi and Devdiwali as well as rituals like *kaul* etc. are performed in this temple. A patch of forest is normally preserved around the temple but certain sanctions with the consent of the villagers. Such sanctions include cutting large timber for renovation and repairing of temple within the sacred groves or village temple school building etc. The rules and regulations for this sanctions and management of sacred groves are not in any written form but are observed meticulously. Gurav is the appointed priest looking after the daily rituals and pooja and has limited sanctions for using the resources within the sacred groves, while Mankaris-specially respected village individuals have special rights over the resources of sacred groves. Mankaris are at a higher position in village hierarchy than the gurav especially during the festivals. Villagers can also use sacred groves resources to some extent. They can collect the dead wood as fuel and dried leaf litter (*pateri*) individually. The leaf litter is collected together for the community. Most of the sacred groves in Sangameshwar Tehsil belong to this category. During the Holi festival the sacred tree of *Maad* (*Caryota urense*) or *Amba* (*Mangifera indica*) can be cut as per the custom from the sacred groves and burnt/ worshipped during the *Holi*.

b. Sacred Groves Preserved for Deity in An Informal Way

Some Sacred groves are maintained for the village deity but without a proper temple and stone idols of the deity. The *Mhasoba dang* of Dewade is a large sacred forest area (30 ha) well preserved and protected for village god *Mhasoba*. such groves or dangs are normally far away from the village settlement. The village may have another sacred grove with proper temple for festivals and rituals. Such groves have a better quality forest than the sacred groves of first category. Very limited sanctions (Sanctions here mean the permissions to use sacred groves and not the punishments for misusing the sacred groves) are allowed and no large timber trees are cut. Only NTFPs like fruit, seed etc. are collected either individually or collectively. Villagers do not visit such groves frequently. They perform simple rituals like offering coconuts to Mhasoba. Dang of Dewade is the only such grove recorded in the study area. (Such groves do occur in the other districts of Konkan viz. Shivapur from Sindhudurg Dist).

c. Groves Preserved as Cremation/Burial Grounds

The practice of preserving forest as a place for cremation/ burial is prevalent in Sangameshwar Tehsil but not as common as first category. Such groves are generally privately owned or form a part of the main village sacred grove. A classic example of such groves is that of Muradpur. This is a small grove of about 0.45 ha. The land is privately owned by a Gawali family. Such groves contain some very old *samadhies* which are now overgrown with ferns moss and *Ficus arnottiana* plants. These are in the memory of the ancestors of the land owners. This grove is used as a site for cremation and specific rituals carried out on the 10th and 13th day after death. Wood for cremation is not collected from the grove. People are reluctant to collect even fuelwood from such groves.

One part of the sacred grove of village Hativ is also used as a cremation and burial ground and protected separately. Groves at Muradpur, Hativ harbour giant tree specimens of *Terminalia bellerica, Lagerstroemia lanceolaria* etc. These cremation groves have a definite role in the traditions of the village. There is no temple in the grove.

Though the sacred groves provide ground for various rituals and traditions, and play an important role in the cultural life of the villages the vegetation maintained for deities around the temples is depleting at a very fast rate. It is clear that faith in the deity has not changed or deteriorated but belief in the concept of preserving forests for deities is vanishing.

The existing diversity within the groves is also under constant threat due to illicit felling, clearing the groves for establishing social forestry plantations, opening up for canal, roads and dams or rehabilitation of oustees or tourism. These developmental activities definitely provide employment to local people. Local people's urgent need is of survival and they are in most of the cases ready to get means of it at the cost of biodiversity.

It is very interesting to see the interventions of government agencies with villagers while implementing the developmental activities.

Conservation of Sacred Groves and the Role of Government Agencies

Traditions of local people are always ignored or described in negative terms by the government planners, extension workers etc. The local

Fig. 1.

Fig. 2.

traditions are regarded as superstitions and reminiscent of primitive past, even as a direct obstacle to development (Gerdan and Mtallo, 1990). Here we would like to discuss three cases of development from Ratnagiri District which are directly affecting the traditional beliefs and neglecting the local people's perception towards them. In these cases the so called modern development is given the utmost importance.

The sacred grove of Kulye is one of the largest Sacred groves recorded from the study area. The Sacred grove occupies the area of 22 acres and located along the hill side near the eastern border of Sangameshwar Tehsil. A small sized irrigation dam is being built near Kulye village on the Gadgadi river. One water canal providing water to far off villages is being built along the length of grove passing almost through the center of the grove. This has divided the grove into two unequal parts of which the smaller one is severely degraded. The remaining vegetation shows relic forest structure which is also disturbed due to opening up of the grove. This grove is on revenue land occupying more than 35 acres before the canal work started. It is managed as village common land. Villagers mainly use this grove to collect dried leaf litter or Patera. This grove harbours endemic species like *Anamirta cocculus* and giant *Ficus benghalensis* and *Ficus racemosa* trees. These lofty trees serve as breeding grounds for great pied hornbills. While designing Gadgadi project, Kulye villagers were not consulted for canal building. Clearing the central part of grove has encouraged the illegal cutting of bigger trees from the grove, in the name of canal. Villagers say that they would have suggested some alternatives to disturbing the grove, if consulted prior to the action.

Another is the case of sacred grove of Masrang near Sangameshwar. This is privately owned sacred grove occupying an area of 16 acres and located along the hill slope. It is rich in species content and storied vegetation structure still exists. Five owners of the sacred grove as well as villagers of Masrang worship the goddess Amba i.e., Deity of sacred grove and carefully follow the rules and taboos constituted by their forefathers. This grove has two wells providing water throughout the year and helps recharging other within the village. The vegetation, especially the trees is well protected. The trees are considered as residences of ancestral spirits as well as some evil spirits.

A small dam is being built near the village Rangav, five km away from Masrang. Government has decided to rehabilitate about 350 oustees of Rangav project on the sacred grove land. Accordingly orders are issued to the owners, of course handsome compensation is offered to them. The owners as well as villagers of Masrang are totally against clearing the sacred grove for rehabilitating the village. Here again the development planners have neglected the value of traditional belief systems. At the same time they are equally ignorant about the ecological value of this well preserved patch of forest. According to the villagers this sacred grove has many functions like recharging wells within the groves, providing large timber for local need (after permitted sanctions from village and owners), honey production resting and feeding place for wild animals and harbouring sacred trees for ancestral spirits. Clearing such grove, villagers believe will definitely be harmful to village as well as owners. Villagers feel other waste land could be made available for the oustees, but the government is considering the sacred grove as non agricultural waste land. Sacred groves owners have decided to fight the case against the government and have already started consultations for the same.

Similarly the intervention of social forestry department is not very encouraging in Ratnagiri Dist. Many sacred groves like Hativ and Chafawali have been cut for social forestry plantations i.e., exotic species like *Acacia auriculiformis* and *Gliricidia sapium* mainly for fuelwood. In Ratnagiri district very little revenue land which can be considered as wasteland is available for social forestry plantations. Only available village common land of sacred groves has been made target for completion of plantation figures, without understanding the value of remaining unique biodiversity of the groves. Similarly role of such an important traditional institution in the social life of local people has been neglected. Many villages are victimised for the idea of renovating existing ancient temple at the cost of cutting sacred groves and plantation of social forestry species.

The solution for conservation of biodiversity lies in empowering village administration and local people's participation for protection and augmentation (Mitra and Pal,1994). It is a long process and local people might continue felling trees but it would be far less than the havoc created by cutting sacred groves for canal building, rehabilitation and Social Forestry plantation.

Of course government intervention is not always negative. Recently the sacred grove of Marleshwar and forest surrounding that grove has been declared as tourist center. However state government has assured that no trees/ forest will be cut for tourism development. As per the plan this development will be completed within five years time. All these cases provide strong evidence that developmental strategies planned without consulting local people are the major causes of threat to traditional institution like sacred grove.

Traditional Institution and Present Legal Framework

Traditional institution of sacred groves as an institutional body has no legal meaning as such in the study area. Most of the sacred groves except one or two are situated on revenue lands. Though they are well preserved forest patches they are totally under the control of revenue department. Villagers still manage them on the basis of traditional management framework. Villagers feel that they have certain rights over sacred groves. Our recent inquiries reveal that the traditional sanctions and decisions taken by the community for management and use of sacred groves are not acceptable to the state administration. In few cases compliants have been lodged against the village Panchayat for cutting trees from sacred groves, though the decision has been taken in the village meeting. At times Panchayat had to pay fines for such acts. On the other hand, matters like illicit felling in sacred groves by an individual without villagers' consent has been neglected or sorted out without any legal action. While in some other cases if villagers come to know the breaking of the rule by individual, such offenses are managed on the village level just by penalizing the offender. The penalty money is used for sacred grove temple management. Due to such confusing scenario, villagers and village Panchayat are normally reluctant to take any responsibility of management and protection of the sacred groves, though they are using it traditionally for generations. Confusions and misunderstandings about the legal status of sacred groves thus pose a serious threat to the conservation and participatory management of sacred groves. Such situation may at times develop conflicts within the village. It is therefore utmost important to formulate the linkage of these traditional institutions with existing legal framework to achieve their protection and conservation by people's participation.

ROLE OF NGOs IN CONSERVATION OF SACRED GROVES

Non governmental organisations and voluntary organisations are very active in the field of rural development in Maharashtra. Their efforts are mainly concentrated on the watershed management, primary health and tree plantations. Few NGOs from study area like Matru Mandir and Siddhi Gram Vikas are actively engaged in activities like rural development, school nurseries, environmental education. But their focus is mainly to involve local people in developmental works that are managed at the village level and beneficial for the village. Sacred groves have been rarely studied and potential of such traditional forest reserves in conservation has not been considered seriously. Local NGOs could be involved in the sacred grove protection programme positively if made aware of the crucial importance of the sacred groves.

LEVEL OF AWARENESS AND FUTURE STRATEGIES

Repeated inquiries for last two years and field observations on the sacred groves revealed that most of the villagers are fully aware of the social dimensions and role of sacred groves in village life but are not aware of the ecological value of these groves. More importance is given to the shrine and deity than to the forest. In this context a phase-wise approach has been designed by Applied Environmental Research Foundation to involve the local people in the augmentation and protection of sacred groves. Details of the approach are discussed below.

Phase I: Understanding Local People's Knowledge of Resources and Their Value

This documentation is a pure research activity. Role of local communities is restricted as informants and positive observers. Such data collections are being done from fifteen sacred groves.

Phase II: Developing and Creating Awareness among Local People about the Resources and Their Value

To create and develop awareness about the resources among the local people is very important and difficult as well. It was found that generally 70% of the local people around the sacred forest areas are quite ignorant about the available resources i.e., locally used medicinal plants, decreasing availability of wild fruits etc. It clearly indicates that the present generation is least bothered and aware about plant resources around them. The two major target groups for awareness generation have been selected i.e., the adults from 20 to 40 years of age and school children from 8 to 15 years. Some knowledgeable village elders above 40 years have been selected as local experts or resources persons. Participation of such local experts has proved very important as the response of local people has increased with interaction of local experts. Awareness generation is a continuous process but it has to be very vigorous in the initial phases of the project. Awareness generation has been tried by using the conventional and non conventional media forms as well as competitions, tours etc. have also been used effectively. Participatory action planning has been initiated through such awareness generation programmes.

Phase III: Preparation of Action Plan for Conservation Protection and Augmentation of Resources

Such participatory planning involves lots of efforts on meetings and discussion to understand the community organisation within the village as well as social hierarchy. In the village clusters in Western Ghats the village level community organisation has become very weak and traditional system of work distribution has been disintegrated. This traditional system has now been replaced by village level political body called Grampanchayat. Other local level institutions like youth sports group of the village, village level watershed management committee (A recent development as a result of participatory approach in rural development) or religious song groups are more respected in the village. Through interacting with such groups we are trying to form the village resource protection committees. Extension work done by active NGOs in the area has proved very important for formation of Resource Protection Committee. With the help of such village resource protection committee or non formal group and local NGO, time schedule has been designed for the activities like ethnobiological survey of resource area (i.e. sacred grove), broad economics of NTFP available etc. The local experts or village resource persons are the key individuals for accurate data collection and analysis. Involvement of children's groups is continued, they are being involved in preparation of charts, posters or specimen collection to support the work carried out by the elder's group. Such activities involving larger group of villagers have increased the curiosity about the Plant Genetic Resources, certain rare or special plants available within the resource area etc. After these concentrated efforts and data analysis with local community, the community gets ready to prepare the action plan for protection of PGR areas and their further augmentation. Such action plan maybe different for different issues and natural resource management strategies prevalent within the area. People from the Ratnagiri District are more interested in augmentation plantation using fuelwood as well as local timber species, rather than providing complete protection. They are interested in collection of NTFPs from sacred groves.Some communities are even interested in developing short term commercial plantations rather than filling the gaps with local timber species.

Phase IV: Involvement of Local People in Protection and Augmentation, i.e. Implementation of Action Plan

Preparation of action plan by the community itself is a time consuming process. Its implementation is also equally complicated and time consuming, as benefit sharing starts from this phase in the real sense. While implementing such action plans, it is necessary to develop a system to organize this implementation. Many a times local NGO as well as researchers have to take initiative to develop such system. Again we have realised that it is necessary to develop a separate system, independent of Grampanchayat where representation from all the sections of the society could be achieved.

Phase V : Collective Benefit Sharing by Community

After 3-4 years of protection and augmentation of resource areas, using of participatory methods, the crucial stage is that of collective and equitable benefit sharing. Certain individuals get monetary benefits during the implementation period. But after 4-5 years harvesting of various NTFPs wild fruits, fuelwood etc. starts. Role of resource protection committees and local NGO is very important as they can formulate a system to regulate the harvesting, processing and marketing of the products as well as benefit sharing. The traditional pattern of community benefit sharing must be considered while designing such system. Benefit sharing pattern must evolved with the consultation of whole community and not by the certain individuals of the community.

Phase VI : Further Management of the Stabilized and Augmented Resource Areas

Stabilised resource area refers to the ability of a system (i.e., sacred groves in this context) to resist the disturbance and to recover to its original state. Through participatory planning of protection and augmentation i.e., to improve it by supplementary plantation of local useful wild species or desired species it is possible to develop stabilized resource area. It is again very long process and is totally dependent on continuos support from the communities. Once local communities realized that the protection and conservation of plant resources is beneficial, it is again very important to continue the process of participation. Management of such developed areas may be given to resource protection committee or village governing body i.e., *Gramapanchayat*. We suggest that it may be given to either of them by rotation. The planning of further management strategies and decision for appointing the management body must be taken by the community. Through this phase self mobilization within the communities should be achieved; which is extremely difficult.

So far we have completed first two phases in case of ten selected sacred groves and in a process of formulating sacred grove Protection committee in village Kundi where local people are keen to involve in the process of sacred groves management in a more formal way.

CONCLUSION

Study of sacred groves from Sangameshwar Tehsil has provided a definite pattern of socio-cultural aspects and changes that are dependent upon geographical locations and ecological status. Most important factor of these traditional factors is the reverence and sanctity attached to it in respect of their management especially use of its produce or to avail their services. Sacred groves are therefore the vegetation area socially designated as common property resource. These are quite flexible in terms of their uses by different villages. The study also revealed that the sacred groves are not only repositories of biodiversity or the means to maintain the pockets of desirable ecosystem or habitat but also forms the focus of cultural and ethical practices developed through Indigenous Knowledge for generations. It is also very important to note that sacred groves are vital for well functioning of the social systems.

There have been rapid changes in the demography, social and economic aspirations of the community at village level and political and administrative changes at policy level within last 20 years. Human intervention in the form of various developmental activities such as settlements roads dams etc. is adversely affecting the ecological balance of the Konkan region as a whole and increasing threat to the very existence and conservation of sacred groves in particular. Therefore the community's perspectives and social dimensions of the relevance of sacred groves have changed. Sacred groves symbolise the dynamic social forces linked with access and control over resources (though it is lacking any legal status).

Role of sacred groves in maintenance of biodiversity is undoubtedly significant. It is very important therefore to revive this traditional institution. and its further conservation. Awareness generation and participatory management are the key aspects for conservation of valuable biodiversity.

ACKNOWLEDGEMENTS

We are grateful to Misereor, Germany for providing financial support to this research project. Thanks are due to Dr. B.G. Kulkarni senior botanist from BSI, Western Circle, Pune. This work could not have been possible without the help of elderly people, school teachers from the villages selected as well as government officials of revenue and forest department of Ratnagiri District.

REFERENCES

Freeman, R. 1994. *Forest and the Folk : Perceptions of nature in Swidden Regimes of Highland Malabar.* Pondy Papers in Social Sciences, No. 15, Institute Français de Pondichéry, Pondichéry.

Gadgil, M. and Vartak, V.D. 1976. The Sacred groves of Western Ghats in India, *Economic Botany* 30: 152-160.

Gerdén, C.A. and Mtallo, S. 1990. *Traditional Forest Reserves in Babati District, Tanzania; A study in Human Ecology.* FTP, Working Paper, 128.Swedish University of Agricultural Sciences, International Rural Development Centre, Uppsala.

Induchoodan, N.C. 1988. *Ecological Studies of a Sacred groves.* M.sc. Dissertation, College of Forestry, Vellanikkara, Trichur.

Kalam, M.A. 1996. *Sacred groves in Kodagu District of Karnataka (South India): A socio-historical study.* Pondy Papers in Social Sciences, No. 21 Institute Français de Pondichéry, Pondichéry.

Mitra, A. and Pal, S. 1994. Besieged the forests of the gods. *Down to Earth, January* 31:21-36.

Nipunage, D.S., Kubhojkar, M.S. and Vartak, V.D. 1988. Studies on Sacred groves of Maharashtra Part 1. Observations on Sagdara grove in Pune district, *Indian Journal of Forest,* 11: 282-286.

Roy Burman, J.J. 1995. The dynamics of Sacred groves. *Journal of Human Ecology,* 6: 245-254.

Sharma , B.D. and Kulakarni, B.G. 1980. Floristic Composition and peculiarities of Devraies (sacred groves) in Kolahapur district, Maharashtra. *Journal of Economic Botany and Taxonomy,*1: 11-32.

Vartak, V.D. and Gadgil, M. 1981. Relic forest pockets of Panshet water-catchment area, Poona district, Maharashtra State, *Biovigyanum* 7: 145-148.

21

Sacred Mangroves in India

A.G. Untawale, S. Wafar and M. Wafar

National Institute of Oceanography, Dona Paula, Goa, 403004, India

INTRODUCTION

Mangrove vegetation is often a prominent feature of estuaries, backwaters, deltas and muddy seacoasts where the land meets the sea. As the mangrove species generally require higher atmospheric and water temperatures for growth and survival, their abundance is more luxuriant in tropical waters, tending through a reduction in density and diversity in subtropical zones to a total absence in temperate zones. The ecosystem constituted by a number of mangrove species, besides others, is referred to as 'mangal' in Goa or west coast of India.

About 60 mangrove species belonging to 41 genera under 29 families are known from Indian coasts. Among them, 58 species are known from the east coast and 37, from the west coast. Species such as *Rhizophora mucronata, R. apiculata, Ceriops tagal, C. decandra, Bruguiera gymnorhiza, Lumnitzera racemosa, Sonneratia apetala, Avecinnia ilicifolius, A. officinalis, A. marina, A. aureum and Excoecaria agallocha* and are most dominant and uniformly distributed along east and west coasts. However, eight mangrove species, among them *Sonneratia caseolaris, Sueda fruiticans* and *Urochondra setulosa*, are known only from the west coast. Other major organisms of economic importance inhabiting the mangals are shrimps like *Penaeus mondon, P. indicus* and *P. merguiensis*, crabs like *Scylla serrata* and fishes like carangids, mullets, hilsa and milkfish. The wild animals like tiger, crocodiles, wild pigs, monkeys, wolves, snakes etc. are also known from mangals (Chaudhury and Chakrabarti, 1973, 1974).

Indian mangals cover an area of 3500 sq. km (Blasco, 1975; Saenger et al., 1983). The largest mangal among them is that of the Gangetic Sundarbans, covering about 2000 km^2, followed by those of Andaman and Nicobar Islands (1000 km^2), that of Godavari and Krishna estuaries (200 km^2) and Mahanadi of Orissa (Untawale, 1985). The west coast of India has fringing mangroves along backwaters, creeks and sheltered areas.

Apart from the fishery yield the mangrove vegetation also have an economic value. The timber and wood products from mangroves find use in house and boat construction, as poles for securing lines including the telegraph lines and as firewood. A managed mangal can yield as much as 3-4 tons of timber ha^{-1} yr^{-1}. Tannin from the bark of some of the Rhizophoraceae such as *Rhizophora mucronata, Bruguiera gymnorhiza* and *Ceriops tagal* is used as a dye. Fruits of *Sonneratia caseolaris* and tender shoots of *Acrostichum aureum* and *Phoenix paludosa* are edible and the leaves of *Avicennia officinalis* and *A. marina* are often used as fodder for cattle. In addition, since ancient times, mangroves have also found a place in traditional medicine. For example, leaves of *Acanthus ilicifolius* are used in curing rheumatic disorders, leaves of *Bruguiera gymnorhiza* are used for treatment of blood pressure, leaves, fruits and roots (oil) of *Cynometra ramiflora* and the latex of *Excoecaria agallocha* is used in curing leprosy and several minor ailments. Other by-products from the mangals are salt from supra-littoral salterns, honey and wax. These by-products could also be of substantial economic value: for example, the Sundarbans mangroves alone produce more than 100 tons of honey every year (CIFRI, 1973).

The ability of the mangroves to trap the sediments with their prop roots render the 'mangal' a zone of strong accretion, counteracting shore erosion by other physical processes. The 'mangals' are also biologically high productive ecosystems and support a substantial capture fishery directly within the mangal and indirectly in the adjacent coastal or estuarine waters by serving as nursery grounds for the larvae and juveniles. The mangrove environment also sustains a large biodiversity including some endemic forms and migratory birds.

CONVENTIONAL CONSERVATION STRATEGY

Gulf of Mannar Biosphere Reserve (Tamil Nadu), Sundarbans Biosphere Reserve (West Bengal), Great Nicobar Biosphere Reserve (Andaman and Nicobar), North Andaman Biosphere Reserve (Andaman and Nicobar) include mangroves as an important component.

Notwithstanding the degree of scientific expertise or the number of possible means for conservation, no effort is going to be successful without peoples' co-operation. Peoples' participation may be envisaged by appropriate conservation of their socio-economic perceptions and the religious importance they attach to the nature. Such pockets are dedicated to forest Gods, Goddesses, saints or deity. The strong faith the people have in the deity translates into a number of religious beliefs, the most important among them being the destruction of the grove will invite wrath. Thus the sacred groves are in a way, traditional sanctuaries.

Quite a few mangrove sites are generally treated as sacred or given some sort of religious significance. Mangroves at such sites experience minimum or no damage at all. Here we present some such examples.

SACRED MANGROVES

Shravan-Kavadia, Rann of Kachchh

Shravan-Kavadia is an isolated temple in the Banni region of the Great Rann of Kachchh (22°-24°N, 68°-71°E). There is a mangrove patch of about 200 *Avicennia marina* trees near the temple. The trees are around 100 years old, with girth (at breast height) ranging from 50-200 cm and an average height of 12-15 m. Apart from this *Avicennia* patch, the entire area is desertic, devoid of any vegetation except for the *Suaeda maritima* species. The mangrove tress are never cut, either for sustenance or for any religious rites like lighting sacred fires. This mangrove patch has been able to withstand the extreme range in temperature (5°C-45°C) and tidal flushing with the Gulf of Kachchh waters. During the last 3-4 years, the tidal waters have rarely reached up to the Shravan-Kavadia temple. As a result, the subsoil moisture appears largely below a depth of about 15-20 cm. Though the mangrove trees have survived the extreme environmental conditions for so many years, the lack of tidal flushing, if it continues, might lead to their degradation.

Pirotan Island, Gulf of Kachchh

Pirotan island off Jamnagar is a small uninhabited island within the Gulf of Kachchh Marine National Park. The island is known for its rich biodiversity and has a good growth of *Rhizophora mucronata, Bruguiera* sp., *Ceriopus tagal and Avicennia marina*. A 'darga' (a religious site for Muslims) in this island is quite popular among the coastal communities who flock to this island for offering the prayers (Urs). The mangrove vegetation around the 'Darga', which is otherwise exploited extensively by these people for firewood and fodder in other islands, is never touched by the pilgrims, even when they camp here.

Khodiyar Mata, Gulf of Kachchh

Long ago there were crocodiles in the mangrove swamps of the Gulf of

Kachchh. *Crocodiles palustris,* a fresh water species but adapted to sea water was common. Because of large scale poaching, this species is almost extinct. However, there is a temple dedicated for this corcodile which is worshipped as a *Khodiyar Mata* (Goddess). Unfortunately, this is a case where even religious beliefs could not help conserve the biodiversity-the temple remains but not the crocodile.

Achra Mangroves, Maharashtra

Achra is a small coastal village in the Sindhudurg district of Maharashtra. A mangrove forest covering about 273 ha lies between the village and the sea coast. The dominant mangrove species are *Avicennia officinalis, A. marina, Bruguiera cylindrica, B. gymnorhiza, Sonneratia alba* and *S. caseolaris*.

The entire mangrove forest area of Achra belongs to the temple trust of Shri. Rameshwar, the deity installed in the local temple. There is no direct religious belief attached to the mangroves but the very fact that they belong to the temple trust is enough to deter people from cutting or otherwise exploiting the mangroves.

Crocodile Conservation in Goa

The crocodile (*Crocodilus palustris*) is a common inhabitant of the mangrove swamps of the Cumbarjua canal connecting the Mandovi and Zuari estuaries of Goa and upstream region of the two rivers.

A popular belief among Goan fishermen is that the crocodiles prey on the predator fish and thus help keep high the yield of the edible fish and prawns from mangroves. Over the years, this popular belief has assumed a religious cover to the extent that the worship of crocodile has become an annual feature. In the Bhoma and Durbhat villages, in the full moon night of March, a figurine of the crocodile is cast from mangrove mud, decorated with shells and flowers and offered animal sacrifices. Curiously, only males from the local community are present at this function, known as 'Mange-Thapani'. The religious significance thus attached to the estuarine crocodiles, helps in their conservation, and through it, the conservation of the mangroves on the whole.

Chidambaram, Tamil Nadu

In the magnificent temple of Chidambaram, Tamil Nadu, there is a stone-carving of the mangrove plant *Excoecaria agallocha*, locally known as 'Thillai. The stone-carving is worshipped everyday even now. In the earlier times the temple pond was surrounded by these plants and there was a belief that if the lepers take a bath in this pond, they would be cured of leprosy. This belief may be because of the medicinal property of the yellowish milky juice of the plants belonging to the family Euphorbioceae. We do not have documented evidences of the effect of extracts from this plant on leprosy, and it might be a worthwhile case to study. In the recent years, *Euphorbia agallocha* plants have disappeared from the temple pond due to the developmental activities.

Sunderbans, West Bengal

The Sunderbans delta is famous for two species: one is the Sundri tree (*Heretiera* spp.) and the other is the Royal Bengal Tiger (*Panthera tigris tigris*). Goddess Durga, the most popular deity in this region, is depicted in the mythology as riding tigers and lions. The associations of tiger with the goddess and its worship serves many purposes.

CONCLUSION

The coastal people, over the ages, have acquired a vast knowledge on the useful properties of various species of plants and animals. To conserve these species from extinction, these settlers knowingly used religious beliefs. Until recent times, the support of folk tales, folk songs, mythological stories etc. was taken to inculcate the spirit of conservation. As a result, the ecological balance was maintained. However, with the rapid urbanization in the last few decades, this concept faded. Because of this, in some areas only traditions have been left behind but not the species. For example, Khodiyar Mata and Chidambaram.

On the other hand, in a civilization that has evolved with assorted traditional religious beliefs, practices and mythologies, conservation written in the language of religious pertinence is a more effective tool of spreading the message. Thus, the association of the message with a religious practice is the surest way of encouraging the local populations to accept conservation measures.

ACKNOWLEDGMENTS

The authors are thankful to Mr. J.K. Tiwari, Wildlife and Environment, Kachchh, for providing valuable information on the mangroves of Kachchh region.

REFERENCES

Blasco, F. 1975. *Les Mangroves de linde (The Mangroves of India).* Institute Francais De Pondichery, Travaux de la section Scientifique et Technique, Rome XIV.

Chaudhury, A.B. and Chakrabarti, K. 1973. Wildlife biology of the Sundarbans forest marine fish. *Sci. Cult.,* 39: 8-16.

Chaudhury, A.B. and Chakrabarti, K. 1974. Wildlife biology of the Sundarbans forests. *Sci. Cult.,* 39: 8-16.

CIFRI Report, 1973. *Fishery Resources of Hooghy-Matlah Estuaries System.* Barrackpore: Bengal.

Saenger, P., Hegers, E.T. and Davie, J.D.S. 1983. Global Status of Mangrove ecosystems. *International Union of Conservation of Nature and Natural Resources.* **3:** 9-18.

Untawale, A.G. 1985. Mangroves of India: present status and multiple use practices. *In: Mangrove of Asia and Pacific: Status and Usage.* pp. 67.

22

Sacred Groves of Meghalaya

B.K. Tiwari, S.K. Barik** and R.S. Tripathi***

*Centre for Eco-development
**Department of Botany, North-Eastern Hill University, Shillong-793014, India

INTRODUCTION

Sacred groves of Meghalaya are public/village forest lands set aside for religious purposes under the traditional land tenure system (Gurdon, 1975). Customarily, it is an offense to cut trees from such groves except for cremation and religious purposes. Three types of forests under traditional forest classification system (and subsequently adopted by the District Councils) viz., *Law Lyngdoh, Law Kyntang and Law Niam* are considered as sacred forests. The sacred groves have been closely interwoven with the social and cultural life of the people and a number of rites, rituals and religious ceremonies have been associated with these forests. Besides, various religious beliefs and taboos, which vary from region to region, and even from one grove to the other, have been attached to the sacred groves and are being passed on from generation to generation. These religious beliefs have been instrumental in protecting these groves in their pristine form since ages.

As the extraction of forest produce from these groves is either minimal or absent the sacred groves are home to a number of plant and animal species which are not found elsewhere (Haridasan and Rao, 1985) and as such, they are very rich in biodiversity. They provide safe sites for reproduction of a variety of floral and faunal species (Darlong, 1995). They help in the maintenance of viable populations of pollinators and predators, conserve germplasm (Khiewtam, 1986) and serve as a potential source of propagules required for colonisation of wastelands and fallows. Besides, due to their dense vegetation often spread over large areas, sacred groves also provide several important ecosystem services.

Despite all their conservational, cultural and aesthetic importance, the sacred groves of Meghalaya are little studied. However, the two sacred groves, one located at Mawphlang and the other at Cherrapunjee, have engaged the attention of a few researchers. Hazra (1975) published a taxonomic account of the sacred grove at Mawphlang. Khan et al. (1986, 1987), Barik (1992), Barik et al. (1992, 1996a, 1996b), Rao (1992) and Rao *et al.* (1990,1997) studied various ecological aspects of this grove, such as community characteristics, gap phase regeneration and regeneration ecology of dominant tree species. Khiewtam (1986) and Khiewtam and Ramakrishnan (1993) studied the vegetation, litter and fine root dynamics, and nutrient flow in a sacred grove at Cherrapunjee.

SACRED GROVES: AN INDIGENOUS KNOWLEDGE SYSTEM

The sacred groves represent a long tradition of environmental conservation based on sound ecological principles being practiced by the three indigenous tribal communities of the State viz., Khasis, Jaintias and Garos since time immemorial. In view of the difficulty to make the common man understand the environmental and conservational values of these forests, the forefathers of these traditional communities, who realised the importance of such groves perhaps devised the simple ways for their conservation and perpetuation by attaching various religious beliefs with them. With the passage of time, it seems, sacred groves became a part of the cultural life of the people and most villages in Khasi hills had sacred groves situated near the summit of hills, composed of oak and rhododendron trees (Gurdon, 1975).

Considering the vast areas under the sacred groves in the State, big size of individual groves (unlike in other parts of the country where they are generally very small), their locations (mostly on the critical sites in catchment areas) and the religious sanctions attached to them, the concept of sacred grove conservation in Meghalaya seems to be an indigenous knowledge system conceived, developed and perpetuated by the indigenous tribal people of the State.

LEGAL STATUS OF SACRED GROVES

Unlike other parts of the country, the sacred forests of Meghalaya enjoy adequate legal support as these are covered under the United Khasi-Jaintia Hills Autonomous District (Management and Control of Forests) Act, 1958 and Garo Hills Autonomous District (Management and Control of Forests) Act, 1961. Both these Acts were passed in pursuance of paragraph 11 of Sixth schedule of the Indian constitution and were extended to all the forest lands of the State except 722.36 sq. km (constituting only *ca* 8% of the total forests of the State), which is directly under the control of the State forest department (Anonymous 1984).

As per the above Act(s), the sacred groves are to be managed by the *Lyngdoh* (religious head) or other person or persons to whom the religious ceremonies for the particular locality or village or villagers are entrusted and in accordance with the customary practice in vogue and the rules framed by the Executive Committee of the District Council(s) from time to time. As per section 7 of the above Act(s), no tree shall be felled in these forests without the previous sanction of the Chief Forest Officer of the District Council or any officer duly authorised by him in writing. Further, section 9 of the Act States, "No tree shall be felled or removed from *Law Lyngdoh, Law Kyntang* and *Law Niam* except for purposes connected with religious functions or ceremonies recognised and sanctioned by the *Lyngdoh* or other persons in accordance with section 4(b)".

Provision for registration of *Law Lyngdoh, Law Kyntang* and *Law Niam* with the District Council has been provided in the United Khasi-Jaintia Hills Autonomous District (Management and Control of Forests) Rules,1960.

STATUS OF SACRED GROVES

Information on the location, extent and status of sacred groves in Meghalaya is too sparse. Although there is a legal provision of registering all the sacred groves with the District Council authorities, no complete and comprehensive list of sacred groves in the State is available. In 1984, the State forest department estimated the area under sacred groves to be about 1000 sq. km (Anonymous, 1984). In 1995, we documented seventy nine sacred forests of the State irrespective of their status and ownership. Seventy nine such groves were located and denoted on a geographical map of Meghalaya. The area of individual grove and the canopy cover were ascertained to know the status of the sacred groves. The area of individual sacred grove varied between 0.01 ha and 900 ha. Of the 56 sacred groves, for which canopy cover was estimated, 7 (12.5%) belonged to undisturbed category (100% canopy cover), 14 (25%) had dense canopy (canopy cover $<100\%$ and $> 40\%$), 11 (20%) were sparse sacred groves (canopy cover $<40\%$ and $>10\%$) and a maximum of 24 (42.5%) belonged to open category (canopy cover $<10\%$). Out of total area of 10,511 ha, 138 ha (1.3%) was undisturbed. About 4420 ha (42.1% of the total area) had dense canopy, 2765 ha (26.3%) had sparse canopy and 3,188 ha with open canopy.

ECOSYSTEM SERVICES OF SACRED GROVES

Protection of Critical Area

Most sacred groves in the State (66 out of 79) covering an estimated area of 10,251 ha are located on the catchment areas of major rivers or rivulets. At least 58 sacred groves with an area of about 9,621 ha are located at the origin of perennial streams while about 38 sacred groves which covered 6,454 ha are on the steep hill slopes. For example, Lum Shyllong-Nongkrim sacred groves are located in an area where from as many as eight streams originate and supply water to a number of human habitations downstream. The other common sites of occurrence of sacred groves are the steep hill slopes (e.g., Trepale Jowai sacred grove). Being located on these critical sites, which are most vulnerable to degradation, sacred groves protect the land and soil from erosion and help in maintaining the quality of water in the streams downhill.

Conservation of Biodiversity

Sacred groves are repository of rich biodiversity and most often represent the climax vegetation of the area. They are home to a number of rare and endangered floral and faunal elements. At least 50 rare and endangered plant species of the State are now confined only to these sacred groves (Haridasan and Rao, 1985). A baseline floristic survey in various sacred groves of the State revealed that as many as 514 species representing 340 genera and 131 families were present in these forests. Orchidaceae is represented by the highest number of species (39) followed by *Poaceae, Rubiaceae* (28 species each) and *Rosaceae* (26 species). Among tree species, *Fagaceae* members are dominant over others in most of the sacred groves. Epiphytic flora is quite abundant. Various types of lianas and ferns are also found in these old growth forests. The vegetation of undisturbed sacred groves is generally very dense and well stratified. It has four major strata viz.; canopy layer, sub-canopy layer, shrub layer, and ground flora. Emergent trees constitute the top canopy layer, while tree species like *Rhododendron arboreum* and *Myrica esculent*a constitute the sub-canopy layer in most groves. Shrub layer is composed of a number of shrubs and saplings of tree species. Ground flora consists of grasses, herbs, ferns and bryophytic species along with the seedlings of the trees.

In general, species diversity in the sacred groves is much greater than the other forests. For example, Rao, et al. (1990) reported the species evenness index, Shannon index of general diversity and species richness index of a sacred grove at Mawphlang were 3.60, 1.00 and 0.96 respectively. On the other hand, in two other forests at upper Shillong and Malki, all the diversity indices were comparatively low (Species evenness index-Upper Shillong, 2.20; Malki, 1.3. Shannon index-Upper Shillong, 0.8; Malki, 0.6, Species richness index-Upper Shillong 0.84, Malki, 0.86).

The regeneration potential in Mawphlang sacred grove was compared with that of the two other forests (Barik, et. al., 1996a). The density of shade-intolerant species like *Schima khasiana* was more in the other disturbed unprotected forests, while seedlings of shade-tolerants like *Quercus* spp. were more abundant in the sacred grove. In general, high seed predation/ disappearance and competition for various resources among the seedling populations, specially for light and space hamper the success of regeneration in the sacred forest (Barik et al. 1996b; Rao *et al.* 1997).

The population structure of tree species in Mawphlang sacred grove was compared with that in a disturbed forest at Upper Shillong by plotting the proportions of seedlings, saplings, small trees and big trees expressed as percentages of the total tree density (Khan et al., 1987). The structure was an upright pyramid in case of the disturbed forest and it was an inverted pyramid in the sacred grove.

SOCIO-CULTURAL ASPECTS

Conservation of sacred groves in the State is closely inter-related with the socio-cultural life of the people. The traditional religions prescribed specific rites and rituals to be organised and performed inside/near the sacred forests. The religious heads (*Lyngdohs*) were made responsible for the performance of such rites and rituals. Rites and rituals varied widely from one sacred grove to another depending upon the socio-cultural set up, religion of the local population, and location of the grove. These rituals are performed with all its rigid procedures in those groves where people still follow the traditional religion(s) and believe in the traditional customs.

The fact that sacred groves of Meghalaya are in existence over a long period of time is evident from the presence of numerous tall monoliths erected in memory of departed elders of the local tribals. Even today, in some areas of the State, the practice of establishing a small grove near the graveyard/ cremation/burial ground is still in vogue and people believe that the souls of their departed relatives would rest in peace in these groves. In the case of most sacred groves local inhabitants believe that *Sylvan* deity would be offended if any product is extracted from these forests. They believe that the gods and spirits who live in the forest look after the welfare of the people and protect them from natural calamities, sickness and invasion of enemies.

The sacred groves are hitherto managed by a committee of nominated members belonging to local community which is chaired by the priest of the community. The priest is also responsible for performance of religious ceremonies and rituals.

RITUALS ASSOCIATED WITH SACRED GROVES

The rituals performed in the sacred grove at Pahempdem village in Ri-Bhoi district of the State have been described here to give some idea about the rituals associated with the sacred groves. The grove belongs to the 'Lyngdoh Syntiew' clan and is managed by the clan council. The belief that the *'U Ryngkew u Basa'*, a deity, lives in the grove, still prevails especially among the non-Christians of this village. They believe that this guardian spirit or *'U Ryngkew u Basa'* which lives in this grove from time immemorial, guards the villages and the people from any harm.

Religious ceremonies are regularly held every year by the 'Lyngdoh Syntiew' clan, who is vested with the right to organise and perform the rites and rituals in order to pay homage and to give thanks to these guardian spirits. Accordingly, every year in the month of April a religious ritual known as *'Ka leh-niam Pyrda'* is performed at a particular spot inside the grove by the 'Lyngdoh Niam' together with about seven or nine male members (in odd numbers) selected from different clans of the village.

The ceremony is performed by sacrificing white cock, dry fish, prawns, and ginger and rice beer (which is specially prepared by a mixture of rice and ginger). After the completion of the ceremony, a feast is organised where the offerings to the spirits are made.

Another ritual associated with the sacred grove is the *'Shad Rah Rynthei'*, which is usually held once in five years in the *Iing Niam* (religious house) of the Lyngdoh. This ritual is held for three consecutive days in the month of December. On the first day of this religious ritual, the ceremony starts just before sun-set near a pond known as *'Ka Pung Kyntang'*. Sand is collected from this pond and carefully wrapped in three different pouches using a white cloth, with the beating of drums accompanied by music. The sand pouches are properly kept in a shed carefully made for the purpose. In the early hours of the next day, a village elder together with the people assemble at the spot where sand pouches were kept, and offer prayer to the Gods and Deities for the well being of the whole village. The sand pouches are then taken by the female members of the village to the *'Iing Sad'* or religious house of the Lyngdoh. On reaching this place the pouches containing sand are opened carefully one after the other for prediction of the future of the village and the people. It is believed that if the sand in the first pouch turns into 'rice grains', then they will get a good harvest from their fields and the village in general will prosper in the coming years. In the second pouch, if the sand turns into 'ants', then they will get poor harvest out of their cultivation and the people will be inflicted with various diseases and will suffer from illness. If the third pouch containing sand does not show any sign of change, but remains as it was, then the general welfare of villagers and their crops will remain as usual.

On the third and final day, the ritualistic ceremonies culminate with a dance and merriment by the villagers in front of the *Iing Niam*. Unmarried male and female members of the community take part in the dance. A pig is sacrificed and afterwards the same is cooked for a feast which is joined by all the villagers. The ceremonies are complete by evening with a vote of thanks from the 'Lyngdoh Niam' where he mainly stresses on the moral rules and obligations of every villager for the upliftment of the tradition, culture and religion.

EROSION OF TRADITIONAL BELIEFS—THE RECENT TREND

Due to several socio-cultural-economic reasons, the traditional beliefs which were hitherto central to sacred grove conservation are now considered as mere superstitions. The traditional value of the groves is gradually being lost with the advent of modernity and education, improved access to once inaccessible sacred groves due to development of road communication, change from traditional religions to Christianity, and increasing aspiration of the people for a better way of life. Over the years, influence of administrative and the judicial authorities of the traditional institutions responsible for the management of groves has considerably diminished due to the establishment of administration and judiciary institutions by the government. All this has brought about an attitudinal change of the traditional societies in many areas of the State resulting in large scale destruction of sacred groves. In most of the areas, people responsible for the observance of rituals seem to have forgotten the rites and procedures for performing rituals. Economic constraints caused by smaller land holdings has resulted in the encroachment of sacred grove areas for agriculture. As a result, a plot of shifting cultivation could be found, in certain cases, even inside the sacred groves which were once totally prohibited for all uses other than the religious purposes.

In a recent study conducted by the authors (Tiwari et al. 1995), it was observed that in the undisturbed sacred groves traditional rituals are still performed in accordance with the customary beliefs. In moderately disturbed groves too where the canopy cover is good, the traditional rituals are performed but not so rigidly as in the undisturbed groves. In the sparse and degraded groves the traditional rituals are not performed.

At present, the rituals are known to very few people, most of whom belong to older generation (age >50). Most persons especially those belonging to younger generation (135 out of 150 interviewed) admitted that the religious belief that was central to sacred grove preservation is now considered as superstition. Only 14% of the people interviewed, seemed to know the significance of sacred grove conservation.

CONSERVATION STRATEGY FOR SACRED GROVES

The religious beliefs and rituals that were responsible for the conservation of sacred groves are now fast dwindling/ languishing and, therefore, these treasure houses of biodiversity can no longer be protected only through the religious belief. Urgent external intervention is essential if these forest patches providing valuable ecosystem services to the local communities were to be saved.

One such external intervention could be in the form of economic incentives to the people who are protecting/managing the groves and also to the people living around the groves. Creation of mass awareness about the intangible benefits and ecosystem services (Cairns and Pratt, 1995) provided by these forest patches and their biodiversity value would be an essential component of any sacred grove conservation plan. It is also required to develop ways and means for limited extraction of produce in order to sustain the interest of the people in preservation of the groves. Protection from fire, cattle grazing and unauthorised product extraction is paramount to any conservation programme and this can be achieved only through people's active participation.

Site-specific conservation/restoration strategies need to be evolved considering the status of the grove and socio-economic conditions of people responsible for its management. For instance, the sacred groves which are largely undisturbed and have dense canopy cover may need level I intervention under multiple tiers/levels intervention programmes suggested by Geller (1992), in which the interventions are least intrusive and involve maximum number of people. The activities under such intervention programme would include behavioural prompting through awareness programmes and some nominal incentives to the people in the form of community development programmes. On the other hand, the disturbed sacred groves may need higher level and more effective intervention processes (Level II under Geller's scheme), which would require increased costs in terms of materials and personnel for their restoration. The activities under this may include any one or combination of the following options: i) immediate rehabilitation of the degraded groves through artificial/natural/ aided natural regeneration programmes; ii) regenerating the adjoining village forests and ensuring their effective management for meeting the bonafide firewood, fodder and non-wood requirements of the villagers so that the pressure on sacred groves on these accounts is kept at minimum; iii) converting them into nature reserves and managing them with people's participation on the principles of joint forest management; iv) constituting them as outdoor recreation areas and environmental research and education centres with a provision to share the revenue accrued with the local populace.

Sacred groves, which are important from wildlife point of view and have potential to act as wildlife corridors may be identified and restored/ managed by the Wildlife Wing of the State/National Government(s). Besides, an institutional linkage involving the local village level traditional institutions, State, national and international organisations needs to be established for launching a joint coordinated sacred grove conservation programme.

REFERENCES

Anonymous. 1984. *Meghalaya Forests*. Chief Conservator of Forests. Meghalaya, Shillong.

Barik, S.K. 1992. *Ecology of Tree Regeneration along a Disturbance Gradient in a Subtropical Wethill Forest of Meghalaya*. Ph.D. thesis, North-Eastern Hill University, Shillong.

Barik, S.K., Pandey, H.N., Tripathi, R.S. and Rao, P. 1992. Microenvironmental variability and species diversity in treefall gaps in a sub-tropical broad-leaved forest. *Vegetatio* 103: 31-40.

Barik, S.K., Rao, P., Tripathi, R.S. and Pandey, H.N. 1996a. Dynamics of tree seedling populations in a humid subtropical forest of north-east India as related to disturbance. *Canadian Journal of Forest Research,* 26: 584-589.

Barik, S.K., Tripathi, R.S., Pandey, H.N. and Rao, P. 1996b. Tree regeneration in a subtropical humid forest: effect of cultural disturbance on seed production, dispersal and germination. *Journal of Applied Ecology,* 33: 1551-1560.

Cairns, J.Jr. and Pratt, J.R. 1995. The relationship between ecosystem health and delivery of ecosystem services. In: D.J. Rapport, C.L. Gaudet and P. Calow (Eds.). *Evaluating and Monitoring the Health of large-scale Ecosystems*. NATO ASI Series, Vol. 128. Springer-Verlag, Berlin, Heidelberg.

Darlong, V.T. 1995. Wildlife preservation and community action. In: B.K. Tiwari and S. Singh (Eds.). *Ecorestoration of Degraded Hills*. Kaushal Publications, Shillong.

Geller, E.S. 1992. *Applications of Behavior Analysis to Prevent Injury from Vehicle Crashes*. Cambridge Centre for Behavioral Sciences, Cambridge, Massachusetts. pp. 147-159.

Gurdon, P.R. 1975. *The Nature Races of India: The Khasis*. Cosme Publication, New Delhi.

Haridasan, K. and Rao, R.R. 1985. *Forest Flora of Meghalaya. Vol.1*. Bishen Singh, Dehradun.

Hazra, P.K. 1975. *Law Lyngdoh (Sacred Grove), Mawphlang*. Government of Meghalaya, Shillong.

Khan, M.L., Rai J.P.N. and Tripathi, R.S. 1986. Regeneration and survival of trees seedlings and sprouts in tropical deciduous and sub-tropical forests of Meghalaya, India. *Forest Ecology and Management,* 14: 293-304.

Khan, M.L.,Rai J.P.N. and Tripathi, R.S. 1987. Population structure of some tree species in disturbed and protected subtropical forests of north-east India. *Acta Oecologica Oeol. Applic.* 8: 247-255.

Khiewtam, R.S. 1986. *Ecosystem Function of Protected Forests of Cherrapunji and Adjoining Areas.* Ph.D. thesis, North-Eastern Hill University, Shillong.

Khiewtam, R.S. and Ramakrishnan, P.S. 1993. Litter and fine root dynamics of relict sacred grove forest of Cherrapunjee in north-eastern India. *Forest Ecology and Management,* 60: 327-344.

Rao, P. 1992. *Ecology of Gap phase Regeneration in a Subtropical Broad-leaved Climax Forest of Meghalaya.* Ph.D. thesis, North-Eastern Hill University, Shillong.

Rao, P., Barik, S.K., Pandey, H.N. and Tripathi, R.S. 1990. Community composition and tree population structure in a sub-tropical broad-leaved forest along a disturbance gradient. *Vegetatio,* 88: 151-162.

Rao, P., Barik, S.K., Pandey, H.N. and Tripathi, R.S. 1997. Tree seed germination and seedling establishment in treefall gaps and understorey in a subtropical forest of northeast India. *Australian Journal of Ecology,* 22: 136-145.

Tiwari, B.K., Barik, S.K. and Tripathi R.S. 1995. *Sacred Groves of Meghalaya: Status and Strategies for their Conservation.* North-Eastern Hill University, Shillong.

23

Status of Orans (Sacred Groves) in Peepasar and Khejarli Villages in Rajasthan

M. Jha, H. Vardhan, S. Chatterjee, K. Kumar and A.R.K. Sastry

World Wide Fund for Nature India 172 B Lodi Estate, New Delhi-110003, India

INTRODUCTION

The sacred groves called '*Orans*' managed by the Bishnoi community of Rajasthan, India are widely known for their conservation ethos of protecting the Khejari trees (*Prosopis cineraria)* and the Blackbuck (*Antelope cervicapra*). However, at present while some of these sacred groves are well protected many others are degraded. In the present study sacred groves, located in Jodhpur and Nagaur Districts, in the villages Khejarli and Peepasar respectively were considered to allow ready comparison of *Oran* functioning between a degraded and well managed groves (Fig. 1).

The villages Peepsar (between latitude 26°25′ and 27°40′ N and longitude 73°18′ and 75°15′) and Khejarli (between latitude 25°0′ and 27°31′ N and longitude 70°55′ and 73°52′) are conspicuous for their extreme dryness, large variations of temperature and highly variable rainfall and low humidity. The normal annual rainfall is reported to be 25-40 mm. The sacred grove at Peepasar is historically important as the birth place of Lord Jambeshwar founder of the Bishnoi sect. Although, the village Khejarli witnessed an unparallel event in history where 363 Bishnoi men, women and children gave away their lives to protect the Khejari trees in 1730 A.D., now the sacred grove at this village is practically non-existent and represents a degraded site, possibly due to its close proximity to the city of Jodhpur.

Fig. 1. *Prosopis cineraria* trees in a sacred grove in village Peepasar, Rajasthan

THE BISHNOIS

The Bishnois, a sect, was founded in 1486 A.D. by Lord Jambeshwarji (born 1452 A.D. at Peepasar to Lohatji and Hansa Devi). He was the 42nd descendant of Maharaja Vikramaditya of Ujjain. The community derives its name from the set of twenty nine (Bis = Twenty and Noi = Nine) precepts that they adhere to. It is an open forum of people belonging to any caste group or religion. Therefore, any one who starts practicing the twenty nine principles can be called a Bishnoi even though he may not suffix the word Bishnoi to his/her name. The sacred grove maintained by the Bishnois are called *Orans* where neither cutting nor axing of Khejari (*Prosopis cineraria*) trees is permitted. Even the young seedlings of this tree whenever they appear in the fields are tended to and protected. The Bishnois twenty nine rules provide protection to animals as well. Due to the efforts of the Bishnois, the Blackbuck (*Antelope cervicapra*) is still surviving in India (Banerjee, 1996). In Bishnoi sites, even the dead are not burnt but buried to protect destruction of biomass. The protection provided by the Bishnoi community at Khichan (Jodhpur) to a migratory bird species Demoiselle Crane (*Anthropoides virgo*), a bird of the Steppe region of Eurasia, which breeds in drylands, is yet another example of this community living in harmony with nature.

The ecological importance of the Khejari tree (*Prosopis cineraria*) is well understood by the Bishnois. The tree withstands the harsh condition of the Indian desert and remains green throughout the year and comes up naturally in fields and wasteland. Having a strong tap root system which goes beyond 40 ft underground, it is not affected by periodic droughts in the region. Being a leguminous tree, it helps in increasing soil fertility by fixing nitrogen and thereby aids increased production in fields. Leaves shed by the tree enrich the soil, lopped dried leaves serve as fodder, small branches and twigs obtained by lopping provide fuel wood, the pods when green are cooked as a vegetable, the wood used in construction of houses and making of agricultural implements. The bark is grinded and mixed with the flour during scarcity of food and famine. *Prosopis cineraria* is a keystone species in the desert ecosystem (S.M. Mohonot, pers. comn.). Many of the Bishnoi Sacred Groves are frequented by faunal species like blackbuck (*Antelope cervicapra*), Jackal (*Canis aureus*), Wolf (*Canis lupus*), Indian fox (*Vulpes bengalensis*), desert cat (*Felis libyca*), Indian desert gerbille (*Meriones hurriane*), and many species of birds. The wetlands in Bishnoi sites attract many migratory avian species.

Every Bishnoi village has within it one or more *Oran*. The word *Oran* comes from the sanskrit word 'Uparanya' which means small forest. The word could also have been derived from the Hindi word *'Auron'* which means for others or not for one's own use. In Rajasthan it is mandatory for every Bishnoi Temple (*Sathri*) to have an *Oran*. Those Bishnois who live on their agricultural fields called *Dhani* make efforts to maintain at least one *Oran* near their Dhani. The ultimate objective of the *Oran* therefore is to allocate a specific area where the birds and in some cases, the wild animals can be fed.

The administration of the *Oran* is undertaken by the Gram Panchayat for the people of the village. The committee of five elders is required to ensure that the *Oran* land is safe from the trespassers. All decision regarding the *Oran* land and its resources is taken by the elders in consultation with the village folk. The management of the *Oran* land is in the hands of the temple committee. All the funds and donations received by the committee are utilized for the upkeep of the temple of the *Oran*, and to meet the needs of the saints or priests looking after the *Oran* premises. The ownership right of the *Oran* land lie with the State Government. Although *Orans* are considered repositories of biodiversity in the desert ecosystem, on the land records of the government, *Oran*, land falls under the *'silvay chak'* i.e., arable land which may be allotted by the Government.

ORANS AT PEEPASAR

Peepasar encompasses an area of 8,000 ha. The focal point of the village is the main village, referred to as Peepasar Gaun- covering an area of 3.52 ha. Surrounding this on all four sides are the personal agricultural fields and pasture lands- this is referred as *'Kankar'* covering an area of 7,996.48 ha. Peepasar has four sacred groves namely *Jambhoji ki Oran* (12.8 ha), *Peepenji ki Oran* (4.8 ha), *Maharaj ki Oran*, (17.18 ha), and *Hanuman ji ki Oran* (2.08 ha) and thus covering an area of 36.8 ha. The village also has *Oodji ki Oran* (0.8 ha) owned by an individual and dedicated to his father. The list of plant species recorded form these sacred groves is placed at Table 1.

It would be interesting to place in this paper an anecdote on the genesis of the concept of *Oran*. In 1916 A.D. a fire disaster (*Agnikand*) racked the small village of Peepasar, virtually bringing life to a standstill. An year later, superstitious and God fearing people dedicated small pieces of land from their personal land holdings to their ancestors in order to invoke their blessings and as a mark of respect to them. They used these lands as pastures (*Gochar*) and no other activities were undertaken by them.

Table 1. Major plant species recorded from the Orans of Peepasar, Naguar district, Rajasthan, India.

1. TREES
   - *Acacia jacquemontii* Benth.
   - *Acacia nilotica* (L.) Del.
   - *Acacia senegal* (L.) Wild.
   - *Balanites aegyptiaca* (L.) Delile.
   - *Maytenus emarginata* (Wild.) Ding Hou.
   - *Prosopis cineraria* (L.) Druce
   - *Tecomella undulata* (Smith) Seem.
   - *Ziziphus mauritiana* Lamk.

2. SHRUBS
   - *Calligonum polygonoides* Linn.
   - *Calotropis procera* (Wild). Dryand ex W. Ait.
   - *Clerodendrum phlomides*
   - *Leptadenia pyrotechnica* (Forsk.) Decne
   - *Lycium europaeum* L.
   - *Saccharam munja* Roxb.
   - *Sericostoma pauciflorum* Stocks. ex Wight
   - *Withania somnifera* (L.) Dunal
   - *Ziziphus nummularia* (Burm.f.) Wight and Arn

3. PERENNIAL HERBS
   - *Aerva Persica* (Burm. f.) Merril
   - *Boerhavia diffusa* L.
   - *Citrullus colocynthis* (L.) Kuntze
   - *Corchorus depressus* (Linn.) Stocks
   - *Crotalaria burhia* Buch. - Ham. ex Benth
   - *Fagonia indica* Burm. f.
   - *Farsetia hamiltonii* Royale
   - *Heliotropium mariafolium* Retz.
   - *Heliotropium subulatum* Hochst ex. DC.
   - *Indigofera linnaei* Ali
   - *Launaea resedifolia* (Linn.) O. Kuntze.
   - *Melothria maderaspatana* (L.) Cogn.
   - *Peristrophe bicalyculata* (Retz.) Nees
   - *Polygala irregularis* Boiss.
   - *Pulicaria angustifolia* DC.
   - *Pulicaria crispa* (Cass.) Benth & Hook. f.
   - *Tephrosia falciformis* Ramaswami
   - *Tephrosia purpurea* (L.) Pers.
   - *Trianthema Portulacastrum* Linn.
   - *Zaleya redimata* (Melville) Bhandari

4. GRASSES
   - *Andropogon pumilus* Roxb.
   - *Brachiaria ramosa* (Linn.) Stapf.
   - *Brachiaria reptans* (Linn.) Gard et C. E. Hubb.
   - *Cenchrus ciliaris* L.
   - *Cenchrus setigerus* Vahl.
   - *Cymbopogon jawarancusa* (Jones) Schult
   - *Cynodon dactylon* (L.) Pers.
   - *Eragrostis tremula* Hochst ex Steud.
   - *Desmostachya bipinnata* (Linn.) Stapf.
   - *Panicum turgidum* Forsk.

Contd.

Table 1. Contd.

5. SEDGES
    Cyperus rotundus Linn.
    Scirpus supinus Linn.

6. INTRODUCED SPECIES
    Acacia tortilis (Forsk) Hayne.
    Albizia lebbeck (L.) Benth.
    Azadirachta indica A. Juss.
    Dalbergia sisso Roxb.
    Eucalyptus terreticornis.
    Ocimum sanctmn.
    Prosopis chilensis (Molina) Stuntz
    Ricinus communis L.
    Tamarix dioica Roxb.

The Jambhoji Ki Oran

This sacred grove is located at the north- north west of the Peepasar. There are only 190 trees in this *Oran* which are Khejari (*Prosopis cineraria*), Kumta (*Acacia senegal*) Kankeri (*Maytenus emarginatus*), Rohida (*Tecomella undulata*) and Bordi (*Ziziphus mauratiana*). The ground vegetation consisted of Khimp (*Leptadenia pyrotechnica*), Massa (*Tephrosia purpurea*), *Boerhavia diffusa*, and *Saccharam munja* etc. Major faunal species observed in this *Oran* were Indian Gazelle, (*Gazella gazella)* Blackbuck (*Antelope cervicapra*), Desert fox (*Vulpes benghalensis*), Desert Cat (*Felis libyca*) etc. Bird diversity was high which included Kestrel (*Falco tinnuncolus*), White winged vulture (*Neophron perenopterus*), Common peafowl (*Pavo cristatus*), Red wattled Lapwing (*Vanellus malabaricus*), Spotted owlet (*Glaucidium radiatum*), Green Bee Eater (*Merops orientalis*), Red vented bulbul (*Pycnonotus cafer*), Indian roller (*Coracias benghalensis*), Large grey babbler (*Turdoides malcolmi*), Indian robin (*Saxicoloides fulicata*), Yellow wagtail (*Motacilla flava*), and tailor bird (*Orthotonue satorius*).

The soil in this *Oran* is compact, with sandy surface soil (*Balu Mitti*) and undersurface soil (*Peeli Mitti*). This undersurface soil is perceived by the villagers as rich in nutrients and hence referred to as *Pukki Mitti*. The survival of the grove rests on rain, the only source of water. This *Oran* serves as a common pasture for the livestock of the villagers. Any kind of extraction is forbidden from this *Oran*. However with increasing needs of the growing population, leniency has crept in and the poor amongst the community are permitted to extract resources of the *Oran*. The Gazelle are a nuisance for the crop of the nearby *Dhabi* and the Desert fox poses a threat to their livestock, specially the sheep and the goat. The villagers are highly aware about the rules and regulations of the Government prohibiting killing of the Gazelle, which is a protected species. Though on sly, some inhabitants indulge in killing the Gazelle for its meat.

The Maharaj Ki Oran

This grove is located at the north-west of the main village at a distance of approximately 3 km from the northern borders of the *Jambhoji ki Oran*. Flanked on the eastern and western sides by sand dunes, the *Oran* rests in a depression akin to a valley. The dunes on both sides help collection of rain water in the *Oran*. Adjacent to the Oran is a village pond (*Gomali Nadi*) which acts as the natural reservoir of the rain water and used by the villagers solely for drinking purpose. The *Oran* is under the charge of the *Maharaj* who acts as the caretaker of the *Saathri*. This ensures that the needs of the *Maharaj* and the other occupants are met and drawal of the resources of *Jambhoji Ki Oran* is not necessary. The *Maharaj* leases out this land to villagers, for agricultural activities, against payment of a certain amount of money. A fifth of the payment is given to *Jambhoji's Saathri*. There are a total of 156 trees of Khejari (*Prosopis cineraria*), Kankeri, (*Maytenus emarginatus*), Bordi (*Ziziphus mauritiana*) and Rohida (*Tecomella undulata*). The ground cover is comprised of Khimp (*Leptadenia pyrotechnica*), Bui Massa (*Aerva persica*), Kali Bui (*Heliotropium subulatum*), Sonali (*Pulicaria angustifolia)* and Dhaman (*Cenchrus ciliaris*) etc. All the faunal species mentioned in *Jambhoji ki Oran* are seen in this oran as well. When not being used for agriculture, the Oran serves as pasture for the livestock of the nearby *Dhani*.

The Peepenji Ki Oran

This grove being surrounded on four sides by agricultural fields is located over the sand dune (*Dhora*). The deity of *Peepanji ki Oran* is Peepenji Bhagat, a *Rajput* and a devotee of Lord Shiva who lived long before the birth of Lord Jambeshwar. The village 'Peepasar' is named in his memory. This *Oran* is the common property of the local villagers and falls under the control of the Revenue department, Tehsil Nagaur. Neither access to the resources and extraction from *Oran* is restricted by any laws or regulations nor villagers pay any revenue for its usage. The *Oran* depends on rain water for its requirement. The nature of the soil is similar to that of the *Jambhoji ki Oran*. There are only nine *Prosopis cineraria* trees, in the *Oran*.

The Hanumanji Ki Oran

This grove is linked to the deity 'Hanumanji' and the control of the land lies with the Revenue department, Nagaur Tehsil and the villagers themselves. This *Oran* too is a common property of the villagers. This *Oran* was earlier referred to and made use of as a *Gochar*. Around 5 years back, however, the area was cleared for the construction of the school and

Fig. 2. A 'Nadi' (village pond) near Khejarli, Rajasthan

sub health centre and the clearing was undertaken by the Government. On account of clearing away of the land, the vegetal cover is lost and there are only 5 Khejari (*Prosopis cineraria*) trees. The area thus has lost the characteristics of an *Oran*.

THE ORAN AT KHEJERLI

In olden days, the present *Oran* land served, as the catchment of the Mansarovar/Khejerli Nadi (pond) and, as the grazing land (*Jord*) for a local land lords' horses. The need for a separate grazing land was due to the fact that, in the event of a war, the Zamindar (land lord) would be required to send to the king of Khejerli, men and horses. These horses therefore, needed proper care and feeding. This land was under the direct control, administration, management and ownership of the Zamindar of the village and encroachment, trespassing on this land was consequently ruled out.

After independence (year 1947), with the enforcement of the Ceiling Act by the then Government, this land with consent of the Zamindar and the villagers was categorised as *Oran*. Coupled with increasing livestock pressure and food requirements, the patronage which Vilayati Babool (*Prosopis juliflora*) received from the then king began the story of the *Oran*'s decimation. (Today the area bears no semblance to an *Oran*. *Prosopis juliflora,* an exotic, is the dominant species in the *Oran* land. There is absolutely no restriction on felling this tree for firewood and timber. With regard to firewood, this tree gives the villagers a daily return of Rs.300/-.

As per the land records available with the village office, the *Oran* in Khejarli was situated in a scattered form to the north-west and north of the village. It was fragmented into six portions by muddy roads (Kuccha Rasta). To the north of the village west was 3.30 ha, and to the north 125.93 ha of *Oran* land. This was exclusive of the ponds which covered a total of 33.37 ha. Within the premises of the *Oran*, on its fringes, are two temples, one dedicated to Lord Shiva and the other to Lord Hanuman.

A scrutiny of the land records of 1954-1973 and 1995-1996 revealed that a total of 19.47 ha of *Oran* land had been allotted through Government permission for various purposes such as, land for a Polytechnic College, land for pasture and, setting up of huts for the victims of the flood of 1979 and establishment of a settlement.

The two communities the Bishnois and the *Raikans* at Khejarli are in conflicts regarding the usage of resources from the *Oran*. The Bishnois being a traditionally agricultural caste group, rear cows and buffaloes due to socio-religio-economic reasons. The *Raikans* on the other hand being a livestock rearing group due to the traditional and economic reasons rear large herds of goat and sheep. Grazing by goat and sheep leave no vegetation for the cows and buffaloes, the Bishnois complain. The Bishnois suggestion of using the community fallow land as pasture by the *Raikans* is not acceptable to latter resulting into a stalemate situation or stand off.

Land shown as *'Oran'* as per the land records is perceived either as *Gochar* or settlement area by the local communities. Table 2 shows the status of *Orans* of Khejarli as on official records, perception of the villagers and field observations. Assessment of people's knowledge regarding the biodiversity of their oran revealed that to 80% of the sample population the word *'Oran'* was unfamiliar. A simple test conducted to test the ability of local people to recall the twenty-nine principles of Bishnois sect, three principles of which relate to protection of nature, revealed that only the age group of 9-20 years and 50-60 years of the sample population could recall the same.

Table 2. Status of Orans in Khejarli , Rajasthan, as on land records, perception of villagers and observations of the Research Team

Sl. No.	Khasra No.	Area (ha)	Land Records (1995)	Peoples Perception
1.	146	16.34	*Oran*	Agor/Gochar
2.	148	16.96	*Oran*	Agor/Gochar
3.	349/1	0.53	*Oran*	* Settlement
4.	330	9.97	*Oran*	** Lions Nagar (Settlement)
5.	323	36.67	*Oran*	Gochar
6.	321	68.40	*Oran*	Agricultural Field
7.	192	7.55	*Oran*	Gochar
8.	189	1.14	*Oran*	Gochar

* Khasra No. 349 is Gochar (Pasture) on land records. However, the area has now been encroached on by new (about 2 years old) residence of the village.

** Settlement established by Lions Club, through community consensus, for the victims of the floods in 1979.

Therefore with practically no *Oran* land, the concept of *Oran* is almost lost in Khejarli. Plants recorded around the temple complex of Lord Jambeshwarji include horticultural and ornamental species like *Punica granatum, Psidium guajava, Parkinsonia aculeata, Eucalyptus* sp. *Cassia siamea* etc. A detailed list of species recorded in the temple complex is placed at Table 3.

Contrary to the belief that *Orans* are units for the in-situ conservation of biological diversity managed by the Bishnoi community, our findings in the study locations indicate that uncontrolled grazing, over exploitation of *Oran's* resources and diversion of *Oran* land for construction activities have resulted in the current dilapidated status of the *Orans*. All the *Orans* of Peepasar are being used as common pasture land. Except *Jambhoji ki Oran*, where usage of its resources is regulated, the remaining three *Orans*

Table 3. List of plants species recorded in the temple complex of Khejarli, Jodhpur district, Rajasthan, India.

1. Trees

Ziziphus mauritiana
Parkinsonia aculeata
Azadirachta indica
Prosopis juliflora
Prosopis cineraria
Albizia sp.
Dalbergia sissoo
Cassia siamea
Pongamia pinnata
Ficus religiosa
Punica granatum
Psidium guajava
Eucalyptus sp.
Salvadora oleoides

2. Shrubs

Tephrosia purpurea
Peristrophe paniculata
Achyranthes aspera
Datura metel
Capparis decidua
Ziziphus nummularia

3. Herbs (including grass)

Phyllanthus fraternus
Dactyloctenium aegyptiacum
Sida sp.
Indigofera sp.
Euphorbia sp.

may be considered as common property resources. In *Peepanji ki Oran*, the control factor is non-existent. As a result of this, it is used in an unrestricted manner by the adjacent *Dhanis*. *Maharaj ki Oran*, is leased out by the *Maharaj* of *Saathri* to the villagers for agricultural purposes. *Hanumanji ki Oran* is nearly decimated following the construction of a school and a sub health centre in its premises. It is likely that the existing pressure on these three *Orans* may shift towards *Jambhoji ki Oran* to meet the unmet demands of the local population. The *Orans* of Khejarli have disappeared, but the *Orans* in Peepasar could still be saved. Although our study indicates that *Orans* do not necessarily serve as repositories of biodiversity, the *Orans* of Peepasar harbors healthy stands of *Prosopis cineraria* and increasingly becoming rare species like *Tecomella undulata* (Rohida) whose felling has been banned by the Government and *Calligonum polygonoides* (Phog), a species heavily exploited for fuelwood and which has almost disappeared from areas close to human settlement. Though few phog shrubs were noticed in an *Oran* it is pre-mature to say that the phogs are found only in an *Oran*. *Leptadenia pyrotechnica* although observed within and outside the *Orans* is currently under exploitation for cordage and for its medicinal value. *Acacia senegal* is increasingly becoming scarce due to over exploitation. (R. Mishra, pers.comn.).

STRATEGIES FOR CONSERVATION OF THE *ORANS*

To ensure the survival of the *Orans*, we therefore suggest the following measures:

1. Develop strategies for revival/conservation of the *Orans* in consultation with all the stakeholders *viz*. Government, local *panchayats* (a village level elected body), civic bodies and the local communities.
2. Management authorities must be approached to review the present practices of converting *Oran* in an *ad hoc* manner to pasture land.
3. A long term sustainable development strategy needs to be adopted by a) local and, b) district administration, to meet needs of local communities during scarcity and drought conditions, so as to maintain the sacred groves unexploited and reserved for long term bio-diversity gains to the user communities.
4. Management authorities needs to identify the age old existence of sacred groves in various villages, especially in the desert terrain, so as to enable the local communities to continue to receive the ecological and biodiversity benefits from the *Oran* as has been received by them hitherto.
5. Urgent measures deserve to be initiated to conserve the Sacred Groves which are severely threatened, some of which are driven to the point of near extinction by way of:
   a) restricting interference by the user community to ensure natural regeneration through the management committee of the *Oran*.
   b) new plantations of native species to be added seasonally, if need be.
   c) revival of cult of sacred groves among younger generation individuals who appear to be less oriented towards the old traditional values of harmonious living with nature.

7. Evolve a meaningful communication package to make stakeholders aware of the ultimate gains from such green lungs.

ACKNOWLEDGEMENTS

We thank Mr. Samar Singh, Secretary General, WWF-India, Prof. Ramakrishnan and Dr. Saxena of Jawaharlal Nehru University for their encourangement and guidance during the study period. Thanks are due to Prof. S. Mishra, Department of Botany, University of Rajasthan, Jaipur, Dr. S.M. Mohonot, Director, School of Desert Sciences, Jodhpur, Dr. Ishwar Prakash, Jodhpur for their valuable suggestions. Mr. Raghuvansh Saxena, Manager, Community Biodiversity Conservation Movement, WWF-India has provided the administrative co-ordination. We thank Mr. G. Sukumar and Ms. V.P. Mallika for the secretarial assistance. Our deepest gratitude goes to the inhabitants of the villages Peepasar and Khejarli who provided us with valuable information making the study possible. Financial assistance for the study through UNESCO, India and the Dutch Embassy in India are acknowledged.

REFERENCE

Banerjee, S.R. 1996. *Protectors of Blackbuck : Bishnois*. Environ., 3: 10-13.

24

Sacred Groves in the Rural Landscape: A Case Study of Shekhala Village in Rajasthan

G.S. Singh and K.G. Saxena***

*Centre for Sustainable Environment and Heritage,
New Delhi-110 022, India
**School of Environmental Sciences, Jawaharlal Nehru University,
New Delhi-110 067, India

INTRODUCTION

Poverty, population growth and traditional resource use practices have been regarded as the major factors causing environmental degradation and loss of biodiversity in the developing countries. This perception led to the establishment of National Parks and Wildlife Sanctuaries where displacement of local communities and/or restrictions on traditional resource use practices was a management imperative. This wildly held view which dominated the mind-set of conservation policy makers for a considerably long period of time has been challenged in the recent years. One of the arguments put forth in favour of appreciation for sustainability and conservation by the traditional societies is the practice of maintaining sacred groves in different parts of the world (Gadgil and Vartak, 1976; Frazer, 1980; Messerschmidt, 1987; Khiewtam and Ramakrishnan, 1989; Ingles, 1994; Ramakrishnan, 1996; Singh et al., 1996 Nair et al., 1997 and Singh, 1997). The role of sacred groves in protection of biodiversity and enhancement of environmental quality by the Bishnois in Indian desert and surrounding arid/semiarid climate zones has received recognition in the recent past. Sacred groves have been portrayed as islands of greenery in the desert landscape. However, an objective analysis of environmental, economic and social functions of sacred groves in a wider context of environment-livelihood-development interphase and changing attitudes towards original religious norms of sacredness is lacking. Quantitative assessment of biodiversity status of sacred groves and other ecosystem types in the village landscape are altogether lacking. This study therefore aimed for a detailed landscape level analysis of biodiversity management in a typical village Shekhala (95 km north-west of Jodhpur) in Indian desert region.

THE RURAL LANDSCAPE

Based on land use-land cover, the village landscape could be classified into five classes viz. irrigated agriculture, rainfed agriculture, private grazing lands, common grazing lands (locally referred to as *Gauchar*) and sacred grove (locally referred to as *Oran*) (Table 1). Spatially, agricultural land forms the central core of the village, common grazing land and sacred grove the periphery and private grazing lands fall in between the two. Houses are also scattered over the agricultural land. Small ponds are more common in the *Oran* than in common grazing land. Ponds in sacred grove and a well in settled area were the sources of potable water until tubewells were established by 1980s. There is a community tubewell established by the government and 4 private tubewells owned by the richer sections of the society. Even now, the economically weaker sections of the society settled, on the village margin and close to *oran* and *gauchar* make use of the ponds. This landscape configuration is a common feature of desert villages except the areas where government canals have reached in recent years. In most of the villages, common grazing lands and *oran* constitute a minor land use-land cover. There are only a few villages where sacred groves are extensive e.g., Kolu Bapuji, Osia, Lohawat-Bhergao, Kokran-Ram Deo oran, Phalodi, Ram Deora and Peepasar. Average land holding size is 9 ha, much larger than the national average.

RELIGION AND CULTURE

The Indian desert had been a territory of the Rajputs, basically a warrior caste/clan of the Hindu religion. In 1485 A.D. a youngman named Jambheshwar propounded 29 tenets and believers in these tenets organized themselves as Bishonois. Prohibition of cutting green trees, castration of bullocks, selling of animals to butchers and non-vegetarian diet are the tenets reflecting conservation norms set through the instrument of religious cult. There seems some sort of incipient hatredness between the Bishnois and non-Bishnois. Non-Bishnois in many instances put forward a different story behind the origin of Bishnoi cult. Rajput warrior married a muslim girl in the historical times and because of this inter-caste marriage they had

Table 1. General Feature of Village Shekhala, District Jodhpur, Rajasthan

Location	95 km north-west of Jodhpur
Total families	175
Total population	1056
Number of castes	21
Land use-land cover (ha)	
Irrigated agriculture	133 (7.5%)
Rainfed agriculture	905 (51.22)
Private grazing land	596 (33.73)
Common grazing land (Gauchar)	27 (1.53)
Sacred area (Oran)	83 (4.70)
Built up area	23 (1.30)
Total Livestock	2475

to face social boycott. To establish their social identity, they established a new cult in the name of Bishnois. Many non-Bishnois argued that the practice of burying dead body is the main feature which distinguishes Bishnois and non-Bishnois. Bishnoi cult could be considered similar to Sufi Movement which gave more importance to humanity rather than specificities of Hinduism or Muslim religion.

ORIGIN OF SACRED GROVES

During the period of princely states, some uncultivated areas were reserved for use by the king and feudal lords. The areas strictly meant for hunting and wood requirements of the kind were called Kailara and the areas meant for sustaining the horses of feudal lords were called Jord. The remaining uncultivated area was open to grazing by livestock of the people and was called as *Gauchar*. With increasing pressure on land partly because of population and livestock pressure and partly because of encroachement in common grazing land by the powerful individuals, carrying capacity of grazing land started deteriorating. Under these situations, isolating an area in the name of a deity or warrior evolved as a mechanism of arresting unsustainable ways of human exploitation. *Lotai Ki Nadi* and *Gogade Ki Oran* are parts of the sacred grove of village Shekhala dedicated to Lotai and Gogade who sacrificed their lives fighting with the enemies. Sacred areas in the name of deities and distinguished departed souls are appreciated both by Bishnois and non-Bishnois. Though each village, irrespective of presence or absence of Bishnois, has its own sacred grove(s), some such as Peepasar and Khejadli are more popular. Peepasar is distinguished for being the birth place of Bishnoi religious supreme Jambheshwar. Khejadli has become popular for being the place of historical incidence of sacrifice of 363 human lives of Bishnoi followers to protest against the felling of *Prosopis cineraria* trees ordered by the king in 1730 AD.

RELIGIOUS AND SOCIO-CULTURAL CONTEXT OF BIODIVERSITY CONSERVATION

Social and religious functions inside the sacred groves are not common. Old generation of Bishnois could recall that they used to perform some rituals decades ago. These were few individuals, generally of *Azadirachta indica* and *Prosopis cineraria* which are worshipped and not the entire grove area. It is believed that the deity or departed soul resides in these individuals and entire grove is their dwelling. The deity and departed souls are believed to enable livelihood in extreme environment prevailing in the desert. Except for water, people believed that sacred area should not be used in any manner directly by the human beings. Indirect uses such as grazing should be undertaken only when there are no other options for sustaining the livestock. People equated *P. cineraria* in desert with *Ocimum sanctum* in other ecological zones of India. Both Bishnois and non-Bishnois perform some rituals on the festival of Dusshehra, the day when Lord Ram (symbolizing the good) got victory over Ravana (symbolic of the bad in Hindu scriptures). For this purpose, all individuals of *P. cineraria* irrespective of its habitat are considered sacred. Ironically, this species has not been planted in artificial sacred grove created around a Bishnoi temple at Khejadli. The artifical grove consists largely of Eucalyptus alongwith a few other species.

BIODIVERSITY STATUS

In the Shekhala village landscape as a whole 59 species were sampled. Species richness was highest in private grazing land (41) followed by sacred grove (39), common grazing land (19) and afforested area (6). Private grazing land was richest in grasses and sacred grove in tree species. All the 4 tree species introduced in afforested area were exotic (*Prosopis juliflora, Acacia tortilis, A. nibica* and *Parkinsonia aculeata*) and 2 grasses (*Aristida adscensionis* and *Eragrostis ciliaris*) regenerated naturally following fencing of the area (Fig. 1). Hardly half of the species were common to two vegetation types (Table 2). The exotic tree *Prosopis juliflora* seems to be a potential invader and has dominated the common grazing lands subjected to uncontrolled grazing. *Acacia nilotica, Azadirachta indica* and *Maytenus emarginata* among trees, *Acacia jacquemontii* and *Lycium barbarum* among shrubs, and *Arnebia hispidisima, Fagonia cretica, Heliotropium ellipticum, Pulicaria wightiana,* and *Trianthema portulacastrum* among herbs were sampled only in sacred grove. *Tecomella undulata*, called commonly as the 'teak of the desert' occurred only in

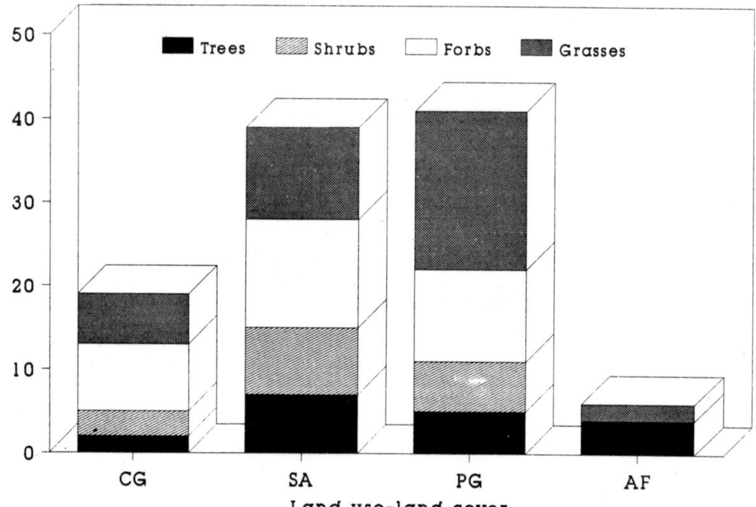

Fig. 1 Life form wise species richness in different land use-land cover types. CG, Common grazing land; SA, sacred area; PG, private grazing land; AF, afforested Area

Table 2. Similarity indices between different land use-land cover types

	Common grazing land	Sacred area	Private grazing	Afforestation land
Common grazing land	1.00	0.55	0.47	0.16
Sacred area		1.00	0.48	0.09
Private grazing land			1.00	0.13
Afforestation				1.00

private grazing land. *Tephrosia purpurea*, a non-palatable species, was more abundant in common grazing lands as compared to private grazing lands. *Tribulus terrestris* and *Dactyloctenium aegyptium* were present in private as well as common grazing lands but absent in sacred grove (Table 3). Nearly all girth classes of trees are present in private grazing land and sacred grove. In common grazing land, both the younger and older stages of trees were absent indicating poor regeneration/excessive wood removal (Fig. 2). Dominance-diversity curves show private grazing lands to be most diverse followed by sacred grove and common grazing lands (Fig. 3).

RESOURCE USES AND BIODIVERSITY MANAGEMENT

P. cineraria seems to be valued more for its food, medicinal, fodder and shade values than for any religious sentiment. It is not cut because it is neither the best timber nor the best fuelwood. *Tecomela undulata* is considered the best timber and *Acacia senegal* the best fuelwood species.

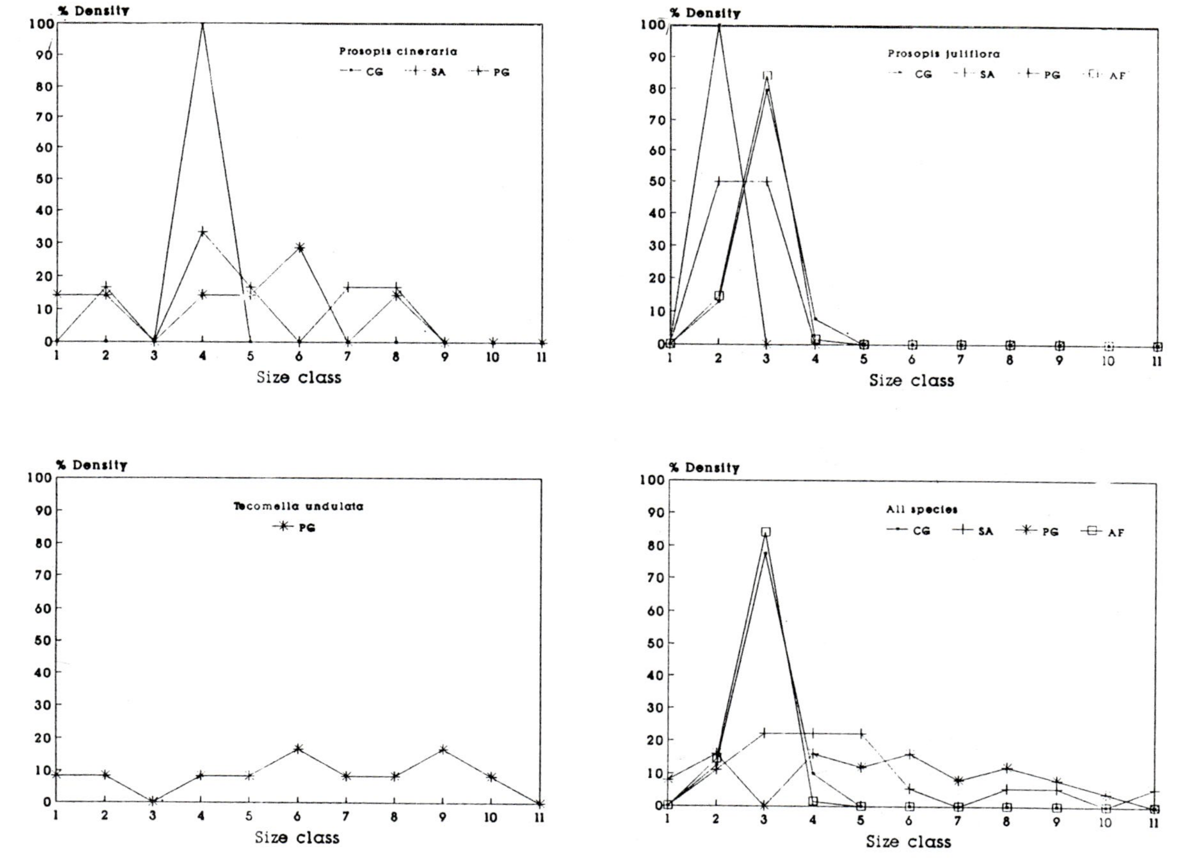

Fig. 2 Girth class distribution in different land use-land cover types. 1, <10 cm; 2, 10-20 cm; 3; 20-30 cm; 4, 40-50 cm; 5, 50-60 cm; 6, 60-70 cm; 70-80 cm; 8, 80-90 cm; 9, 90-100 cm; 10, >100 cm diameter at breast height; other details as in fig. 1.

Table 3. Importance value indices in different land use-land cover types

Species	Common grazing land	Sacred area	Private grazing land	Afforestet area
Trees				
Acacia nibica	-	-	-	65.52
Acacia nilotica	-	6.88	-	-
Acacia senegal	-	12.17	11.30	-
Acacia tortilis	-	-	-	98.70
Azadirachta indica	-	11.41	-	-
Maytenus emarginata	-	4.89	-	-
Parkinsonia aculeata	-	-	-	9.55
Prosopis cineraria	2.56	17.77	13.25	-
Prosopis juliflora	45.00	3.04	0.85	13.47
Tecomella undulata	-	-	30.40	-
Zizyphus mauritiana	-	1.73	2.46	-
Shrubs				
Acacia jacquemontii	-	1.47	-	-
Aerva persica	-	3.11	3.07	-
Calotropis procera	25.59	4.69	7.45	-
Capparis decidua	2.35	5.86	3.06	-
Crotalaria burhia	-	8.49	3.77	-
Leptadenia pyrotechnica	-	11.18	1.23	-
Lycium barbarum	-	0.64	-	-
Zizyphus nummularia	1.97	4.87	3.24	-
Forbs				
Arnebia hispidisima	-	1.62	-	-
Boerhavia diffusa	3.49	3.78	3.41	-
Citrullus colocynthis	-	-	2.42	-
Citrullus lanatus	-	-	5.55	-
Convolvulus microphyllus	1.68	2.33	1.92	-
Corchorus aestuans	3.49	3.64	-	-
Cucumis callosus	-	-	1.13	-
Fagonia critica	-	1.31	-	-
Heliotropium ellipticum	-	1.45	-	-
Indigofera cordifolia	27.12	17.00	11.90	-
Indigofera linifolia	7.10	10.16	4.76	-
Pulicaria wightiana	-	1.16	-	-
Pupalia lappacea	7.85	3.04	-	-
Tephrosia purpurea	59.45	26.24	18.31	-
Tephrosia uniflora	-	10.85	7.23	-
Tephrosia wallichii	-	-	3.21	-
Trianthem portulacastrum	-	4.92	-	-
Tribulus terrestris	11.84	-	3.38	-
Grasses				
Aristida adscensionis	-	32.01	9.00	38.40
Aristida articulata	20.84	15.20	8.99	-
Aristida funiculata	42.73	10.71	-	-
Aristida mutabilis	-	-	3.52	-
Brachiaria ramosa	-	-	7.23	-
Cenchrus biflorus	-	3.92	20.19	-
Cenchrus ciliaris	-	-	15.44	-
Cenchrus setigerus	-	3.33	30.02	-

Contd.

Table 3. Contd.

Species	Common grazing land	Sacred area	Private grazing land	Afforestet area
Dactyloctenium aegyptium	15.49	-	3.21	-
Dactyloctenium sindicum	-	26.94	6.90	-
Digitaria adscendens	5.52	-	-	-
Eragrostis ciliaris	12.72	6.83	9.19	74.28
Eragrostis pilosa	-	-	3.13	-
Eragrostis viscosa	-	-	6.59	-
Eremopogon foveolatus	-	3.34	4.99	-
Heteropogon contortus	-	-	4.10	-
Lasiurus sindicus	-	-	10.39	-
Panicum antidotale	-	-	3.83	-
Panicum turgidum	-	2.32	3.77	-
Perotis indica	-	-	2.76	-
Sporobolus tenuissisimus	-	6.24	3.43	-
Tragus racemosus	3.20	3.49	-	-

P. cineraria is considered to be a species which adds to soil fertility and does not compete with agricultural crops for soil resources. On the other hand, *T. undulata* is believed to reduce food crop yields. These perceptions in indigenous knowledge could explain dominance of *T. undulata* in private grazing lands and *P. cineraria* in farm lands. Uprooting of non-palatable herbs and shrubs is common to check natural regeneration of species poor in respect of fodder, fuelwood and timber uses in private grazing lands. Maximum pressure on utilization of biomass is in common grazing land. There is practically no social or legal control on grazing, lopping and cutting in the commons and this could account for their low biodiversity status.

Religious norms have promoted dominance of trees in sacred area but these areas as at present are subjected to biomass removal. Survey of people's perceptions in village Shekhala revealed that, at present, 10% of total requirement of fuelwood, 20% of livestock feed and 40% of medicinal plants, wild edibles and other non-timber forest products of the village are met from sacred grove (Table 4). Tree cutting is becoming increasingly common particularly by the outside-village community. Availability of water was the most valued attribute of sacred area followed by protection of private lands from erosion and other direct uses.

ECOSYSTEM DEGRADATION

People perceive exogenous factors the ultimate and main factors behind the diminishing significance of traditional conservation practices (Table 5).

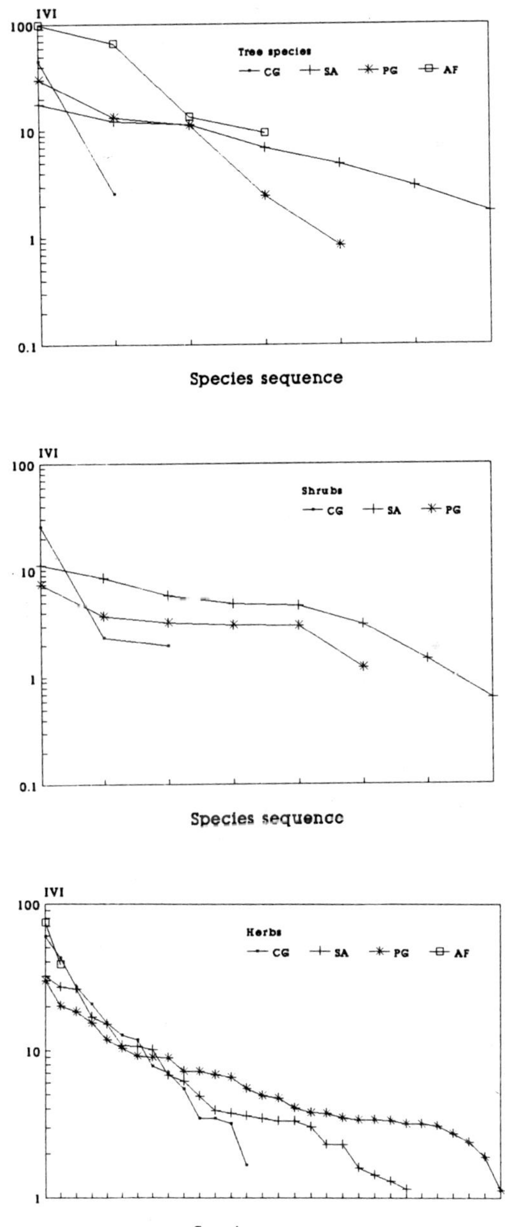

Fig. 3 Dominance-diversity curves of different land use-land cover types; othe details as in fig. 1.

Table 4. Direct benefits (% of total) from different land use-land cover in village Shekhala

Land use-Land cover	Common grazing land	Sacred area	Private grazing land	Agricul-tural land	Afforestet area
Fuelwood	40	10	46	2	2
Timber/small wood	20	0	80	0	0
Fodder/grazing	30	20	38	10	2
Medical plants and other plant products	5	40	40	15	0
Water	0	15	0	0	0

Table 5. People's perceptions on factors causing erosion of traditional conservation ethos

Factor	% response
Population growth	5
Livestock population growth	35
Increasing attention to monetary economy	90
Disregard of traditional land rights system	90
Government resource management and development machinery	87
Fragmentation of the society	92

Socio-cultural-religious norms fostering conservation are fast changing under a variety of forces. The resource use and conservation norms in traditional societies seemed to be adapted to prevailing ecological constraints, property regimes, subsistence economy and small isolated societies. In people's perceptions, property regimes imposed through the State, increasing preferences for monetary economy, development interventions promoting neglect/exploitation of traditional conservation practices and politicization of conservation attitudes of traditional societies for vested interests are the major factors causing ecosystem degradation. People are concerned not only for the degradation threats to sacred groves but also for the degradation of common grazing lands as livestock are the backbone of livelihood in the deserts.

State commitments for providing drinking water facility at the door step eroded traditional practice of avoiding the human disturbances around ponds in sacred groves. Allotment of land to landless in/around sacred grove has been a common political factor causing dissolution of cultural-religious sentiments. Historical incidence of human sacrifice by the Bishnois at Khejadali seems to have been exploited more for political mileage than for promoting conservation and sustainable utilization of the resource base. There is considerable income from donations for religious sentiments to the temple at Khejadli but this is being invested for non-conservation purposes like construction of buildings, motorable transport and felicitation of political leaders and expansion of temple complex. Plantation of Eucalyptus, an exotic and practically of no value to local human and wildlife needs around the shrine at Khejadli and a variety of exotic species within sacred grove in village Shekhala are indicative of the extent of loss of religious appreciation for indigenous species. Traditional worship of the deity in nature is being replaced by worshipping the deity in idols in artificial rooms. Natural tree falls in sacred grove area are now auctioned to generate income for village council or shrines. However, even the green trees are cut at the time of removal of auctioned dead wood. Village community has no power to penalize the outsiders.

Government efforts have concentrated on agricultural intensification in private lands and infrastructural development to integrate the small scale subsistence oriented village societies into larger society and market economy, resulting in marginalization of local socio-cultural controls for protection of grazing lands and the sacred areas. Competition for maximizing monetary gains over a short period and vested political interests caused fragmentation of traditional cohesive village society and erosion of collective initiatives to retain and enhance traditional conservation norms.

CONCLUSIONS

The important tentative conclusions emerging from this study are: (a) practice of putting least biomass use pressure in sacred areas is common to both Bishnoi and non-Bishnoi communities (b) conservation and sustainable utilization of biodiversity needs to be looked into at the scale of village landscape comprising sacred grove, common grazing land, private grazing land, agricultural land and water bodies (c) biodiversity in sacred groves could be enhanced/improved through improvement in productivity of other ecosystems by way of improvements in traditional resource use practices (d) the objective of biodiversity conservation can be enhanced only when the local people realize economic benefits from conservation in the present circumstances. These tentative conclusions need to be confirmed and elaborated through further studies.

ACKNOWLEDGEMENTS

We thank to Prof. P.S. Ramakrishnan of Jawaharlal Nehru University, Ms. Sudha Mehndiratta, UNESCO Regional Office, New Delhi, TSBF (Nairobi) and Dr. Rajendra Parihar of Jodhpur University for support.

REFERENCES

Frazer, J.G. 1980. The Golden Bough, MacMillan, London.

Gadgil, M. and Vartak, V.D. 1976. The sacred groves of Western Ghats. Economic Botany, 30, 152-160.

Ingles, A.W. 1994. The influence of religious beliefs and rituals on forest conservation in Nepal, Discussion paper, nepal Australia Community Forestry Project, Kathmandu.

Khiewtam R.S. and Ramakrishnan, P.S. 1989. Socio-cultural studies of the sacred groves at Cherrapunji and adjoining areas in the north-eastern India. Man in India, 69, 64-71.

Messerschmidt, D.A. 1987. Conservation and society in Nepal: Traditional forest management and innovative development. In: *Land at Risk in the Third World: Local Level Perspectives.* Edited by P.D. Little, M.M. Horowitz and A.E. Nyerges, 373-397, Westview Press, Colorado.

Nair, G.H., Gopikumar, K., Krishnan, P.W., and Kumar, K.K.S. 1997. Sacred groves of India—vanishing greenery. Current Science, 72, 697-698.

Ramakrishnan, P.S. 1996. Conserving the sacred: from species to landscapes. Nature and Resources, 32, 11-19.

Singh, G.S., Saxena, K.G., Rao, K.S. and Ram, S.C. 1996. Traditional knowledge and threats of its extinction in Chhakinal Watershed in north western Himalaya. Man in India, 76, 1-17.

Singh, G.S. 1997. Sacred groves in western Himalaya: An eco-cultural imperative. Man in India, 77, 247-257.

25

Conservation through 'Socio-cultural-religious Practice' in Garhwal Himalaya: A Case Study of Hariyali Sacred Site

B. Sinha and R.K. Maikhuri***

*School of Environmental Science, Jawaharlal Nehru Univeristy, New Delhi, India
**G.B. Pant Institute of Himalayan, Environmental and Development, Srinagar (Garhwal), Uttar Pradesh, India

INTRODUCTION

Garhwal Himalaya in India commonly referred to as Dev Bhumi (land of the Gods) houses many important religious shrines like Badrinath, Kedarnath, Yamnotri and Gangotri etc. besides the sacred confluence of five tributaries of holy Ganga. It is interesting to note here that many a times an entire landscape represented by a variety of species and ecosystems had been considered sacred and conserved as such in pristine condition by forbidding the use of any resource from it. This strategy seems to be quite analogous to the present days concept of species conservation through sanctuaries, national parks and biosphere reserves. A number of sacred groves are also reported from different parts of Garhwal and they include Tarkeshwar, Kot, Nandisain, Paabo, Dewal, Chapdon and Hariyali (Fig. 1). The present study aims to identify and enlist the 'socio-cultural religious' valued species and also to report the existence of the sacred groves in this region. Hariyali sacred forest is discussed here as a case study from Garhwal.

SACRED SPECIES OF THE GARHWAL HIMALAYA

The history of conservation of plant species by the traditional societies of

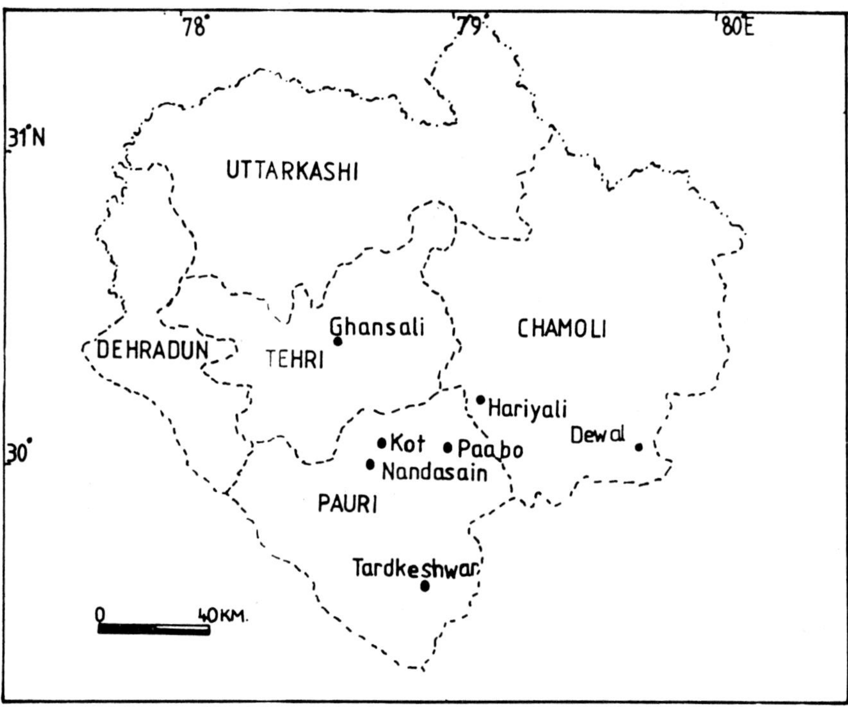

Fig. 1 : Map showing distribution of sacred forests in Garhwal Himalaya. (O) indicates the location of sacred forests

Garhwal Himalaya dates back to millennia. There are many plant species of different economic importance which have been conserved by the locals based on intuitive knowledge which has ecological significance too. By evolving a unique strategy of species conservation by attaching sacred value to them, many are conserved from excessive exploitation (Table 1).

HARIYALI SACRED FOREST

Hariyali sacred forest (Fig. 2) is situated above Kodima village which is at a distance 32 km from the nearest town Gauchar (enroute Badrinath shrine) in Chamoli district (Garhwal Himalaya of Uttar Pradesh). Three villages Kodima, Jasholi and Pavo form the boundary of sacred landscape as they are directly linked through the rituals of the Goddess. The summit of the hill which abodes the goddess is 2850 m from mean sea level and the sacred forest cover 5.5 sq. km in area (according to the boundary of the forest which encircles totally prohibited area of forest from human interference). Demarcation of sacred forest (protected zone) from the

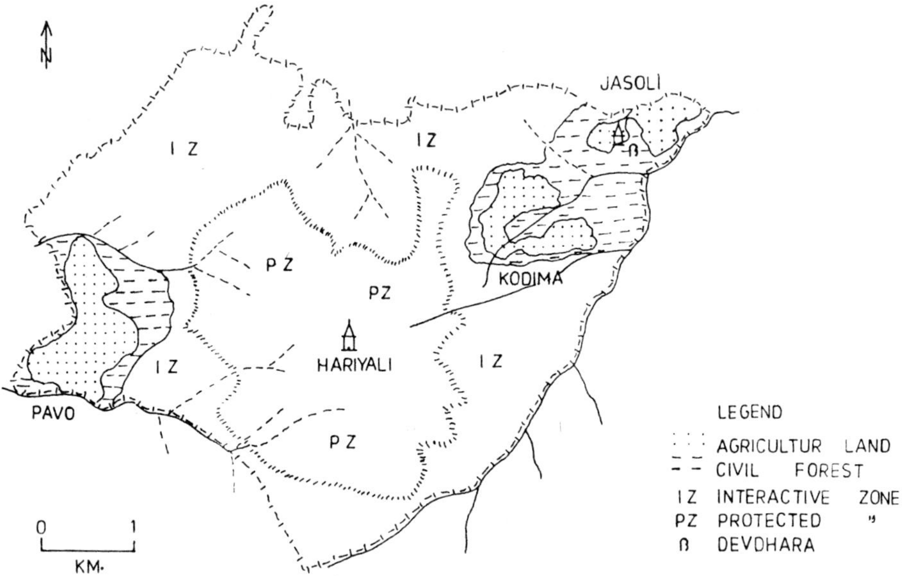

Fig. 2 : Hariyali Sacred forest in Garhwal Himalaya.

interactive zone has been done by taking four spots from four different directions namely Banso, Lathwanakhal, Kakhotu, Hitgaja where people of different villages put off their shoes considering the above forest as completely sacred.

The cultural landscape boundary extends from 30°13' to 30°19'N and 79° to 79°7'E (Fig. 3). The cultural landscape boundary includes villages which worship Hariyali as their saviour and participate in the religious rituals like Doli Yatra etc., encompasses about 15 villages and inhabited by more than 6,000 people belonging to three different caste groups i.e., Brahmins, Rajput and Schedule Castes.

The status of the forest is reserve forest as it falls under the jurisdiction of Forest Department. This forest is dominated by *Quercus* and *Rhododendron* spp. which fits in 'Himalayan moist temperate forest' type according to Champion and Seth's classification. The herbaceous layer underneath mainly comprises of *Thalictrum foliolosum, Strobillanthus dalhausinus, Hedychium Spicatum, Geranicum walichianum, Reinwardtia indica, Swertia chiraita* etc.

This sacred grove can be considered as one of the largest sacred groves in India covering an area of 5.5 sq. km. A wide variety of forest types including oak mixed pine forests and broad leaved forests of oak-*Lyonia* and *Rhododendron* are seen here (Plate 1).

Table 1. Sacred species of the Central Himalaya

Plant species	Local Name	Material use	Religious cultural value	Ecological features
Pinus roxburghii	Kulain	Fuel, timber & manure	Worshipped on the occasion of marriage	Early successional stress tolerant fire adapted
Cedrus deodara	Devodar	Timber, oil of medicinal value	Planted in temple premises	Susceptible to fire
Ficus religiosa	Pipal	Fodder	Temple premises	Open non-forest areas
Aegle mormelos	Bel	Edible fruit	Leaves offered to Lord Shiva	Dry bare slopes with *Sapium* Adina & Mallotus
Ficus bengalensis	Bar		Worshipped on a particular new moon day	Open areas
Opuntia	Nagphani	Edible fruit	Protect from evil spirit	Degraded areas
Euphorbia royleana	Sullu	Fencing	— do —	— do —
Zanthoxylum spp.	Timru	Teeth cleaner & spices	— do —	Wide ecological amplitude
Dandrocalamus hemiltonii	Bans	Buidling material handicrafts	Dead body lifted out	Degraded open areas
Ziziphus numularis	Ber	Edible fruits, fodder & fencing	Offered to Lord Shiva	Degraded forest
Datura stramonium	Datura	Opium	— do —	Scrubs
Cannabis sativa	Bhang	Edible seed & fibre	— do —	Grow above 1500 masl
Ocimum sanctum	Tulsi		Worshipped as a Goddess	
Searamura indicum	Til	Oil & medicinal value	Offered to sacred pyre	Kharif season crop
Amarathus sp.	Marchha	Edible grain	Offered to Lord Ganesh	Kharif season crop
Fagopyrum tatairicum	Phafar	Flour	Chappatis/puris offered to Goddess	Kharif season crop
Cynodon dactylon	Doob	Fodder & medicinal	Used to offer water to God	Open areas
Saclarum spp.	Babla	Fodder, thaching material	Rope used as pure for water pots for offerings	Exposed calcarios rock
Prunus cerasoides	Payan	Fuel	Leaves & wood, used in sacred pyre	Degraded forest on village margin
Cedrella toona	Ton	Timber	Planted in temple premises	An associated species on sandy soil
Bauhinia vahalii	Malu	Fodder & edible seed	–	Shaded areas at altitude <1000m
Quercus spp.	Baj	Fodder, fuelwood & timber	–	Dominant climax species at elevation 1400-2200 m

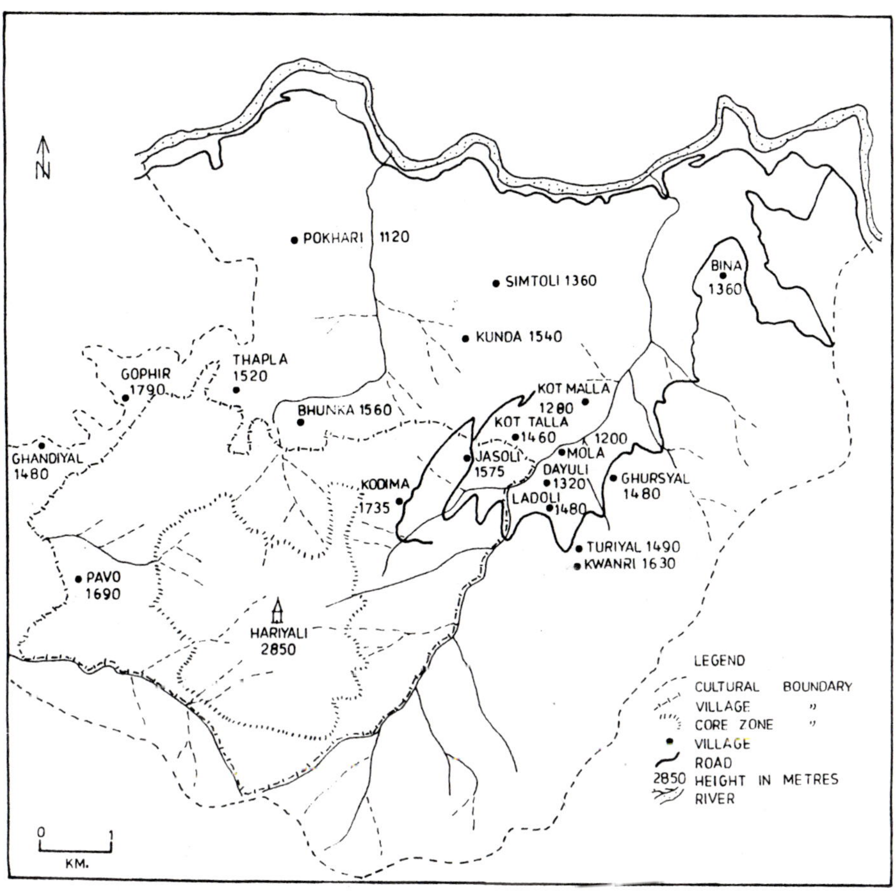

Fig. 3 : Cultural landscape of Hariyali sacred forest.

Myth of Hariyali Devi

The myth which prevails is that according to the *Bhagwat Puran*, *Yogmaya* was the sister of Lord Krishna, and who replaced him in the cell of his parents when Kansa threw her against the wall, she turned into lighting and came to 'Hari Parvat' to make her abode. Since then she came to be known as 'Hariyali Devi' and the adjoining forest is called as 'Hariyal'.

Three other Gods considered as guards associated with the Hariyali Devi referred to as **Hit** located at different places surrounding the Hariyal and for whom small stony temples have been built.

A myth most widely believed is that related to the location of the sacred site, i.e. the abode of the Hariyali Devi was known when a cow from the adjoining village **Pavo**, suddenly used to disappear at night; the

Plate 1. Hariyali Devi Temple

owner could not understand why the cow was not lactating in the evening. So one night he followed the cow and found her lactating upon a stone. The same night, in his dreams, the goddess dictated the procedures and the rituals that has to be followed. As the cow belonged to the **Pavo** village which is west of Hariyali temple, village came to be regarded as Goddess's 'Mayaka' (parental home). Another village **Kodima** northeast of Hariyali temple is Goddess's Sasural (in-law's home). The villagers are solely responsible for carrying mother's planquin and finally the responsibility for carrying out the rituals has been delegated to 'Maithani' Brahmins (and secondly to 'Chamoli' Brhamins in absence of the former) of the village Jasholi. Since the site was quite formidable to climb daily, it was decided that a temple be constructed for the Goddess in the village itself. During 'Mahayajna' ceremony for the recital of 'Vedas' and the 'doli yatra' (the annual pilgrimage to the sacred forests), different rights of the religious rituals are performed by different castes of adjoining villages.

The principal structure (deity) dedicated to the Goddess is the sacred black stone on which the scratches of the tiger's claw are engraved (as Devi rides on the tiger). The stone forms a part of the large rock mass buried within, which measures one and half feet in height and approximately one feet in breadth.

Rituals, Taboos and Folklores

* Women are strictly prohibited from entering the sacred forest due to the belief that they are impure (menstrual cycle being the most commonly cited cause).
* The lower caste people are not allowed to go above Kodima village, i.e. in interactive zone.
* Fetching fodder or fuelwood from this forest is strictly prohibited; the myth prevails that use of tools in any form (knife, sickle etc.) on the plants or animals will be a step to hurt the sentiments of Devi. The forest fairies in turn are angered and their wrath is exhibited in the form of making the person speechless, mad or deformed but it can also lead to disaster in the family of the offender.
* If someone making a pilgrimage comes across a snake in midway, then he not only has to abort the journey but will have to restart only after worshipping the 'hit' God after an interval of a week.
* One week before, the villagers stop taking onion, garlic, egg, meat and so forth. The same holds true if anyone wants to make a trip to the sacred forest.
* Milk of a cow fed upon the leftovers or having physical deformities is forbidden to be offered to Hariyali Devi.

Plate 2. Drinking water source in Hariyali Devi Sacred forest

* Earlier people used to leave their slippers way behind in the fringes of the village boundaries but with attenuation in faith with the beginning of modern era, the villagers take off their shoes or chappals in the fringes of the sacred forest (Protected zone). Any thing which is made out of leather is to be left behind.

Socio-economic and Ecological Role of the Sacred Grove

Inhabitants are primarily farmers raising a variety of traditional crops which include grains, millets and pulses etc. All households entirely depend on forest for fuel, fodder, timber and leaf litter for organic manure. Variety of non-timber forest products are collected from the forests to meet their sustenance.

From the view of resource utilization pattern, two villages are crucial as far as the Hariyali is concerned (Table 2). However, it doesn't imply that other adjoining villages do not exert pressure on the sacred forest. The interactive zone of sacred grove plays a vital role in supplying fodder and fuelwood to the inhabitants of two villages of the sacred landscape thereby reducing the pressure on the protected zone. Five major perennial streams which originate from the grove supply water for irrigation and drinking to a number of villages before they finally drain into Alaknanda (Fig. 4).

Table 2. Resource collection (ton/yr) by two villages from different sources.

	Village			
Sources of fodder and fuelwood	Jasholi		Kodima	
Interactive zone (Hariyali's fringes)	10.55	(-)	151.57	(9.2)
Agroforestry land	54.07	(11.96)	103.93	(15.33)
Other forest	192.87	(137.62)	177.55	(128.7)

Values within parentheses indicate the fuelwood collection.

Species diversity of tree stratum was observed slightly lower in sacred forest (protected zone) than the non-sacred forest which may be due to human induced disturbances in the latter forest leading to invasion by some of the early successional species e.g. pine. However, the density and basal cover of the non-sacred forest (interactive zone) was recorded significantly lower than the sacred forest indicating anthropogenic pressure on the former (in the form of fuel wood and fodder extraction). The dominant tree species in the protected forest are *Rhododendron arboreum, Quercus semecarpifolia, Q. leucotrichophora* and *Betula alnoides,* while the non protected (interactive zone) forest was dominated by species like *Quercus leucotrichophora, Rhododendron arboreum* and *Pinus roxburghii* (Tables 3 and 4).

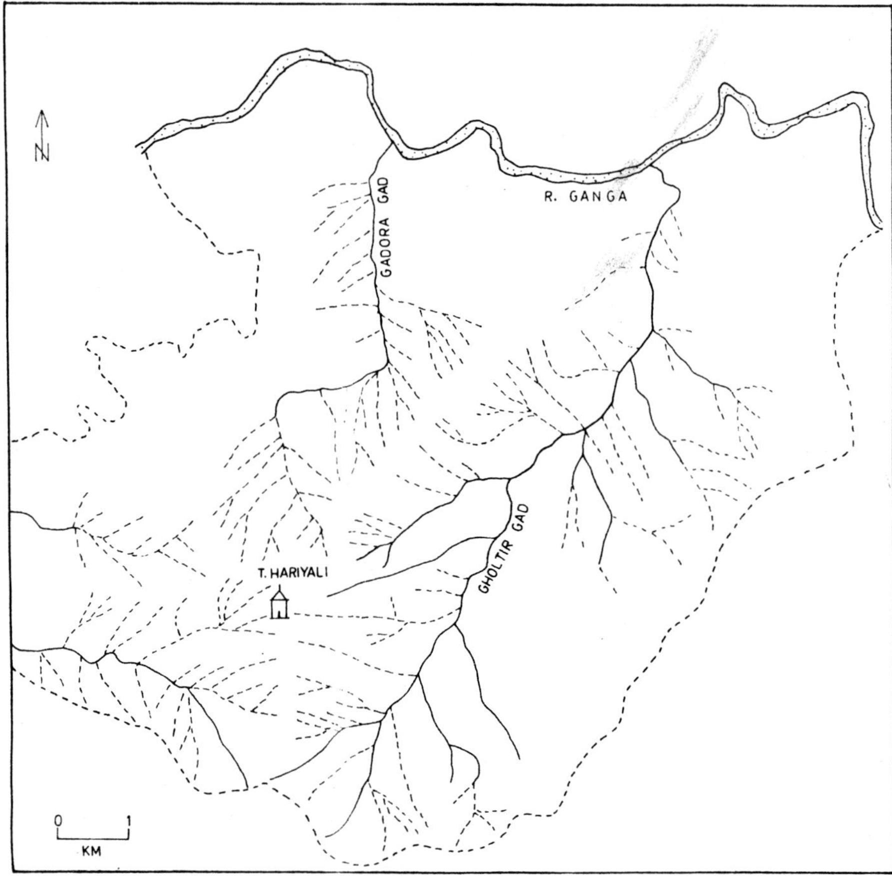

Fig. 4 : Streams originating from the Hariyali sacred forest.

From the protection of wildlife point of view, this forest is very significant as their habitats are undisturbed and hunting is also prohibited due to belief. Therefore, it nurtures rich fauna e.g., deer, wild boar, Himalayan bear, leopard, porcupines and forest birds etc.

The rituals associated with the grove foster social integration, equality and conviviality in that region. Conservation values assigned by the traditional societies through cultural-religious institutions had scientific rationale behind maintaining sustainability. However, due to direct conflict between ever increasing human population and limited natural resources such traditional ways of resource management are becoming non-functional. Priority thus needs to be given to strengthen traditional systems of conservation of natural resources.

Table 3. Phytosociologial attributes of sacred forest (core zone) at Hariyali, Garhwal Himalaya

Species	Density/ha	Total basal cover (m^2/ha)	IVI
Alnus nepalensis	9.00	0.044	3.366
Betula alnoides	200.00	1.160	27.257
Ilex spp.	36.00	0.252	10.996
Lyonia ovalifolia	145.00	1.300	20.989
Pyrus pashia	9.00	0.056	3.391
Quercercus leucotrichophora	182.00	5.127	36.939
Q. semecarpifolia	300.00	23.76	95.054
Rhododendron arboreum	509.00	15.57	98.048
Lauraceae member	9.0	0.321	3.948
Total	1399.00	47.59	

Table 4. Phytosociologial attributes of non sacred forest (interactive zone) Hariyali Devi.

Species	Density/ha	Total basal cover (m^2/ha)	IVI
Alnus nepalensis	8.00	0.050	3.515
Betula alnoides	16.00	0.092	4.37
Cotoneaster sp.	16.00	0.086	4.348
Lyonia ovalifolia	58.00	0.614	17.881
Myrica esculenta	16.00	0.344	7.940
Pinus roxburghii	83.00	11.33	62.568
Pyrus pashia	16.00	0.107	7.058
Quercus leucotrichophora	541.00	7.389	103.73
Q. semecarpifolia	16.00	0.179	7.326
Rhododendron arboreum	333.00	6.381	68.641
Symplocos spp.	33.00	0.182	8.823
Lauraceae member	8.00	0.123	3.786
Total	1144.00	26.877	

ACKNOWLEDGEMENTS

Support from the Director and the Scientist In-charge, Sustainable Development and Rural Ecosystems, G.B. Pant Institute of Himalayan Environment and Development, Kosi-Almora is gratefully acknowledged. Dr. R.L. Semwal, Mr. Sunil Nautiyal and Mr. S.D. Tiwari helped a lot during field work.

26

Eco-cultural Analysis of Sacred Species and Ecosystems in Chhakinal Watershed, Himachal Pradesh

G.S. Singh, K.S. Rao** and K.G. Saxena****

*Centre for Sustainable Environment and Heritage, New Delhi, India
**G.B. Pant Institute of Himalayan Environment and Development, Kosi-Katarmal, Almora, India
***School of Environmental Sciences, Jawaharlal Nehru University, New Delhi, India

INTRODUCTION

Conservation and sustainable use of biodiversity to the benefit of both present and future generations, an ethical concern, has now a global political and legal sanction in the form of Convention on Biological Diversity signed by almost all countries. Yet, irreversible loss of biological diversity and inequitable distribution of resource base remain the serious problems. While deficiencies of State driven conservation approaches have been brought out, proven examples of perfect conservation approaches are lacking. With increase in population pressure and biological resource demands set by fast economic growth ambitions, threats to protected areas are getting more and more serious. There is a need for new approach to biodiversity conservation (Borrini-Feyerabend, 1996) where conservation emerges as a people's initiative rather than a State enforcement.

Sacred species, groves, forests and other ecosystems which have been conserved for religious and cultural reasons offer an opportunity to build on an age old traditional ethos of conservation. Descriptive accounts of origin, religious and cultural practices and attitudes related to sacred groves/forests/ ecosystems/ landscapes have been extensively published in recent years (Gadgil and Vartak, 1976; Frazer, 1980; Messerschmidt, 1987; Khiewtem and Ramakrishnan, 1989; Bachmann, 1992; Anonymous, 1996; Singh *et al.*,

1996; Ramakrishnan,1996; Nair *et al.*, 1997). Much more limited is the quantitative information on biodiversity status of these traditionally protected areas as a component of landscape, people's perceptions on tangible and intangible benefits from the biodiversity and, opportunities and constraints in improving the biodiversity resource base and its sustainable utilization in the present scenario (Chandrakanth and Romm,1991; Ingles,1994; Roy Burman, 1995).

Himalaya is a unique geo-ecological system. It is the youngest mountain system of this planet, and it is extremely rich in biological, ethnic and cultural diversity. At the same time, Himalaya is considered to be sacred in one way or the other by both local and regional communities. This paper deals with social, cultural and ecological dimensions of sacred groves, forests and pastures at landscape scale, considering a microwatershed as a landscape unit, in the Western Himalaya.

SACRED AREAS IN THE ECO-CULTURAL LANDSCAPE

Area covered in District Kullu of the State of Himachal Pradesh in Western part of the Indian Himalaya is well known among the followers of Hindu religion as 'Dev Bhoomi' (the land of Gods). The density of sacred sites in this part of the Himalaya is generally much higher as compared to the other parts of the Himalaya. However, the size of sacred sites and their distribution in the landscape could tremendously vary depending upon ecological conditions, human settlement patterns and intensity of external impacts. In Chhakinal watershed, there are nine hamlets, six on south facing slope and three on north facing slope. There are 322 households with a population of about 2340. The average size of land holding is about 0.84 ha. Total annual rainfall recorded here is 1100 mm. While the minimum temperature may vary from 2-16°C, maximum temperature may vary from 9-24° C. Out of the total area of 45 sq km, 21.5% area comprises sacred area, 6% area as agricultural land and remaining 72.5% as forests/ pastures/permanent snow areas subjected to varied intensity of resource use/ disturbance regime. There are many sacred sites below tree line but only one in the alpine zone. Each hamlet has its own sacred site near the settlement (Table 1). These sacred sites comprise a house dedicated to the deity surrounded by a few trees. In addition, Nagoni sacred forest far away from the settlements is revered jointly by all the hamlets. Above tree line (3500 m amsl.), an area of about 5 ha is considered sacred by all. This alpine meadow area is considered sacred only for the purpose of performing some rituals when people camp here to graze livestock during summer period and to collect medicinal plants, and when they cross the highest peak of the watershed by the community. Unlike sacred groves and sacred

Table 1. Sacred groves in the Chhakinal watershed

Name of the hamlet	Location in relation to main settlement	Associate deity	Area (ha)
Baregra	Middle of the village	Jamlu	0.1
Puling	Middle of the village	Jamlu	0.1
Saran	Middle of the village	Jamlu	0.1
Rumsu	Towards one end of the village	Narayani	0.4
Raman	Towards one end of the village	Basukinag	0.1
Baltha	Towards one end of the village	Basukinag	0.2
Cherang	Towards one end of the village	Basukinag	0.1
Kumarhatti	Middle of the village	Basukinag	0.3
Kumarhatti	About 250 m away from the settlement	Narayani	2.8
Nagoni	1 km away from the settlement	Basukinag	5.5
Alpine meadow	10-16 km from settlement	Basukinag	5.0

forest where resource uses are permissible only for performing the rituals and maintenance of the deity house, grazing is permitted in sacred alpine meadow after the rituals are over. Higher density of small size sacred sites and lower density of large size sacred sites observed in the present caste is similar to observation of Ingles (1994) based on an extensive survey of 26 sacred forests in three districts of central Nepal.

ORIGIN OF SACRED GROVES

Two stories were narrated by the people in response to the question when and how the sacred sites came in existence. According to one story, there was a miraculous youth named Basukinag living in the watershed area long ago. Once while relaxing on the hill top he happened to look at a beautiful girl sitting on another hill top far away near Bashishta. The girl was attracted by the handsome personality of Basukinag. Before she could reach the area where Basukinag was sitting to express her desire of marrying him, Basukinag had disappeared. She conceived as a result of thought of marrying Basukinag and disappeared after giving birth to 36 Nagas (cobra). All these snakes dispersed in the landscape. Each snake is revered as a hamlet deity and the deity is believed to reside in the sacred site near the settlement. Nagoni sacred forest is believed to be the place where Basukinag is residing along with the girl who thought of to marry him. Nagoni sacred forest is thus considered to be the place of more powerful deity while the groves near settlements to be of less powerful deities.

The other story says that son of a king named Kulang Thakur started going to Nagoni forest along with a cow though the king forbade him to do so. The child observed that a snake used to suck the milk of cow and told this to his parents. Once, the queen followed the child. When she saw the snake sucking the milk and the child sitting nearby, she rushed to protect the child. Suddenly the snake turned a man, introduced himself as Basukinag. He told that the sites where his offsprings are living (i.e., sacred site near each hamlet) should not be treated as the property of king, otherwise the king would be penalized. Later he disappeared. The king then requested some priests to take care of these sites.

Folklore evolve based on a continuous thinking process. The stories behind origin of sacred sites and traditions of protecting these sites from all sorts of human disturbances bring out the importance attached to humans as a part of natural ecosystem.

DIVERSITY-DOMINANCE IN SACRED AND NON-SACRED ECOSYSTEMS

Comparison of species richness and dominance-diversity curves of sacred forests and pastures with that in other seven forest/pasture ecosystem types in the watershed is shown in Figs. 1 and 2. Species richness and evenness did not differ significantly between the sacred and non-sacred alpine meadows because here sacredness meant just some rituals and not protection from grazing and other human disturbances like collection of medicinal plants and wild edibles. Nagoni sacred forest, the most protected sacred site in the areas, had highest species richness followed by *Aesculus indica* and *Quercus semecarpifolia* dominated forests. In all, 13 tree species, 6 shrubs and 34 herbs were sampled in Nagoni sacred forest compared to 3-8 trees species, 2-7 shrubs and 11-34 herbs in other forest ecosystems. This species richness of an economically important group like medicinal plants was correlated with total species richness (Fig. 3).

Different ecosystems existing in the watershed differ in terms of community similarity. Nagoni sacred forest had highest level of similarity with *Abies pindrow, Aesculus indica* and *Quercus semecarpifolia* dominated forests with nearly 40% of species present in the former being also present in the latter (Table 2). No species was found exclusively in the Nagoni sacred forest. Thus, conservation value of sacred forest lies in respect of presence of a large number of species over a small area and not in terms of presence of unique species. Too small sacred areas like the sacred groves near settlements, may not sustain viable population of many species.

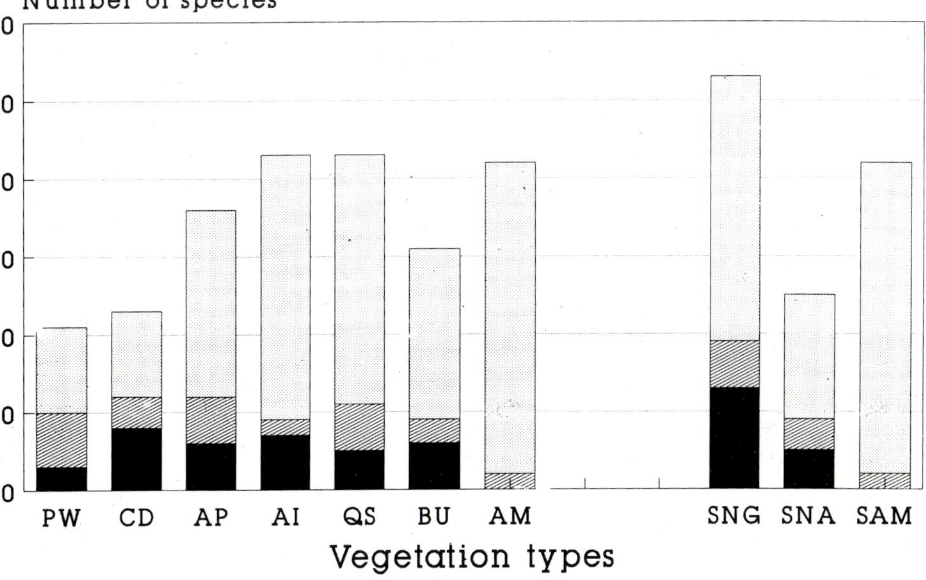

Fig. 1. Species richness of sacred forests/meadows and other ecosystem types. PW, *Pinus wallichiana* forest; CD, *Cedrus deodara* forest; AP, *Abies pindrow* forest; AI, *Aesculus indica* forest; QS, *Quercus semecarpifolia* forest; BU, *Betula utilis* forest; AM, Alpine meadow; SNG, Sacred forest of Nagoni; SNA, Sacred forest of Naryani; SAM, Sacred alpine meadow

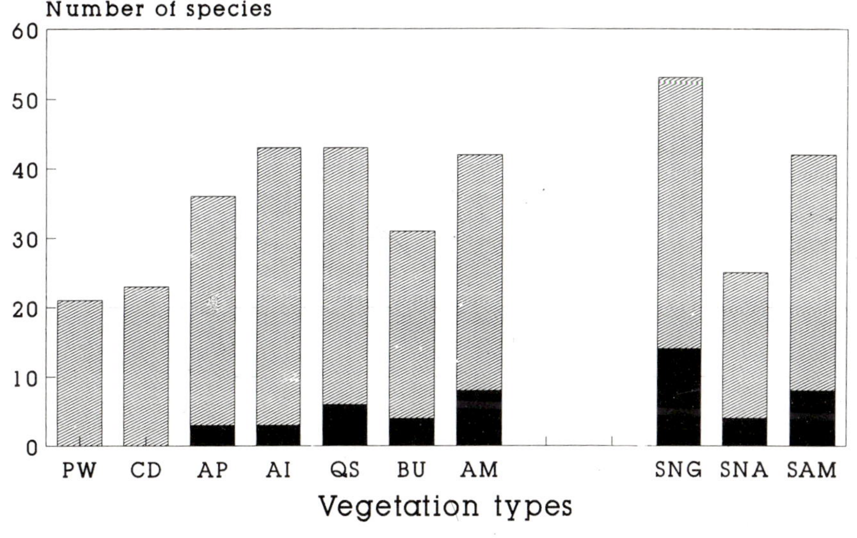

Fig. 2. Richness of medicinal and other species in different ecosystems. Other details as in Fig. 1.

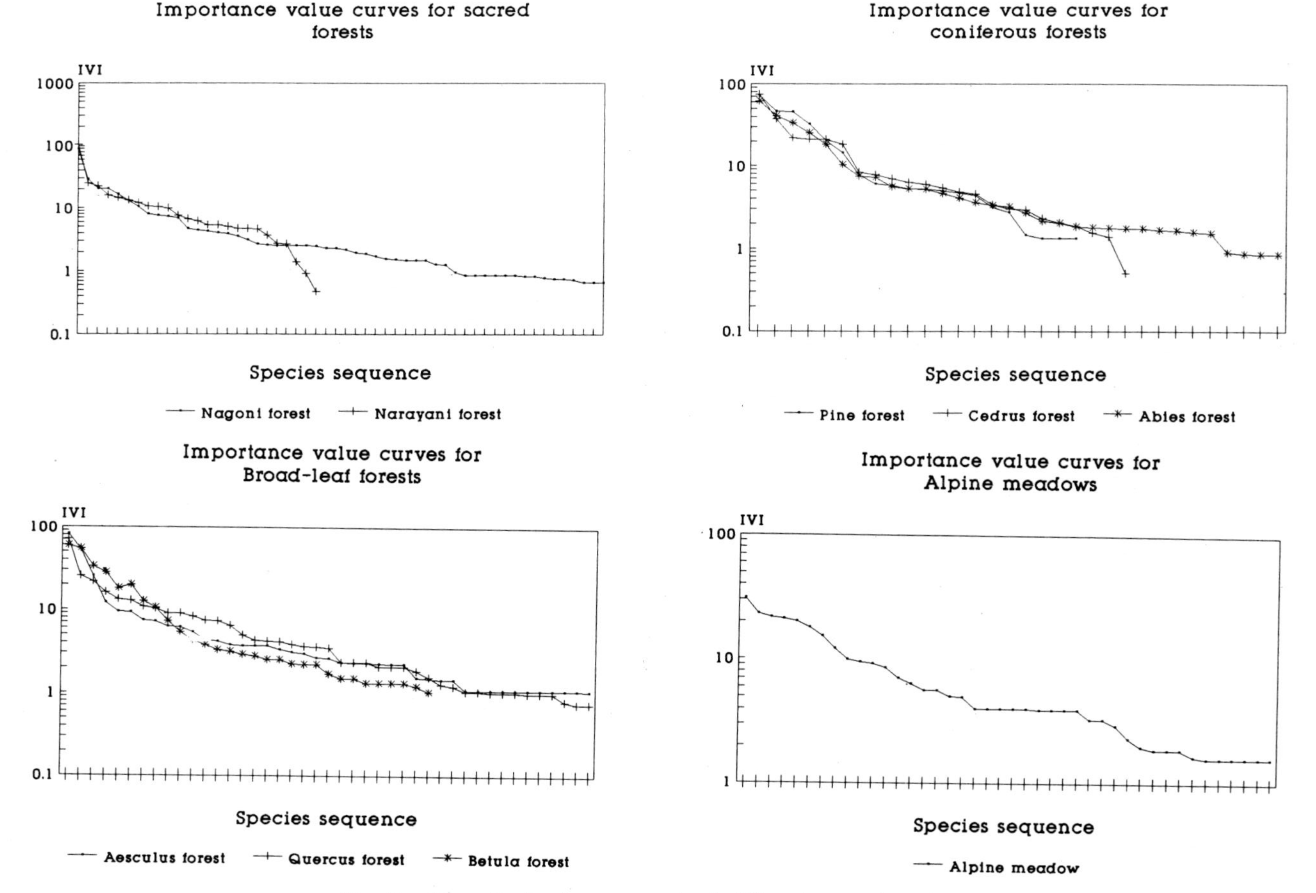

Fig. 3. Diversity-dominance curve of different ecosystem types.

Table 2. Similarity index between sacred ecosystems and other ecosystem types

	PW	CD	AP	AI	QS	BU	AM	SNG	SNA	SAM
PW	1.00	0.55	0.18	0.16	0.19	0.04	0.10	0.27	0.52	0.10
CD		1.00	0.20	0.15	0.21	0.07	0.22	0.29	0.33	0.22
AP			1.00	0.43	0.33	0.18	0.17	0.41	0.30	0.17
AI				1.00	0.47	0.38	0.35	0.42	0.15	0.36
QS					1.00	0.51	0.38	0.35	0.24	0.38
BU						1.00	0.41	0.29	0.14	0.41
AM							1.00	0.31	0.21	1.00
SNG								1.00	0.41	0.31
SNA									1.00	0.21
SAM										1.00

PW, *Pinus wallichiana* forest; CD, *Cedrus deodara* forest; AP, *Abies pindrow* forest; AI, *Aesculus* indica; QS, *Quercus semecarpifolia* forest; BU, *Betula utilis* forest; AM, *Alpine meadow*; SNG, Sacred forest of Nagoni; SNA, Sacred forest of Naryani; SAM, Sacred alpine meadow

Apart from the area effect, the groves near settlements are highly disturbed because of grazing and timber extraction for maintenance of the deity's house. Two hypothesis have been proposed to explain the patterns of diversity-dominance: niche preemption hypothesis assumes that the most dominant species occupies a fraction 'k' of the resource hyperspace, the second dominant a fraction approximating 'k' of that not occupied by the first and so on (Whittaker,1972); random niche boundary hypothesis assumes that the boundaries between species hypervolumes in resource space are set at random i.e., species are limited by competition at randomly located boundaries (Vandermeer and MacArthur, 1966). Diversity-dominance plots exhibit a geometric series if niche pre-emption hypothesis is true and is observed in highly perturbed situations characterised by low species richness. As the number of species increases, the number of factors governing their relative importance increases and dominance-diversity plot shows a lognormal series. Lack of a geometric trend in diversity-dominance curves even in ecosystems other than the sacred Nagoni forest indicates the cultural appreciation of promoting diversity all through the landscape.

LOCAL PERCEPTIONS OF SACREDNESS

Apart from religious and cultural values attached to the sacred sites, people do realise their ecological, economic and social values (Table 3).

Socio-cultural-religious Context

Protection from earthquakes and abnormally high or low precipitation/temperature are the major concerns fostering protection of sacred sites. People believe protection from catastrophic events is provided by the supernatural powers residing in the sacred sites. The history of sacred sites

Table 3. People's perceptions on values of sacred forest

Perceived values	% of total response
Respository of species diversity in the wild	92
Monetary benefits from resources in the grove	5
Rendering sustainable yields of food, fuelwood, fodder and other needs	70
Enabling availability of irrigation water of desired quality	70
Enabling slope stability of private land	20
Enabling availability of pure drinking water	85
Ability to prevent natural catastrophic events	90
Source of strength to survive catastrophes	92
Source of water preventing/curing diseases, within the grove	92
Production of new species/cultivars	0

remind people that economic or political power wielding individuals are not endowed with the power enabling survival following catastrophes.

Apart from spatial connotation of sacredness, people consider many species sacred. Seven species are used as incense on various religious-cultural functions, six species are valued for offering their flowers to the deity and 12 species are believed to be preferred by the deity all across the landscape (Table 4). Many of these species such as *Valeriana jatamansii, Abies pindrow, Cedrus deodara* and *Juglans regia* have potential of providing economic benefits to the local community. Religious and cultural importance of these species is a factor promoting their sustainable utilization as well as conservation.

Management activities critical to the survival of the people such as sowing, harvesting, migration to alpine meadows for grazing and medicinal plant collection are undertaken following religious functions in and around sacred sites. Though, each village has its own deity, all deities are equally respected. Goats and sheep are sacrificed to please the deity. Higher caste families who also happen to be large holders offer their livestock and the meat is shared by the entire community including the poorer low caste people. On a few occasions community feasts are organized with all food items contributed by the high caste families. Low caste people own the responsibility of collecting fuelwood from the forests excluding sacred forests and cleaning.

Ecological and Economic Context

While people do not perceive any ecological value of the sacred groves located nearby settlements, they believe that Nagoni sacred forest located on steep and rugged slope towards ridge regulates flow of water and sediment to the settlements and agricultural land downslope. People know that sacred forest is a repository of a variety of useful species but only a few feel that this potential should be exploited for direct monetary benefits.

Table 4. Species of religious values

I. Species used for incense	
1. *Artemisia parviflora*	Leaves/floral tops
2. *Betula utilis*	Bark
3. *Juniperus communis*	Young twigs and leaves
4. *Jurinea macrocephala*	Roots and leaves
5. *Selinum tenuifolium*	Roots
6. *Valeriana hardwickii*	Leaves
7. *Valeriana jatamansii*	Leaves
II. Species used in rituals and prayers	
1. *Primula denticulata*	Flowers
2. *Rhododendron campanulatum*	Flowers
3. *Ranunculus laetus*	Flowers
4. *Rosa macrophylla*	Flowers
5. *Rosa moschata*	Flowers
6. *Viola odorata*	Flowers
III. Species considered to be liked by the deity	**in areas other than sacred forests also**
1. *Abies pindrow*	Whole plant
2. *Aesculus indica*	Whole plant
3. *Cedrus deodara*	Whole plant
4. *Cupressus torulosa*	Whole plant
5. *Ficus palmata*	Whole plant
6. *Ficus religiosa*	Whole plant
7. *Juglans regia*	Whole plant
8. *Picea morinda*	Whole plant
9. *Pinus wallichiana*	Whole plant
10. *Quercus dilatata*	Whole plant
11. *Quercus incana*	Whole plant
12. *Quercus semecarpifolia*	Whole plant

The majority attaches more importance to the indirect benefits in terms of reduction in erosive force of water, control of mass movement from upper reaches, availability of water of desired quality and natural dispersal of seeds of useful species (Table 3). Thus the existence value of sacred forest for religious or cultural reasons is blended with the indirect values of biodiversity in terms of its ecological functions in the landscape.

Productivity in private farmlands covering 6% of the area is sustained with the inputs from the forests and pastures covering 94% of the area, of which sacred elements constitute a minor component, and accounts for 94% of household income potential (Table 5). A large number of medicinal plants whose pharmaceutical values and economic potential are well known to the people are harvested for both subsistence and cash but from the areas other than sacred groves and forest. Cultural control on consumptive use of biological resource in the sacred areas could be viewed as a mechanism of maximizing indirect benefits from these areas. The advantages of maintaining sacred

Table 5. Monetary values (Rs) of annual production and income by sale of the produce in an average household (values in parentheses are % of total annual production or income; based Dobariyal et al., 1997).

Production system	Monetary value of annual production	Annual income by sale of the produce
Agricultural crops	14847 (21.28)	3599 (6.76)
Horticultural crops	47742 (68.43)	46310 (87.02)
Livestock	6493 (9.31)	2622 (4.93)
Medicinal Plants*	689 (0.99)	689 (1.29)
Total	69771	53220

*Monetary value of medicinal plants collected for self-consumption is not included.

forests from the point of facilitating natural process of speciation and as source of germplasm for research to the benefit of wider global community are not perceived by the local community.

Carbon sequestration is an important intangible function of biodiversity in sacred forest and needs to be integrated with the more common concern for loss of species. Based on basal area as an indicator of standing biomass, carbon stock in vegetation component of sacred groves is significantly higher compared to other forest ecosystems. Soil carbon stock is high in well protected Nagoni sacred forest but is depleted in Narayani sacred forest subject to disturbances of grazing and litter removal (Table 6).

Table 6. Soil organic carbon (%), pH and basal area (m^2ha^{-1}) of different ecosystem types.

Ecosystem types	pH	Organic carbon	Basal area
Pinus dominated forest	6.08	1.70	33.16
Cedrus dominated forest	6.39	2.32	61.64
Abies dominated forest	5.72	2.24	113.60
Aesculus dominated forest	6.25	6.56	62.48
Quercus dominated forst	5.59	4.29	34.82
Betula dominated forst	6.20	6.83	39.43
Alpine	6.12	3.45	–
Sacred grove: Nagoni	6.15	8.45	230.14
Sacred grove: Narayani	6.22	2.68	205.34
Sacred alpine area	6.10	3.30	–

Institutional Context

Traditional informal institutions coexist along with formal government supported local institutions in Chhakinal watershed unlike the more common scenario of replacement of the former by the latter (Table 7). Management of sacred areas is looked after by Deota committee which comprises 12 male members including one head priest, one member from the priest family and one member from each hamlet selected by the village

Table 7. Status and operational functions of different village institutions

Name of the institutions	Status	Existence	Operational functions
Village committee	Informal/traditional with 8-10 selected individuals	since historical times	Conflict resolution within village, natural resource management maintenance of common facilities
Deota (deity) committee	Informal/traditional with 12 members; 2 from Head priest family and others selected from the villages	since historical times	All affairs related to the deity and the sacred forests; psychological/ spiritual treatment
Gram Sabha	Formal consisting of 11 elected members since 1960s	since 1960s	Getting government grants for infrastructure development, helping individuals for loans/ grants
Mahila Mangal Dal	Formal comprising all women who want to become members, three office bearers	since 1991	Promoting education and employment of women
Forest Protection Committee	Formal comprising selected 17 villagers and 3 forest department officials	Since 1994	Protection and regeneration of forests excluding sacred groves

community. This committee is empowered to impose fine on adults found violating the cultural norms. The fine realized in terms of cash is spent for religious functions. Head priest is believed to foresee the future and is revered as a link between the deity and the common man. The priest as well as all the members of Deota committee are non-Brahmins. This committee is regarded as the most powerful institution in that the members have spiritual capacity to influence the deity. Apart from the rituals needed to appease the deity, religious functions serve as occasions for community gathering of each village independently as well as for all the villages in the watershed.

FACTORS CAUSING EROSION OF TRADITIONAL CONSERVATION CULTURE

Analysis of people's perceptions revealed that marginalization of traditional informal institution of sacred area management following establishment of formal institutions of Gram Sabha, Mahila Mangal Dal and Forest Protection Council, increasing economic interests of the society at large including the members of sacred area management committee have caused erosion of cultural value for conservation rather than increase in human or livestock population pressure (Table 8). People think that many sacred groves are now desecrated because the power cable coming from low caste households passes through the sacred areas. Protection of sacred areas for their value as a natural source of propagules of a variety of species is getting weakened with introduction of government programmes making propagules/seedlings available free of cost/at subsidized price. Cultural valuation of protected

Table 8. Factors causing degradation of sacred groves/forest and associated cultural values

Factors	% of total response
Increase in population pressure	4
Increase in livestock pressure	8
Development of infrastructure viz. road and electrification	64
Introduction of new crops/cultivars	84
Increasing economic interests of priest families	36
Increasing economic interests of society at large	94
Immigration of aliens	8
Marginalization of traditional institution due to establishment of new formal institutions by the government	96

areas for hydrological balance is vanishing with emergence of government water supply services. Improvement in accessibility brought in through construction of motorable road has brought in temporary migration of outsiders who do not believe in local deity and sacredness. Government grants to Forest Protection Council, Gram Sabha and Mahila Mangal Dal but not to sacred area management committee is leading to demoralization and erosion of cultural responsibility in the latter. Though violation of cultural norms were common, there were no cases of penalty in the recent past indicating callous attitude of Sacred area management committee at present.

BUILDING ON TRADITIONAL CONSERVATION ETHOS

Government attention to environmental conservation in the Himalaya dates back to 1890s when all uncultivated areas were notified as 'forest land' in state ownership by the colonial rulers. As at present, degree of protection from human disturbances or biomass uses varies. National Parks have been accorded the highest level of legal protection to the extent that it is mandatory to resettle all indigenous settlements elsewhere following notification of an area as National Park, followed by Wildlife Sanctuary,

Reserve Forests, Protected Forests, Village Panchayat/Community Forests and Civil Forests. Yet, there are many threats to the goal of conservation set for the protected areas, including those from the local communities (Uniyal et al., 1996). Necessity of local participation in protected area management is getting more and more recognition (Borrini-Feyerabend, 1996) for a variety of reasons. These include people's movement for local empowerment and restoration of traditional land rights, growing scientific evidences supporting conservation values of traditional management practices, failure of distant government machinery to control illicit actions in inaccessible remote habitations, emerging financial crisis in developing countries making conservation projects conceived as investment activities with negligible returns financially unviable. Local knowledge should not be presumed to be perfect in all respects and locations, but should be comprehensively analysed in the dynamic socio-cultural, economic and political context (Maikhuri et al., 1997). Conservation ethos related to sacred ecosystems needs to be capitalized upon for promoting the cause of conservation and sustainable utilization of biodiversity in the Himalaya to the benefit of local and global community through appropriate scientific management innovations.

ACKNOWLEDGEMENTS

Support from Director, G.B. Pant Institute of Himalayan Environment and Development and TSBF (Nairobi) is thankfully acknowledged.

REFERENCES

Anonymous 1996. *Sacred and Protected Groves of Andhra Pradesh*. WWF State Office, Hyderabad, Andhra Pradesh.

Bachmann, V.K. 1992. *Mythen*. Geo Wissen, Hamburg.

Borrini-Feyerabend, G. 1996. *Collaborative Management of Protected Areas Tailoring the Approach to the Context, Issues in Social Policy*. IUCN, Gland (Switzerland).

Chandrakanth, M.G. and Romm, J. 1991. Sacred forests, secular forest policies and people's actions. *Natural Resources Journal*, 31: 741-756.

Dobriyal, R.M., Singh, G.S., Rao, K.S. and Saxena, K.G. 1997. Medicinal plant resources in Chhakinal Watershed in the north-western Himalaya. *Journal of Herbs, Spices and Medicinal Plants*, 5:15-27.

Frazer, J.G. 1980. *The Golden Bough*. MacMillan, London.

Gadgil, M. and Vartak, V.D. 1976. The sacred groves of Western Ghats. *Economic Botany*, 30: 152-160.

Ingles, A.W. 1994. *The Influence of Religious Beliefs and Rituals on Forest Conservation in Nepal*, Discussion paper, Nepal Australia Community Forestry Project, Kathmandu.

Khiewtem, R.S. and Ramakrishnan, P.S. 1989. Socio-cultural studies of the sacred groves at Cherrapunji and adjoining areas in the north-eastern India. *Man in India*, 69: 64-71.

Maikhuri, R.K., Semwal, R.L., Rao, K.S. and Saxena, K.G. 1997. Agroforestry for rehabilitation of degraded community lands: a case study in the Garhwal Himalaya, India. *International Tree Crops Journal, 9:* 91-101.

Messerschmidt, D.A. 1987. Conservation and society in Nepal: traditional forest management and innovative development. In: P.D. Little, M.M. Horowitz and A.E. Nyerges (Ed.), *Land at Risk in the Third World: Local Level Perspectives.* Westview Press, Colorado. pp. 373-397.

Nair, G.H., Gopikumar, K., Krishnan, P.G. and Kumar, K.K.S. 1997. Sacred groves of India—vanishing greenery. *Current Science, 72*: 697-698.

Ramakrishnan, P.S. 1996. Conserving the sacred: from species to landscapes. *Nature and Resources,* 32: 11-19.

Roy Burman, J.J. 1995. The dynamics of sacred groves. *Journal of Human Ecology,* 6: 245-254.

Singh, G.S. 1997a. Sacred groves in Western Himalayas: An eco-cultural imperative. *Man in India, 77*: 247-257.

Singh, G.S. 1997b. Socio-cultural evaluation of sacred groves for biodiversity conservation in north-western Himalaya. *Journal of Hill Research* 10: 43-50.

Singh, G.S., Saxena, K.G., Rao, K.S. and Ram, S.C. 1996. Traditional knowledge and threats of its extinction in Chhakinal watershed in north-western Himalaya. *Man in India, 76*: 1-17.

Uniyal, V.K., Pandey, S. and Sawarkar, W.B. 1996. Buffer zones—key to the success of protected areas. In : P.S. Ramakrishnan, A.N. Purohit, K.G. Saxena, K.S. Rao and R.K. Maikhuri (Eds.), *Conservation and Management of Biological Resources in Himalaya.* Oxford & IBH, Delhi. pp. 313-327.

Vandermeer, J.H. and MacArthur, R.H. 1966. A reformulation of alternative (b) of the broken stick model of species abundance. *Ecology,* 47: 139-140.

Whittaker, R.H. 1972. Evolution and measurement of species diversity. *Taxon,* 22: 213-251.

27

A Study of Indigenous Community Based Forest Management System: Sarna (Sacred Grove)

S. Patnaik and A. Pandey

Indian Institute of Forest Management, Bhopal, India

INTRODUCTION

Sarnas are the sacred groves found in the eastern part of Madhya Pradesh. These groves are believed to be the abode of local deity, holy and ancestral spirit and also the place where religious rites and rituals are performed by indigenous communities of the area. They are the sacred relics which are protected on the strength of traditional beliefs.

This paper is an outcome of the study conducted on sacred groves (Sarna) in Jaspur Forest Division of Raigarh district in eastern Madhya Pradesh. The major area of the Division is covered by Sal (*Shorea robusta*) forest. The study area is predominantly inhabited by tribals (66%). They include Oraon, Birhors, Kawnar, Korwa, Hill Korwas, Munda. Oraon and hill Korwa are considered as primitive tribal groups. The high lands are inhabited by Oraon and Korwas. The Korwas generally confined to hill tops and slopes and practice shifting cultivation whereas Oraon usually cultivate in valley. The plains, is inhabited by Roytia, Kawnar, Nagesia and Mahakuls who practice sedentary agriculture. Besides agriculture, the communities are also dependent on forest areas for their livelihood amongst which collection of non timber forest produce (NTFP) such as sal seeds and leaves, leaves of Bauhinia, Mahua (*Madhuca latifolia*), Harra (*Terminalia chebula*) etc. are the major activities.

Sarnas are characterised by the vegetation with a cluster of sal (*Shorea robusta)* trees as dominant species along with other tree species which includes: Peepal (*Ficus religiosa*), Saja (*Terminalia tomentosa*), Mahua (*Madhuca latifolia*), Harra (*Terminalia chebula*), Bahera (*Terminalia belerica*) etc.

Sarnas in the study area are found in wide range of locations, from mountain top to plateau, and are found to vary in size from 0.02 ha to 21 ha. The individual trees in the Sarna may be as old as 200 years, tall with an average height of 20-30 m. and girth ranging from less than 1 meter to 3 meters. The Sarnas can easily be identified in a tribal rural landscape, which look like an island of forests in human settlements.

BELIEFS, DEITY AND RITUALS AND THEIR RELATION TO CONSERVATION OF SARNA

The indigenous communities living in the area believe that nature is sacred and need to be worshipped. Nature and natural phenomena occupy the central place in ritual festival and customs of indigenous communities and sacred groves serve the functions of cultural and religious sanctum in tribal life. Tributes and prayers are offered for good rain, good harvest and good health. Interestingly the community rituals are often synchronised with blossoming of flowers of forest trees and different agricultural operations. These rituals reveal the intimate sense of harmony that exists between nature and indigenous societies. These ritual celebrations of social occasion including song and dance are generally a communal activity.

The indigenous communities identify Sarna as a religion. To them not only the trees within the Sarna are important but the whole patch/micro habitat (living and non-living components) is sacrosanct.

The indigenous community believe that the deity along with various other spirits resides in the Sarna and they protect the community from evil forces, epidemics and calamities. These deities and spirits are feared greatly and it is believed that if they are displeased they can bring sorrow to individuals and the community. To pacify them appropriate rituals and sacrifices are offered by individuals as well as the communities.

Some deities are also believed to be malevolent. If the deities are provoked by breaching or disobeying the taboo in the Sarna, it is believed that implication could be mild to dangerous depending on the action of provocation.

The traditional deities of the Sarna are: Sarna Mata, Sarna buhria, Mahadani, Pat, Beru, Dharmesh etc. However, due to influence of Hinduism the traditional communities believe that the Sarna are abode of Hindu gods and goddess like Bhagwan, Mahadev, Hanuman etc.

TYPES OF SARNA

In the study area, four types of sarnas were found. They are: Sarhul Sarna,

Kadamara Sarna, Mahadani Sarna and Phool Sarna. Generally, a village has at least one Sarna and sometimes named after a village or a hill on which or near which it is situated.

Each of the above mentioned sacred groves have specific deities associated with specific power (assigned to fulfil certain needs of the people) and a set of rituals.

Sarhul Sarna

This sarna is specifically meant for the celebration of the Sarhul festival. The Sarhul puja is celebrated in the month of March-April, when sal trees are in full blossom. The day of sarhul festival is generally fixed after having discussion amongst village elders and consultation with Baiga (Tribal Priest) in the village council. Earlier onset of the flowering of Sal used to act as a bioindicator to decide the time of celebration of the festival but now it is celebrated on the new year day of the Hindu calendar i.e. in the Baisakh month every year. This might have been adopted by a process of acculturation of tribals since they have been living in close proximity of non-tribal communities for a very long time.

The Sarhul puja celebration is an elaborate process carried out by the community with the guidance of village Baiga or Pahan (Traditional priest) to please the Sarna deities so that the coming year should bring happiness and prosperity to individual households and village as a whole. The Baiga brings three new pots filled with water and covers it with Sal leaf plate. Sal flowers are collected from Phool Sarna for the rituals. After the ritual is completed the Baiga distributes the Sal flower offered to the deity to individuals to be kept in the house the year long to bring prosperity. The Baiga studies the water level in the pot and predicts the nature and quantity of rain in the coming year. After the puja, cultural celebration takes place on community basis for two to three days in the Sarna.

-Baiga of Ghamariya Village

The size of Sarhul Sarna is approximately found to vary from 0.5 ha to 2 ha and is located near the village. The deities associated with this Sarna are: Sarna mata, Beru, and Bhagvan. Apart from Sarul puja, the Sarna is also used for recreational purposes like community dance and celebration of marriages.

Kada Mara Sarna

The Kada Mara Sarna is named after "Kada Mara" which means "*killing of a buffalo*" or sacrifice of a buffalo to appease the deity. The sacrifice of the buffalo is carried out once in 12 years, or whenever the village is under a threat of any calamity/epidemic or any other similar problem.

People of Echikela village in Jaspur range narrated an incidence to substantiate their belief:

Few years back a huge herd of elephants entered from east direction of village from Orissa side and was heading towards their village. The elephant herd was destroying everything on their way. The villagers requested the deity in Kada Mara Sarna to prevent the herd of elephant from entering the village for which they would sacrifice a buffalo. It seems that the herds' of elephants turned back from the outskirts of the village without causing any damage to their area for which they sacrificed a buffalo as per the request made earlier.

Kadamara sarnas are generally situated at a distance of 200 m to nearly 2 km from the village. The deities associated with this Sarna are Andheri pat, Budadev, Khuts, and spirits. Basically, they are considered to be ferocious and harmful in nature. Care is taken not to displease them by either causing harm to the vegetation in the Sarna or any other similar activities which are considered harmful. Women who have attained puberty age are strictly prohibited from entering the Sarna.

The size of the Sarna ranges from 0.5 ha to 14 ha. Earlier collection of wood and non-wood products in the sarnas were prohibited. But at present, collection of NTFPs takes place. The reason given for this change in usage pattern of Sarna is that there has been decrease in forest area which was available for NTFP collection. At present they are collecting NTFP from this Sarna under compulsion but by appropriately easing the deity through some offerings.

Mahadani Sarna

This Sarna receives its name from the word "Mahadan " which means the "biggest sacrifice" made to the deity. Roy (1928) reported that human sacrifice were given in this Sarna to appease the deity, so that they would be benevolent to the community but gradually the practice of human sacrifice was stopped and now substituted by animal sacrifices. The size of this Sarna varies from 2 ha to 21 ha. This Sarna is located at a distance from the village, generally near the forest boundary.

The deities associated with this Sarna are Mahadev, Bhuria Khut, Budhadev and other local spirits. Traditionally all deities and spirits living in this Sarna are believed to be so frightful that the villagers would not dare to go inside the Sarna except during rituals. Presently, disturbances in terms of collection of dried twigs, NTFPs and usage as pathways and trespassing is common.

Phool Sarna

This Sarna as the name signifies harbours flower bearing plants and trees. Sal is found to be the predominant species. If a village does not have a Sarhul Sarna then Sarhul rituals are performed in Phool Sarna. These sarnas are generally found near the village and the size of Sarna varies from 0.2-1.5 ha. The deity associated with this sacred grove is 'Sarna Mata' considered to be benevolent but punishes those who cause harm to the Sarna. This Sarna is mostly used for recreational purposes such as play ground for children and marriage ceremony.

THE CHANGING TREND IN USAGE PATTERN OF SARNA

As mentioned earlier, Sarnas have been used for selective purposes i.e., for religious rites and practices. Taboos and beliefs including certain social norms were designed for permission and restriction for various usage, which led to effective conservation of biodiversity of Sarna in the area.

Presently, the sarnas are being used for religious, recreational as well as economic purposes, amongst which grazing encroachment and collection of NTFPs are common.

Earlier, exploitation of these groves for individual benefits was prohibited. Only the Baiga was allowed to collect the dead fallen wood. However there has been considerable change in the usage of Sarna with the passage of time. For example, in certain sarnas the trees are distributed amongst the villagers for collection of Sal seeds whereas in others, villagers are allowed to collect certain quantity of Sal seeds.

FAITH AND ITS IMPLICATION FOR THE CONSERVATION OF SARNA

Belief in deities, supernatural powers and indigenous values are now changing amongst the indigenous community in the area and undergoing a transitional phase i.e., a phase of cultural oscillation. There are three category of believers in the area.

a) Ardent Believers of Sarna Faith

This group includes the people who belong to the indigenous community and still continue to have faith in their traditional values and practices. These groups continue to maintain strong belief in Sarna and practice strictly all the norms, rituals and taboos associated with the Sarna.

b) Believers but not Practicing the Sarna Faith

This includes the category of people (i) who have converted to Christianity from the indigenous core (ii) Hindu communities who have migrated to this area.

Inspite of conversion to Christianity, people are yet to adopt to new religious value and culture of single deity worship in church. Therefore they still persist in believing the old belief system but socialize their children in the new faith.

The migrated Hindu communities have adopted the Sarna faith on a marginal basis because belief in spirits and supernatural powers is equally prevalent in Hindu communities. They also participate in ritual celebration carried out by the indigenous communities.

c) Do not Believe and Do not Practice

This consists of the subsequent generation of the converted indigenous communities i.e. strict followers of Christianity who consider indigenous religion as primitive. Neither they believe in Sarna faith nor do they practice.

REVIVALISTIC MOVEMENT OF SARNA FAITH

In recent past, the area is witnessing a revivalistic movement of Sarna faith. This movement encourages the revival of the old cultural practices of the indigenous communities. The main role players in the movement are the two key political parties of the region and a local Non Government Organisation (NGO).

For the last four years the two major political parties separately organised sarhul festival as a major event. The NGO too encourage the indigenous communities to practice beliefs in their old custom, practices and advise them to continue to follow the traditional form of religion.

Development Programmes Influencing Sarna

Of late there has been increased interest in sacred groves because of the belief and values; therefore efforts have been made to ensure its conservation and management. In this forest division, the government has initiated a pilot project in 1993-1994 under Employment Assurance Scheme (Srivastava, 1994). Under this, the forest department identified 275 sarnas in the forest division and demarcated the boundary. In some sarnas cattle proof trenches have been dug out, and trees planted in blank areas.

CONCLUSION

Presently the Sarnas are a fairly degraded landscape with over-matured Sal trees with hardly any regeneration because of grazing, NTFP collection and various other biotic pressures. Sarnas which are closer to human habitation have been encroached upon for agriculture as well as infrastructure development of the village such as school, hospital etc.

The sarnas were protected over centuries due to belief and fear of the wrath of deities. The communities in the area are experiencing transitional changes in the cultural and economic life style. Therefore to ensure better conservation of sarnas with well developed management strategy is essential. The sacred groves that are well protected, and those that are adjacent to forest area could be protected as a priority. In addition to that, rehabilitation of degraded sarnas with local peoples involvement is essential.

ACKNOWLEDGEMENTS

We would like to thank Director, Indian Institute of Forest Management, Bhopal for his encouragement and guidance and Regional Centre for Afforestation and Eco development Board, Bhopal for providing financial support.

REFERENCES

Roy, S.C. 1928. *Oraon Religion and Customs*. Gyan Publishing House, Delhi.

Srivastava, M.K. 1994. *Hill Korwa : Past, Present and Potential*. Sri Mudran and Publication, Raipur.

28

Structure and Functions of Sacred Groves: Case Studies in Kerala

U.M. Chandrashekara and S. Sankar

Kerala Forest Research Institute, Peechi - 680 653, Kerala, India

INTRODUCTION

The practice of assigning a patch of forest as the abode of Gods or Goddesses is not new. The societies of Greece, Roman, Asia and Africa had long preserved sections of the natural environment as sacred groves to Gods and Goddesses (Gadgil and Vartak, 1975; Khiewtam and Ramakrishnan, 1989; Hughes, 1994; Ramakrishnan, 1996). In Kerala, there are 1000s of such sacred groves and many of them are still rich in biodiversity (Induchoodan,1992). It is estimated that about 500 ha forest area is under sacred groves (Prasad and Mohanan, 1995) contributing only 0.05% of the total forest area of the State. However, these sacred groves preserve more than 800 species of angiosperm (20% of total flowering plants recorded from the State). Out of which 150 plants are medicinal and 40% are rare and endangered. Apart from helping in the conservation of biodiversity, watershed protection and moderation of the local climate, sacred groves also form the binding force of the people of an area (Unnikrishnan, 1995). Thus, enmeshed in the landscape of agricultural and urban societies, the sacred groves, also represent the cost-effective community-based models for the conservation of the culture as well forests and its biota (Nayar, 1997). In the State, regardless of whether the responsibility of managing the sacred grove is under one or few families or is fully assigned to a statutory agency for temple and sacred grove management, it has been a fact that many stakeholders have an interest and role to play for ensuring effective management of such systems. Thus in reality, different stakeholders together form the institution for the management of sacred groves. However, the strength of these institutions have been eroding due to changing socio-economic conditions and landuse systems. Thus many sacred groves are now threatened and altered both in terms of size, vegetation structure and species composition. In this paper, we discuss the influence of different management systems on the vegetation structure, the present strengths and weaknesses of all important stakeholder groups and the strategies to be adopted for effective conservation and management of sacred groves.

CASE STUDIES

In Kerala, based on management systems, sacred groves can be broadly categorised into three types; sacred groves managed by individual families, by groups of families and by the statutory agencies for temple management (Devaswom Board). Three sacred groves namely Sri Bhagavathi Kavu at Iringole (here after Iringole Kavu), Sri Shangukulangara Bhagavathi Kavu at Sree Narayana Puram (here after S.N. Puram Kavu) and a Sarpa kavu at Ollur (here after Ollur Kavu) which are located within a radius of about 40 km were selected. While the Iringole Kavu is managed by a temple trust in association with the Devaswom Board, the S.N. Puram Kavu is managed by a committee comprising of a group of families and the Ollur Kavu by an individual owner. The area of Iringole Kavu is about 20 ha and that of S.N. Puram Kavu and Ollur Kavu are about 2 ha and 1 ha respectively. These sacred groves lie between 10°10' and 10°43' N latitude and 76°15' and 76°53'E longitude. The average annual rainfall in these sites is between 2500 mm and 2680 mm. May and October are the wet months while November to April is relatively dry. Relative humidity is always greater than 55% and attain 100% during rainy season. Mean maximum temperature is between 25°C to 30°C while mean minimum temperature is about 18°C. The soil is sandy loam to laterite and acidic (pH 4.8 to 5.2).

Vegetation Structure and Composition

In Iringole Kavu, a total of 51 tree species are recorded with *Hopea ponga* and *Artocarpus hirsutus* as the dominant species in all three stages namely, mature phases (individuals with girth <10.0 cm and height <1 m), sapling phase (individuals with 10.1-30.0 cm gbh) and seedling phase (individuals with girth <10.0 cm and height <1 m) (Table 1). In S.N. Puram Kavu, while *Hopea ponga* and *Memecylon umbellatum* are the dominant species in mature tree phase and sapling phase but *Memecylon umbellatum*, *Artocarpus hirsutus* and *Hopea ponga* in seedling phase. In case of Ollur Kavu, *Strychnos nux-vomica* is the dominant species in mature phase and seedling phases while *Tabernaemontana heyneana* is dominant at sapling stage. Total number of tree species recorded in S.N. Puram and Ollur Kavu are 17 and 24 respectively (Table 2).

Table 1. Tree species and their Importance value index values in Iringole Kavu, S.N. Puram Kavu and Ollur Kavu, in Kerala, India

	Iringole *Kavu*	S.N.Puram *Kavu*	Ollur *Kavu*
Mature trees			
Adenanthera pavonina L.	0.6	———	———
Adenanthera triphysa (dennst.) Alston	———	———	5.9
Alstonia scholaris (L.) R.Br.	———	———	19.6
Anacardium occidentale L.	———	———	8.7
Areca catechu L.	———	———	9.0
Aporosa bourdillonii Stapf.***	3.8	———	———
Aporosa lindleyana (Wt.) Baill.		1.6	———
Artocarpus hirsutus Lamk.***	64.7	9.8	———
Borassus flabellifer L.	———	———	16.6
Bridelia retusa (L.) Spreng.	———	———	4.7
Celtis timorensis Spanoghe	0.7	———	———
Cinnamomum malabathrum (Burm.f.)Bl.**	0.6	———	———
Cocos nucifera L.	———	———	6.7
Eriodendron pentandrum Kurz.	———	———	4.6
Ficus benghalensis L.	———	4.7	———
Ficus microcarpa L.f.	———	———	48.1
Ficus mysorensis Heyne	0.7	———	———
Garcinia gummi-gutta (L.)Robs.	———	8.1	———
Holigarna arnottiana Hk.f.***	7.7	3.0	———
Hopea parviflora Bedd.***	39.0	———	———
Hopea ponga (Dennst.) Mabber***	83.0	166.1	———
Hydnocarpus pentandra (B.-H.)Oken***	———	1.5	
Litsea laevigata (Nees) Gamble***	4.1	———	———
Macaranga peltata (Roxb.) M.-A.	2.4	———	———
Mammea suriga (B.-H. ex Roxb.)Koster.	5.7	———	———
Mangifera indica L.	———	———	40.8
Memecylon umbellatum Burm.f.	———	67.1	———
Mesua ferrea L.***	21.4	———	———
Morinda tinctoria Roxb.ex DC.	———	———	9.0
Moringa oliefera Lamk.	———	———	4.3
Myristica malabarica Lamk. ***	1.2	———	———
Naringi crenulata (Roxb.) Nicolson	———	———	14.1
Olea dioica Roxb.	———	2.0	———
Pajanelia longifolia (Willd.) K.S.	0.6	———	———
Plumeria alba L.	———	———	4.5
Polyalthia fragrans (Dalz.)Bedd.***	10.9	———	———
Mature trees (cont'd)			
Prunus ceylanica (Wt.) Miq.	0.7	———	———
Quassia indica (Gaertn.) Nooteb.	———	3.1	———
Schleichera oleosa (Lour.) Oken	———	———	4.2
Strychnos nux-vomica L.	0.6	———	75.2
Syzygium caryophyllatum (L.) Alston	———	14.5	———
Syzygium cumini (L.) Skeels	4.2	———	10.2

Contd.

Table 1 (Cont'd).

Syzygium rubikundam Wt. & Arn.	1.6	——	——
Tamarindus indica L.	——	——	4.8
Trema orientalis (L.) Bl.	5.6	——	——
Vateria indica L.***	37.3	1.5	——
Vitex altissima L.	2.4	1.6	——
Xanthophyllum flavescens Roxb.	——	15.4	——
Tree saplings			
Adenanthera pavonina L.	0.9	——	——
Aglaia elaeagnoidea Juss.*	0.5	——	——
Antiaris toxicaria (Pers.) Lesch.**	0.9	——	——
Antidesma menasu Miq.ex Tul.	0.5	——	——
Antidesma zeylanicum Lamk.	1.3	——	——
Aporosa bourdillonii Stapf.***	27.4	——	——
Aporosa lindleyana (Wt.) Baill.	0.2	12.8	——
Artocarpus hirsutus Lamk.***	82.5	29.4	——
Canthium umbellatum Gamble	0.2	——	——
Caryota urens L.	0.3	——	——
Cassia fistula L.	——	——	13.9
Celtis timorensis Spanoghe	0.2	——	——
Cinnamomum malabathrum (Burm.f.)Bl.**	3.1	7.5	——
Dichapetalum gelonioides Engl.	——	1.6	——
Elaeagnus kolaga Schlecht	0.2	——	——
Ficus tsjahela Burm.f.	0.2	——	——
Ficus virens Ait.	0.2	——	——
Flocourtia montana Grah.	0.8	——	——
Garcinia gummi-gutta (L.)Robs.	——	12.1	——
Holigarna arnottiana Hk.f.***	25.5	10.9	——
Hopea parviflora Bedd.***	2.0	——	——
Hopea ponga (Dennst.) Mabber***	83.0	119.8	——
Hydnocarpus pentandra (B.-Ham.)Oken***	0.3	6.9	——
Ixora brachiata Roxb.	0.9	——	——
Litsea laevigata Nees) Gamble***	6.0	——	——
Macaranga peltata (Roxb.) M. -A.	3.4	——	——
Mallotus tetracoccus (Roxb.) Kurz.	0.2	——	——
Mammea suriga (B.-H. ex Roxb.)Koster.	22.0	——	——
Memecylon umbellatum Burm.f.	——	91.6	——
Mesua ferrea L.***	0.8	——	——
Mimusops elengi L.	0.5	——	——
Morinda tinctoria Roxb.ex DC.	——	——	29.3
Myristica malabarica Lamk. ***	4.8	——	——
Naringi crenulata (Roxb.) Nicolson	——	——	31.3
Olea dioica Roxb.	——	3.4	——
Polyalthia fragrans (Dalz.)Bedd.***	5.2	——	——
Prunus ceylanica (Wt.) Miq.	1.9	——	——
Santalum album L.	——	——	25.8
Streblus asper Lour.	0.2	——	——
Strychnos nux-vomica L.	——	——	76.9
Syzygium caryophyllatum (L.) Alston	——	2.6	——
Tabernaemontana heyneana Wall.**	1.0	——	125.3
Trema orientalis (L.) Bl.	0.3	——	——

Contd.

Table 1 (Cont'd).

Vateria indica L.***	22.2	——	——
Vitex altissima L.	0.3	——	——
Wrightia tomentosa Roem. & Schult.	0.7	——	——
Xanthophyllum flavescens Roxb.	——	10.1	——
Zanthoxylum rhetsa (Roxb.) DC.	0.3	——	——
Tree seedlings			
Adenanthera triphysa (Dennst.) Alston	——	——	29.2
Antidesma menasu Miq.ex Tul.	0.7	——	——
Antidesma zeylanicum Lamk.	31.8	——	——
Aporosa bourdillonii Stapf.***	7.8	——	——
Aporosa lindleyana (Wt.) Baill.	0.2	——	——
Arenga wightii Griff.**	0.5	——	—
Tree seedlings			
Artocarpus hirsutus Lamk.***	43.1	49.3	——
Borassus flabellifer L.	——	——	33.8
Canthium umbellatum Gamble	3.2	——	——
Caryota urens L.	2.8	——	——
Cinnamomum malabathrum (Burm.f.)Bl.***	34.5	14.6	——
Garcinia gummi-gutta (L.)Robs.	0.3	29.0	——
Holigarna arnottiana Hk.f.***	8.3	10.8	——
Hopea parviflora Bedd.***	11.9	23.4	——
Hopea ponga (Dennst.) Mabber***	52.8	35.2	——
Hydnocarpus pentandra (B.-Ham.)Oken***	0.2	18.8	——
Ixora brachiata Roxb.	19.2	——	——
Litsea laevigata Nees) Gamble***	0.5	——	——
Mallotus tetracoccus (Roxb.) Kurz.	0.5	——	——
Mammea suriga (B.-H. ex Roxb.)Koster.	25.1	——	——
Memecylon umbellatum Burm.f.	——	83.2	——
Mangifera indica L.	0.3	——	——
Memecylon sp.	23.5	——	——
Mimusops elengi L.	0.2	——	——
Myristica malabarica Lamk. ***	0.2	——	——
Naringi crenulata (Roxb.) Nicolson	——	——	47.4
Nothopegia beddomei Gamble **	3.9	——	——
Olea dioica Roxb.	1.5	5.5	61.0
Persea macrantha (Nees) Kosterm.	1.2	——	——
Photinia integrifolia Lindl.	0.2	——	——
Polyalthia fragrans (Dalz.)Bedd.***	21.9	——	——
Quassia indica (Gaertn.) Nooteb.	——	3.6	——
Strychnos nux-vomica L.	——	——	128.6
Tabernaemontana heyneana Wall.**	0.2	——	——
Theobroma cacao L.	0.2	——	——
Vateria indica L.***	2.4	——	——
Wrightia tomentosa Roem. & Schult.	0.2	——	——

* Endemic to peninsular India;

** Endemic to Western Ghats, India;

*** Endemic to South-western Ghats, India.

Among these three sacred groves, the Iringole Kavu and S.N. Puram Kavu are comparable to other evergreen formations in the Western ghats of India in terms of species richness, stem density and basal area distribution of trees (gbh more than 10 cm). For example, in the lowland and mid-elevation evergreen forests of the Western Ghats of India (Pascal, 1988: Chandrashekara and Ramakrishnan, 1994), the range of stem density is from 663 ha^{-1} to 2140 ha^{-1} while the range of basal area is from 32.67 m^2 ha^{-1} to 83.83 m^2 ha^{-1}. The stem density value obtained for Iringole Kavu (3341 ha^{-1}) is higher than for the other forests of the region, whereas values for basal area of trees in S.N. Puram Kavu (53.15 m^2 ha^{-1}) and Iringole Kavu (37.37 m^2 ha^{-1}) are within the range reported in other studies. Similarly, the number of tree species per ha in Iringole Kavu (42) and S.N. Puram Kavu (16) is also within the range (from 12 to a maximum of 56) recorded for other evergreen forests of the Western Ghats of India. The mature tree species diversity index in the Iringole Kavu and S.N. Puram Kavu is 3.102 and 2.140 respectively (Table 2). These values are lower than those recorded in tropical rainforest in Silent Valley, India (4.89: Singh et.al., 1981) and in Nelliampathy, India (4.0: Chandrashekara and Ramakrishnan, 1994). On the other hand, the values obtained for the concentration of dominance for the tree layer (0.163 for Iringole Kavu and 0.3643 for S.N. Puram Kavu) are higher than the range (0.06–0.14) recorded for tree layers in the Silent Valley (Singh et. al., 1981) and Nelliampathy (Chandrashekara and Ramakrishnan, 1994). Higher values for the concentration of dominance and lower values for the species diversity index indicate that these sacred groves are essentially mixed type of tropical forest with little dominance by a few species.

Inventory of tree species in three sacred grove also showed that out of seventy three species recorded thirteen are endemic to south-western Ghats, three are endemic to Western Ghats and one is endemic to peninsular India. *Myristica malabarica, Nothopegia beddomei* and *Antiaris toxicaria,* are rare and threatened species, while *Aporosa bourdillonii* is a vulnerable species (Ahmedullah and Nayar,1987; Nayar, 1996). They are also recorded in these sacred groves. Thus these sacred groves can be regarded as the treasure houses of rare and endemic species.

Vegetation structure and species composition of a sacred grove is greatly influenced by the level of disturbance. Therefore, here we made an attempt to compare the three sacred groves using phytosociological data. It may be mentioned here that earlier studies indicated that in relatively undisturbed evergreen forest patches both the number and density (individuals ha^{-1}) of primary tree species are comparatively more than those of late secondary and early secondary species (Bazzaz, 1984; Brokaw, 1985; Chandrashekara and Ramakrishnan, 1994). However, if the forest is disturbed, species number and density of late and early secondary species would increase and the the magnitude of this increase will depend on the intensity and frequency of disturbance (Whitmore, 1984; Pascal, 1988; Chandrashekara and Ramakrishnan, 1994). This kind of information, are useful to assess whether a given forest patch has been subjected to disturbance or not. However, such information are too general and inadequate to investigate differences within and between floristic types of forest, and to relate these to canopy disturbance, and also to explore variations through time. In this context, the following new method of estimating Ramakrishnan Index of Stand Quality (RISQ) considering the life history, density, basal area and distribution pattern of all constituent species (Chandrashekara, 1998) is used to determine the stand quality i.e. whether the stand is disturbed or not and if disturbed to measure the level and frequency of disturbances.

$$RISQ = S \{(ni \times \text{pioneer index})/N\}$$

where, RISQ = Ramakrishnan Index of stand quality (name is given in honour of Prof. P.S. Ramakrishnan, School of Environmental Sciences, Jawaharlal Nehru University, New Delhi, India); ni and N being the same as in the Shannon index of general diversity; Pioneer Index is 1 for the species whose seedlings establish in closed canopy area but need small canopy gaps to grow up, Pioneer Index is 2 for the species whose seedlings establish in small gaps but need small to medium size gaps to grow up, and Pioneer Index is 3 for the species whose seedlings need larger canopy gaps for both establishment and growth.

In the sacred grove at Ollur, higher RISQ values are recorded for all three tree layers (Table 2). This could be attributed to the vegetation changes due to extensive canopy disturbance both in the distant and recent past at Ollur. The RISQ value in Iringole Kavu and S.N. Puram Kavu is between 1.319 and 1.61. This is due to the recruitment of some of the species belonging to pioneer index 2 and 3 along with those belonging to pioneer index 1 due to natural canopy gap formation. For example in Iringole Kavu, late secondary and early secondary species such as *Adenanthera pavonina, Celtis timorensis, Macaranga peltata, Trema orientalis* and *Vitex altissima* are recorded. In already disturbed sites, as in the case of the Ollur Kavu, higher 'RISQ' values for saplings and seedlings than for mature trees mean further disturbance or failure of shade-bearers to establish in such sites. Presence of stumps of trees of both larger and smaller girth classes and trampling of seedlings by cattle are the clear evidence of repeated man-made disturbance to the vegetation of the Ollur Kavu. Similarly, the size class distribution of trees in Iringole Kavu and S.N.Puram Kavu showing a typical negative exponential curve with a clear

Table 2. General characterisitcs of the tree strata of Iringole Kavu, S.N.Puram Kavu and Ollur Kavu in Kerala, India

Parameters	Location		
	Iringole Kavu	S.N. Puram Kavu	Ollur Kavu
Number of species			
Mature trees	23	14	20
Saplings	37	12	6
Seedlings			
Density (trees ha^{-1})			
Mature trees	657	665	367
Saplings	2682	1178	130
Seedlings	108536	21000	26678
Basal area (m^2 ha^{-1})			
Mature trees	31.6	51.4	45.5
Saplings	5.7	1.8	0.09
Seedlings	41.2	15.3	10.1
Species diversity (H)			
Mature trees	3.102	2.140	3.579
Saplings	3.117	2.529	2.206
Seedlings	3.640	3.017	2.093
Species dominance (C)			
Mature trees	0.163	0.364	0.120
Saplings	0.178	0.126	0.270
Seedlings	0.102	0.152	0.272
Ramakrishnan Index of Stand Quality (RISQ)			
Mature trees	1.319	1.404	2.265
Saplings	1.439	1.648	2.731
Seedlings	1.407	1.610	2.368

Seedlings: GBH< 10.0 cm and height <1m ; Saplings: GBH, 10.1-30.0 cm ; Mature trees: GBH > 30.1 cm,

preponderance of stems of small girth classes indicates the better regeneration potential of the sites. On the other hand, absence of such a sharp 'L' shaped size class distribution curve in the case of Ollur Kavu reveals the poor regeneration of trees in the site. Therefore, the Ollur Kavu is degenerating further instead of recovering from the past disturbances. Adoption of proper management strategies for the eco-restoration of this sacred grove is needed.

Stakeholder Analysis for Sacred Grove Management

As mentioned earlier, although the responsibility of managing the sacred grove is under one or few families or is fully assigned to a statutory agency such as Dewaswom Board, it is a fact that many stakeholders have an interest and role. Thus in reality, different stakeholder groups together form the institution for the management of sacred groves. In the sacred groves discussed here, the local people, owners of the sacred groves (in case of single family or group of families owned sacred groves), trustees and Devaswom board (in the case where sacred groves are managed by the statutory like Devaswom Board), priests, shop owners, tourists, youth clubs, schools, forest department, municipality and panchayat are the major stakeholders. Among them, in general local people are the major stakeholders. This is because associated with faiths, taboos and beliefs over years have developed a strong affinity of local people towards the temple and the sacred grove. The local people of each sacred grove in general also believe that their livelihood, security and cultural existence are due to the blessings of their deity. However, the dominance of the local people over other stakeholder groups in three sacred groves differ significantly. For example, although the Iringole Kavu is managed by the Devaswom Board, local people are the major stakeholder. This is because here the local people strongly feel that they are enjoying benefits and at the same time they are contributing man-power, money etc. with dedication and commitment. Thus the Iringole Kavu may be considered as the community-based property. The S.N.Puram Kavu and Ollur Kavu are either single owner property or the property of a few families. Of course, in case of the S.N.Puram Kavu also local people are allowed to participate in all festivals connected to the sacred grove and they are also allowed to visit the grove throughout the year. However, the management or the owners of the grove are reluctant to receive any financial contribution from other sources, even from local people towards the protection and maintenance of the system. The only source of financial contribution from other groups are in the form of offerings to the temple by the local people and tourists. The man-power and other contributions for the maintenance are contributed exclusively from the family members of the owners. They do not encourage other group member to involve in the management activities. This is the case in many other sacred groves of Kerala (Prasad and Mohanan, 1995).

It may be pointed out here that, at present, the youth clubs, schools, Forest Department, Municipality and local Panchayat appear to be less important stakeholder. This is mainly because their contributions for the management of the sacred grove is poor and their present attitude towards the conservation and management of the grove is either negative or neutral.

Collection and removal of any material from the sacred grove is prohibited and this is strictly enforced in the case of Iringole Kavu and S.N. Puram Kavu. This strong positive attitude of the managers which is supported by the local people is responsible for the conservation of the forest associated with the temples. Similar observation has been made for several other sacred groves in Karnataka and Kerala in south India (Gadgil, 1987; Unnikrishnan, 1995) and in northeastern States of India

(Khiewtam and Ramakrishnan, 1989). However, in case of the Ollur Kavu, biomass in the form of green manure, leaf litter, naturally fallen wood and branches as well as timber harvested by the owner of the sacred grove are removed. This kind of repeated disturbance of the sacred grove by its owner and the local people is responsible for its degradation as also recorded by Prasad and Mohanan (1995). It is also clear from the opinion by the stakeholder groups that through well managed sacred groves local people and other stakeholder groups who are proxime to the sacred groves enjoy indirect benefits of the dense vegetation. According to them, the presence of sacred grove associated with dense forest has various ecosystem functions such as amelioration of microclimate, wind shelter-belt effect, protection and regulation of local hydrology, maintenance of visual quality and air purification. Enhancement of mental peace and happiness due to visit to the temple are also appreciated by them. These benefits are felt clearly and more prominently by the stakeholders of the well managed sacred groves such as Iringole Kavu and S.N. Puram Kavu than those by stakeholders of poorly managed Ollur Kavu.

Discussion with stakeholder groups also indicated that better management of the sacred grove by the stakeholders can offer them direct monetary benefits even without the extraction of biomass of the grove. For example, in the case of Iringole Kavu and S.N.Puram Kavu, several stakeholders such as temple trustees, priests, Devaswom Board and shop owners mentioned that they are getting direct benefits because of their involvement with the sacred grove in one way or the other. The economic or material benefits are coming through the tourists, devotees and visitors. Thus, further increased contributions by the stakeholder groups in terms of man-power, financial and other contributions and change in their neutral or negative attitude towards sacred grove into more positive attitude would lead to conservation and management of this natural resource. This kind of shift in their attitude can also increase the benefits they are going to get due to involvement with sacred groves.

STRATEGIES FOR EFFECTIVE CONSERVATION AND MANAGEMENT OF SACRED GROVES

Stakeholder analysis indicated that there is a necessity of better coordination of all stakeholder groups for a common goal of conservation and sustainable management of sacred groves. Subsequent discussion with the stakeholder groups also highlighted the fact that the participation of groups in management may be of different types. For example, in the sacred grove, self-imposed complete ban on removal of biomass, as was done in the

S.N.Puram, would certainly help in revitalisation of ecosystem and its importance in the locality. Misuse or abuse of the land under sacred grove by the local people should be discouraged. In fact, some sacred groves are protected considerably well due to associated faith, beliefs and interests shown by majority of the stakeholder groups. Even in such situations a fear is developing among the managers and people that due to changing socio-economic and political conditions, protection of sacred groves in future is difficult if the area is not fenced. However, the type of fencing suggested by different stakeholders of a given sacred grove or the managers of different sacred groves may often vary. For example, in some cases, a mere social fencing by developing awareness among local people and other stakeholder groups may be quite effective. In this context, it is worth to mention here that the young and active members among the owners of the S.N.Puram Kavu have formed a committee to protect the sacred grove. Recently they organised a Workshop involving local people and resource persons from outside for creating awareness about the functional role and importance of sacred groves. This strategy of holding awareness campaigns can help to jointly chalk out plans to conserve the existing systems from possible threats of encroachment and habitat destruction. However, in many sacred groves, the situation has reached a stage where both awareness campaign and protection measures such as live fencing, compound wall etc. should go hand in hand.

During stakeholder analysis different stakeholder groups made an attempt to identify their future role/responsibilities. For example, youth clubs of the locality felt that they can become an active stakeholder by organising people and conducting seminars, exhibitions and coaching/ training for the conservation of the sacred groves. Similarly, local schools identified their role as organisation of excursions and field trips for children from other areas for creating awareness about the importance and management of sacred groves. Agencies such as Forest Department, Municipality and Panchayat recognised their role as participation in activities such as awareness campaigns etc. which will also help in framing effective a management strategies in future. This will also be a helping hand for the local committee members, in bringing their issues and problems at the decision and policy making authorities. However, by becoming the active stakeholders any one group should not, directly or indirectly, adversely affect the ecosystem. Therefore, careful analysis of present management system and development of future strategies which will not adversely affect the interest and role of different stakeholder groups but help in conservation and management of these natural resources are required.

In Kerala, as elsewhere in the tropics, while some sacred groves are relatively well preserved, others show different degree of degradation.

Restoration of such degraded sacred groves based on the ecological knowledge gained from well preserved sacred groves should be one strategy for conservation of sacred groves of the State. Species present in the system as well as in comparatively less disturbed sacred groves of the same ecological zone may be used to restore degraded systems. However, while doing so, autecological aspects such as regeneration pattern, successional status, architectural features, site requirement of the species should be considered. At the same time, due importance should also be given to social, cultural and religious importance of such species.

ACKNOWLEDGEMENTS

We are grateful to Dr. K.S.S. Nair, Director, KFRI for his keen interest and encouragement. Prof. P.S. Ramakrishnan and Dr. K.G. Saxena of School of Environmental Sciences, Jawaharlal Nehru University, New Delhi are gratefully acknowledged for their support and valuable suggestions. Thanks are due to Mr. P.C. Anil, Mr. Sathian P. Joseph and Mrs. V. Sreedevi for their assistants in the field works and data analysis. The study is supported by the UNESCO, India.

REFERENCES

Ahmedullah, M. and Nayar, M.P. 1987. *Endemic Plants of the Indian Region*. Botanical Survey of India, Calcutta.

Bazzaz, F. A. 1984. Dynamics of wet tropical forests and their species strategies. In: E.Medina, H.A. Mooney, and C. Vazquez-Yanes (Eds.), *Physiological Ecology of Plants of the Wet Tropics*. W.Junk Publishers, the Haque, pp. 233-243.

Brokaw, N.V.L. 1985. Gap phase regeneration in a tropical forest. *Ecology, 66*: 682-687.

Chandrashekara, U.M. 1998. Ramakrishnan Index of Stand Quality (RISQ): an indicator for the level of forest disturbance. In: A.D. Damodaran (Ed.), *Proceedings of the Tenth Kerala Science Congress*. State Committee on Science, Technology and Environment, Thiruvananthapuram. pp. 398-400.

Chandrashekara, U.M. and Ramakrishnan, P.S. 1994. Vegetation and gap dynamics of a tropical wet evergreen forest in the Western Ghats of Kerala, India. *Journal of Tropical Ecology,* 10: 337-354.

Gadgil, M. 1987. Diversity: cultural and biological. *Trends in Ecology and Evolution,* 2: 369-373.

Gadgil, M. and Vartak, V.D. 1975. Sacred groves of India: a plea for continued conservation. *Journal of Bombay Natural History Society,* 72: 314-320.

Hughes, J.D. 1994. *Pan's Travail: environmental problems of the ancient Greeks and Romans*. The Johns Hopkins University Press, Baltimore.

Induchoodan, N.C., 1992. *Ecological Studies on the Sacred Groves of Kerala*. Final Report Submitted to WWF, India.

Khiewtam, R.S. and Ramakrishnan, P.S. 1989. Socio-cultural studies of the sacred groves at Cherrapunji and adjoining areas in northeastern India. *Man in India,* 69: 64-71.

Nayar, M.P. 1996. *Hot Spots of Endemic Plants of India, Nepal and Bhutan*. Tropical Botanical Garden and Research Institute, Palode, Kerala, India.

Nayar, M.P., 1997. Biodiversity challenges in Kerala and science of conservation biology. In: P. Pushpangadan and K.S.S. Nair (Eds.).*Biodiversity and Tropical Forests: The Kerala Scenario*. The State Committee on Science, Technology and Environment, Kerala. pp.7-80.

Pascal, J.P. 1988. *Wet Evergreen Forests of the Western Ghats of India: ecology, structure, floristic composition and succession*. French Institute of Pondicherry, Pondicherry, India.

Prasad, G.A. and Mohanan, C.N. (1995). The sacred groves of Kerala and biodiversity conservation. In: P.K.Iyengar (Ed.). *Proceedings of the 7th Kerala Science Congress*.State Committee on Science,Technology and Environment, Kerala pp. 125-126.

Ramakrishnan, P.S. 1996. Conserving the sacred: from species to landscapes. *Nature and Resources*, 32: 11-19.

Singh, J.S., Singh, S.P., Saxena, A.K. and Rawat, Y.S. 1981. *The Silent Valley Forest ecosystem and Possible Impact of Proposed Hydroelectric Project*. Reports on the Silent Valley Study. Ecology Research Circle, Kumaun University, Nainital, India.

Unnikrishnan, E. 1995. *Sacred Groves of North Kerala : An Eco-folklore Study*. Jeevarekha, Thrissur, Kerala, India. (in Malayalam).

Whitmore, T.C. 1984. *Tropical Rain Forests of the Far East (2nd edition)*. Oxford University Press, Oxford.

29

Conservation and Management of Sacred Groves in Kerala

S. Chand Basha

Principal Chief Conservator of Forests (Retd.)
B-503, Kanchanjunga Apartments, Palarivattom, Kochi-25, Kerala, India

HISTORY AND PRESENT STATUS

Sacred groves are the 'Gardens of Gods' containing natural vegetation of the locality associated with the supernatural power and got preserved over a period of time. These are as old as starting of settled agriculture some where in the 6th Century AD (Gadgil and Vartak, 1976). This system had probably originated during the time of aboriginals or forest dwellers who had started agriculture in the form of shifting cultivation or on a permanent basis. While clearing land for agriculture invariably they used to keep a patch of forest growth where they had established their 'God' for protecting their cultivation and settlements. They believed that the 'forest god' would protect them and their cultivation from the evil spirits and wild animals. They found supernatural power in the 'stones' or some other form of structures which they placed in the groves. This practice of worshipping stones which are mostly of different shapes (not of human or animal shapes) are still prevalent among same of the primitive tribal groups of Kerala.

These blocks of natural forest vegetation which are available in different parts of Kerala State have undergone considerable changes both in extent and quality with the result that what we see today are groves with considerable reduction in extent and degraded vegetation of more drier type. Side by side with this change there was also change in the mode of worship and belief of the people of the supernatural being placed inside the grove.

In the beginning, the supernatural power was attributed on the pieces of stones established in the open mostly at the base of trees. Along with these there were also stones representing the ancestors of the forest dwellers. These original forest dwellers were driven away during the invasion of the civilised society. The aborigines took shelter in the deep forests situated especially in the high ranges and restarted activities like shifting cultivation.

The usurpers continued their cultivation and developed the land by further deforestation. In the process, the extent of the original groves was reduced to a great extent. These civilised people gave a new dimension to the mode of worship to Gods which originally existed in the groves and to the beliefs which were practised. The stone gods which had no specific shapes originally were gradually replaced by human shapes in most cases and became the Gods of Hindu Pantheon. Side by side the Gods were placed within enclosures which by passage of time became well built temple structures; where in the idols were placed. More importance was given for the development of the temples by giving specific architectural designs. These structures were put up mainly based on Vasthu (vedic architecture).

Almost all these sacred groves came to the hands of families and remained as family owned properties. In course of time the families were divided into smaller units and while dividing the family property the groves also got divided. Those who got parts of the grove without the worship places converted them to agricultural fields or home gardens. Invariably the eldest male member retained the grove with the worship place, locally called the "Kavu" and continued daily rituals of worship like lighting of lamp which became the responsibility of this family. Once in a year the all family members got together in the 'Tharavadu' (joint family) and conducted the festival in the grove to propitiate the 'God'.

Many of such old families which were unable to practise the Kavu rituals regularly due to disintegration and the grove came to the hands of organisations like Religious Boards or local temple committees. Thus majority of the groves ceased to be family properties, while there are still a few which are maintained by individual families.

The fragmentation process of the original grove across different generations led to the disappearance of larger parts of the original vegetation and now only small patches of original vegetation housing the temples or worship places exist. Some of the groves are completely devoid of vegetation and only the temples exist although still known as 'Kavus'. There were many instances where even the worship places were removed by a process called *'Avahichu Kudiyiruthal'* (absorb the spirit and rehabilitate) where in the temple deity is transferred to a new place after conducting special rituals or 'Poojas'. Thus the grove was freed from the power of the God and the vegetation cleared for conversion to alternate land use.

During the survey of the Travancore region alone Ward and Conner (1894) came across the existence of thousands of groves. During the beginning of the 19th century there were more than 30,000 groves in Kerala (Velupillai, 1940; Logan, 1951). A recent survey conducted by Induchoodan (1996) revealed presence of only 761 sacred groves of which 399 (52.17%) were of less than 0.02 ha in extent. Only 362 groves were larger than 0.02 ha.

Of the 362 groves having an area of more than 0.02ha, 286 (79%) are below 0.5 ha, 24 (6.6%) between 0.5–1.0 ha, 41 (11.4%) between 1.0 and 5ha and 11 (3.02%) above 5 ha. The total extent of the sacred groves in Kerala is about 440 ha. The district-wise extent and number of sacred groves are given in Table 1.

Table 1. District -wise distribution and extent of sacred groves in Kerala.

Sl. No.	District	Number	Total area (ha)	Av. area (ha)	% of toal
1.	Thiruvananthapuram	43	1.7790	0.0414	0.40
2.	Kollam	44	6.6740	0.1517	1.52
3.	Pathanamthitta	33	115.2740	3.4932	26.20
4.	Alappuzha	49	18.5630	4.3788	4.22
5.	Idukky	3	6.5400	2.1800	1.49
6.	Kottayam	10	2.1800	0.2180	0.59
7.	Ernakulam	7	12.2200	1.7457	2.77
8.	Thrissur	17	3.4840	0.2049	0.79
9.	Palakkad	3	5.1600	1.7200	1.17
10.	Malappuram	11	7.3520	0.6684	1.67
11.	Kozhikode	23	27.8780	1.2121	6.34
12.	Wyanad	5	109.2520	21.8504	24.83
13.	Kannur	54	100.2080	1.8557	22.27
14.	Kasargode	60	23.440	0.3907	5.33
	Total	362	440.0040	1.2155	100.00

Poonkavanam of Pathanamthitta District and Pakshipathalam of Wyanad District are the two bigger groves having an extent of more than 100 ha and are parts of the Reserved forests lying contiguous to forest areas. These two are well protected by the forest laws. Though Kasargode District contains more number (60) of groves, Kannur district with 54 groves excels in extent with 100.28 ha . In many districts the groves are small with poor floral diversity. Side by side with the decline of biological diversity an increase in cultural diversity (totems and rites) is observed in a large number of groves.

Broadly the vegetation of sacred groves can be classified into two types viz., the evergreen type and the most deciduous type. Evergreen type includes four sub-types namely the Swampy type, High moist type, Low moist type and disturbed type. These groups are biologically rich containing a rich flora. Majority number of groves support the above sub-types. In sacred groves with moist deciduous type of vegetation are mostly dry with open canopy and poor plant diversity.

There are about 722 plant species belonging to 474 genera and 129 families in the sacred groves of Kerala. Out of 722 species 154 species are endemic to the Western Ghats and 459 species contain one or more medicinal properties. The groves also harbour many wild relatives of cultivated plant species (Induchoodan, 1996).

Faunistically the sacred groves are very poor due to small extent of natural vegetation and the impact of the surrounding dense population. In the two groves viz., Pakshipathalam and Poonkavanam (Sabarimala) which are larger in extent and contiguous with natural forests, the faunistic and floristic diversity is very high. In other groves only smaller animals, birds, insects etc. are found in appreciable numbers.

Due to varied beliefs of the people living around the groves, the cultural diversity in the groves is very high and such a varied diversity probably unique to Kerala groves, is not found anywhere in the world.

In most of the groves the presiding deity is the local goddess. The presiding deity may be one in most cases while in a few cases more than one. In many groves there are sub-deities the number of which varies and goes up to even 16 sub-deities, all of which are worshipped as seen in Thirur Palliyara Bhagavathy Kavu in Kollam district. The subdeities are called by different names and there are as many as 133 sub-deities as could be seen from a detailed study of 63 sacred groves in Kerala. The number can increase, if detailed studies are conducted in all the groves.

The most common god worshipped is the Snake God which is a presiding deity in only about 30% of the groves, but they are largely worshipped as subdeity in the majority of the groves (60%). There are also groves without any deity as in Oachira Kavu in Kollam district. Here 'Parabrahmam' (the universal creator or lord and father of all creatures) is worshipped. In the Kanayi Bhagavathy Kavu dedicated to Kanayi Bhagavathy in Kanathodam and the Muthappan Kavu of Kanjangad dedicated to Sastha, there are no idols or deities in the worship place. In the Bhayankavu of Malappuram district, people worship Bhadrakali but before a mirror which represents God. There are many such varied phenomena which remain to be investigated.

The festivals in these groves are varied and highly interesting from the cultural and religious points of view. In this respect each grove differs from the other. The description of even a few is beyond the scope of this paper. It is suffice to state that cultural diversity of groves is fantastic and unique to Kerala.

CONSERVATION OF SACRED GROVES

Conservation of sacred groves is of great importance in order to conserve the biological diversity and retaining the cultural importance of conservation depends mainly upon two important matters viz. ownership of the groves and beliefs of the people on supernatural power of the Gods or deities established in the groves.

Ownership of the Sacred Groves

More than 99% of the sacred groves are privately owned. The privately owned ones can be further grouped into four groups:

1. Family owned, where management will be looked after by a single family as in Mannarassala Sacred of Alappuzha District dedicated to Snake God or managed by a group of families like the family trust as in Pambumekattumana Grove of Thrissur District again dedicated to the Snake God.
2. Establishment owned where, the management is entirely left to the hands of Dewaswam Boards or Hindu Religious and Charitable Endowment (HR & CE). Here the establishments provide annual maintenance grants to the temples situated inside the groves. Most of the groves located in the southern parts of Kerala are under the Dewaswam Board and many situated in the northern parts of Kerala are supported by HR & CE.
3. Community owned where some groves are under the ownership of certain communities as in Cheermakavu dedicated to Sri Kurumba Bhagavathy in Neeleswaram of Kasargode District. Here the 'Asari' (carpentor) community is managing the affairs.
4. Local trust owned, where many groves are managed by the local committees or trusts. These committees have come into existence as a result of disintegration of the families which originally owned the groves. Pramancheri Kavu in Kannur District is one such groves managed by the 'Pramancheri Kavu Samrakshana Samithi'. There are a few more under this type of management.

Ownership is a decisive factor in formulating plans for conservation as many groves need protection in order to conserve the biological diversity and to perpetuate the existing varied and interesting cultural activities which are mostly based on faiths. Through varied cultural festivals, the faiths are kept alive in most of the groves.

In the present set up it is found that in general conservation is very effective in single family owned groves as owners receive sufficient income from the groves especially in the form of offerings to the Gods. With regard to conservation of groves owned by many families there is always problem in the effective conservation due to difference of opinion among the individual families constituting the committee. With regard to the conservation of groves owned by the establishments, many difficulties are being experienced viz., the organisation are mostly interested in the income from offerings from the worship places of the groves. There are also cases where the produces from the existing vegetation like timber etc., are used to supplement the income to conduct the temple rituals. Conservation status of groves belonging to community or local committees is better as they mostly depend upon the offerings and supplement any shortage of income through local collections. Only in rare cases the produce is taken out by destroying the existing vegetation.

The groves under private ownership do not enjoy any legal protection. But the groves under the Government ownership are under rigid protection due the acts and rules prevalent in the State. The Kerala Forest Act and Rules and the Kerala Wildlife Protection Act provide sufficient protection to the forest areas.

One of the factors which is mostly responsible for the degradation of the vegetation is the festival conducted in most of the groves regularly. There may be one or more festivals in a year depending upon the cultural importance or significance of the deity or deities housed inside the groves. One of the redeeming feature of such festivals is the assemblage of large number of people from the adjacent villages or from the different parts of the State and India. The number depends upon the significance or importance of the God/Gods located in the worship places/temples in the groves. Irrespective of the caste, creed and religion people take part in cultural functions. Due to movement of large crowds the ground vegetation which comprises regeneration of the species prevalent in grove is entirely destroyed making the ground barren. This being an annual feature the continuity and growth of the regeneration are lost. This is one of the reasons for the degradation of the groves.

Belief of the People

The other most important factor which helps in conservation of the sacred groves in Kerala is the ardent belief which people have on the deity or on the supernatural power existing in the grove. The people consider the grove or the vegetation as the garden of the deity. Different kinds of beliefs prevail according to the deities reigning the sacred grove. In many cases people believe that the God/deity not only protects the grove but also protects the people of the locality from bad spirits and natural catastrophes.

People dare not to violate the old traditions or practices, prevalent for generations, due to the fear of receiving the wrath of the God. Violation of the practices and rituals leads to inflicting punishments of different kind including death of the person involved. There are certain sacred groves where the surrounding people believe that the wrath of god for violation of rituals not only falls on the individual but also on the family as a whole.

In order to perpetuate the rituals and practices prevalent for generation regular festivals are held in the worship places/temples of the sacred groves, where people of all faith and religions congregate to participate in the same. The congregation of more people means more offerings from devotees which constitute the major revenue used to perform the regular daily rituals of the deity/deities.

Due to financial constraints administrators of many sacred grove find it difficult to organise the festivals even in a smaller way. Thus the cultural rituals are fading out and in course of time these cultural practices are forgotten and consequently the groves disappear.

In many groves there exist powerful deities and tantric rituals are carried out by the learned Brahmin priests to recharge the power of the deities.

For the violation of the traditional practices, the priests or the 'Theyyams' or 'Velichappadu' decide the punishments to be inflicted on the disbelieved. The utterances of these 'Theyyam' or 'Velichappadu' in trance are considered as commands of the presiding deities. There are also cases where the temple committees decide the penalty for violation of rituals.

There is a gradual erosion in the beliefs of the people in the power of the deities reigning the groves. Such erosion has lead to the gradual decline in the extent of the groves and their gradual disappearance.

There are many instances where the belief on the presiding deity has became so strong that people construct wholesome temples to house the deity/deities and fully neglect the groves. In such instances there exist only the temples and the surrounding groves have been cleared for the improvement of the temples. Thus we come across many 'Kavus' without the groves.

It is a fact that so long as the beliefs are strong among the people, the groves will exist and the present consciousness of the people about the conservation of natural vegetation surrounding the groves adds to the same.

STRATEGIES FOR CONSERVATION AND MANAGEMENT OF SACRED GROVES

As already pointed out conservation of the sacred groves is important on two counts:

1. To conserve the biological diversity of the natural vegetation
2. To perpetuate the cultural festivals which are continued for generation.

When the above two aspects are considered, the conservation of most of the groves becomes a paramount necessity. Conservation of biological diversity will depend upon the extent of the groves, more extent more the diversity. As already pointed out there are only eleven (11) groves having an extent of more than 5 ha. Many of the small groves also need protection not only from the point of view of high biological value but also for perpetuating cultural values, which are unique to each grove.

Some of the important aspects which need attention for the conservation of the groves are the following:

1. According complete protection to the grove by putting strong fencing to facilitate the natural process of vegetation regeneration and succession.

   Limiting the cultural or religious festivals to periphery of the grove or in one portion of the grove so that the natural regeneration is not periodically destroyed and the natural process of vegetation development is maintained.
2. Supplementing the natural process by mixed seeding in the degraded areas so that the vegetation development takes place. The seeding should be done by collecting the seeds of the indigenous vegetation existing in the locality. No exotic species should be introduced.
3. It is necessary to avoid use of grove lands for public purposes as this activity will reduce the belief of the people on the deity (the deity is not able to protect its own property as the garden of the God). It is found absolutely essential to parcel part of the grove for public purpose, which should be spared only after undertaking appropriate tantric or religious rituals so that people can be under the impression that the God has willingly given part of the grove for the purpose of the devotees.
4. It is advisable to publicise the beliefs prevalent among the people of the locality so that the community is interested to protect the grove.
5. It would be necessary that a good local NGO is involved in the conservation of the grove by imparting necessary training to it about the ecological importance and biodiversity value and also the economic and cultural benefits.
6. It is better not to enforce any Government legislation on the groves as it becomes almost a no man's property and people will loose the beliefs on the deity of the grove.

7. The groves should not be brought under the control of the Forest Department as it will be difficult to protect the groves delinking the temple or worship place attached to it.
   It would be necessary to avoid political interference into the administration of the groves as it will create a sense of disbelief among people and people will gradually loose interest in the conservation of the grove.
8. It would be advisable to promote eco-pilgrimage to the groves having high cultural values. This can improve the income of the grove through the offering of devotees and the grove can be maintained properly . Eco-pilgrimage should get priority over eco-tourism.

It is not enough that attempts are made to conserve the groves. The purpose will be served only it the groves are managed properly and efficiently so that they retain their pristine condition and remain for ever. Therefore management of the groves is of paramount importance.

Effective and efficient management of the groves depends upon the availability of sufficient funds for running the day to day rituals connected with worship of the deity/deities of the groves. Many of smaller sacred groves are under the Dewaswam Board, HR & CE and similar organisations which give a small grant annually to run the worship places. This along with the offerings of the devotees in most cases fall short of the requirements. With the result the groves with smaller extents are disappearing. Along with the natural vegetation the cultural values also getting diluted.

The land hunger due to the high density of population and the soaring land value speed up process of disappearance of the groves in many parts of the State. The process becomes easier when the tantric rituals also came handy. The 'Avahichukudiyiruthal' process of the Gods reigning the Sacred Groves adds fillip to use of grove lands for alternate land uses. The encroachments of grove lands by the adjacent home gardens in many places reduce the extent of groves.

Fragmentation of grove lands in the case of groves under family ownership in some cases results in the partitioning of the property wherein the heirs who get the land without the worship place convert it to other uses or homegardens.

There are also instances where collection of produce becomes excessive. This includes even removal of trees for construction purposes or for sale to add revenue to the management. The excessive collection of the produce by the stake holders becomes in most cases unscientific leading to deterioration of the original vegetation.

Keeping all the above factors it is necessary to plan suitable management strategies with the co-operation of the local villagers or devotees. Since the activities of the groves are mainly connected with religious rituals it will be difficult to frame an entirely new management strategy. It would be better to follow the present management system with some practical and innovative changes acceptable to the management. Further fragmentation of the grove property will have to be stopped by paying adequate compensation to the claimants. It is necessary to restrict excessive collection of the produce by the right holders and villagers by developing appropriate scientific harvesting methods.

Enhancement of the grants presently given to the grove management to necessary levels so that along with the money from the offerings received is sufficient enough to run day to day rituals of the grove worship places. Side by side, it would be essential to give appropriate publicity about the supernatural power of the presiding deity/deities so that more devotees visit the worship places in the grove. This will help to add revenue to the management. The management should take interest in organising the yearly festivals with the co-operation of the local villagers so that it becomes an attraction to people from other parts to visit. Though this proposal may look odd, it would be necessary to have a gathering of the people of all faiths and religions to develop communal harmony. Moreover, this will also help to enforce some sort of discipline among people to refrain from destroying the groves.

In many cases it would be advisable to acquire the adjacent fallow lands, which were once the grove lands, and add up to the grove in order to extend the grove.

For effective management it is absolutely essential to monitor the financial accounting of the management especially in the case of groves under the Dewaswam boards, HR & CE, local committees and trusts so that the funds obtained from different sources are suitably spent for the purposes which are absolutely essential for running rituals of the worship places of the grove. The funding organisations should impose restrictions on overuse or abuse of the grove vegetation.

The management strategy should be to develop both the worship place and the grove simultaneously so that the natural vegetation of the grove can remain permanently and the worship place developed to attract more devotees. Unless some innovative modern management techniques are introduced with full involvement of the grove managers and the people living around there will not be any future to the groves especially the smaller ones.

CONCLUSION

Ownership of the groves and the beliefs of the people on the God/Gods "living" inside the groves are two decisive factors which decide the conservation of the sacred groves in Kerala. The fact that about 79% of them are small i.e. below 0.02 ha. in extent should never be taken as a criterion to neglect the groves. By appropriate management practices many of them can be developed or at least kept in their existing condition with a possibility for improving the vegetation. Neglecting the smaller groves will lead to the disappearance of both vegetation and cultural diversity. Therefore, it is necessary to encourage conservation even by appropriate funding for development as there are the last resorts of existing relicts of natural vegetation in the unforested areas of Kerala.

REFERENCES

Gadgil, M. and Vartak, V.D. 1976. Sacred Groves of the Western Ghats of India. *Economic Botany, 30:* 152-60.

Induchoodan, N.C. 1996. *Ecological Studies of the Sacred Groves of Kerala.* Ph.D. Thesis submitted to the Central University of Pondicherry.

Logan, W. 1951. *Manual of Malabar District.* Madras Government Press.

Vellupillai, T.K. 1940. *The Travancore State Manual.* 4 volumes. Government of Travancore

Ward and Conner 1894. *Memoir of the survey of the Travancore and Cochin States.* Government of Kerala.

30

Biodiversity Conservation through Sacred Groves (Deorais) in Maharashtra: Retrospect and Prospect

A.G. Raddi

809, Bhandarkar Institute Road
Pune-411 004, Maharashtra, India

INTRODUCTION

It sounds incongruous and improbable. But nevertheless it is true. Even in present times of environmental disruption and stress, there actually exist in Maharashtra a number of small pockets of relict vegetation. They stand out as green oasis amidst the enveloping barrenness. They are sacred groves or Deorais, traditionally venerated by the villagers as abodes of village deities. As a consequence of the sanctity bestowed on them they have remained relatively inviolate and harbour within them relict ecological communities and biological species. Noting the widespread destruction of natural habitat all around, such islands of green relict vegetation and their associated faunal elements have assumed great significance. Their Biodiversity has not only reference value but significant biotechnological potential. Gadgil and Vartak(1976) through their pioneering work in Maharashtra focussed attention on this priceless natural heritage of the state and highlighted the need for their protection and further study. Ramakrishnan (1996) has also reviewed the theme with particular reference to east India.

CURRENT PROBLEMS

The sacred groves are presently confronted with many survival problems. the more important of these are discussed on following page.

Erosion of Sanctity and Values

It is obvious that the element of sanctity which had ensured their survival so far is fast getting eroded. The younger generations in the villages are often oblivious or casual to the groves. While some of the sacred groves on account of their relative remoteness and inaccessibility and sentiment continue to be well wooded, others are not so fortunate. With the spread of road groves get denuded. It is also not uncommon for farmers owning lands in proximity of the sacred groves, to fell trees within sacred groves trees resulting in management apathy and neglect. Instances have also been heard of where in the temple committee set up by the villagers for management of the sacred grove have sold parts of it ostensibly to raise funds for building a temple of the village deity in the sacred grove. The price of large sized timber being very high, the large sized trees in sacred groves are of particular interest to timber dealers. For the villager economics is easier to understand than ecology. There is therefore strong temptation on the part of some villagers to clear at least part of the grove against substantial monetary returns to the temple management or to some unscrupulous individuals. Instances have also been noticed where in removal of dead wood from the grove has been permitted. This has serious implications. After all illicitly grilled trees do eventually die and turn into dead and can then be removed with a clean conscience by interested parties as 'dead wood'.

Legal Limitations

It is generally observed that most of the sacred groves are noted in government records as temple property and are either on private lands or land in charge of government departments other than the forest department. Thus most of them have not been notified as forests and as such have been neglected by the forest department. Its staff is presently overstretched protecting the government forests and wild life therein, as well as executing numerous developmental works in its area of operation. This keeps it thoroughly occupied and often harassed. Consequently, it has refrained from voluntarily taking upon itself additional responsibilities in area so far considered as out side its domain e.g. sacred groves. Further the sacred groves are small in size and occur in a loosely scattered manner in the state. Administering such small and diffused units is a difficult proposition. These factors may have contributed to a situation where in the sacred groves in the state are left by government agencies in a state of benign neglect leading to their continued attrition.

Imbalances in Observational Studies

The studies so far on the sacred groves are basically botanical. They generally consist of inventories listing the dominant and uncommon tree species found in them, the size attained by them and celebrating the fact of their long term survival on account of the sanctity of the groves. Attention however also needs to be paid (to highlight its full biodiversity value) to the herbaceous and shrub flora, Propagation methods of the species of plants found in the sacred groves need to be worked out and standardised. Such information is available today only in respect of some plant species of economic value and not for others which may be of biodiversity value. Also observation on the faunal elements of the sacred groves are conspicuously absent, particularly the smaller vertebrates and various invertebrates. Management and sociological studies also need to be undertaken to chart an effective approach for sustained management of the natural heritage represented by the sacred groves.

MANAGEMENT OPTIONS

Against the above stated background and the ground realities of the situations it is essential to chart out a definitive course of management and protection of this rich natural heritage, putting to an end the current directionless drift. In this context the following possibilities are evaluated.

Notify Sacred Groves as Forest Areas

Well meaning botanist and naturalists often talk of this as the right solution. In theory, after recording them as "Forest", it is possible for the forest department to notify under provisions of the Indian forest act its intention to declare the sacred groves as reserved forest. Although the actual procedure of converting them into reserved forests can be lengthy, the provisions of the Indian forest act become applicable to the area intended to be declared as reserved forests right from the date on which the intention is notified to the public in the government gazette, facilitating forest department intervention. It must also be realised that once this is done, the forest department will treat these areas as forest areas. As a consequence of the above, the Forest conservation act 1982 will also then automatically apply to these areas. The provisions of the forest conservation act are very stringent and specifically prohibit the undertaking of any non forest activity in these areas without the express approval of the government of India. Any social and cultural activities pending clearance from the govt. of India will create in its wake confrontation with the forest department and generate public ill will, since the forest department has to implement the act as it is and has no discretionary powers in this regard. Thus while the Forest conservation act is well meaning and helps in the protection of our depleting government forests it may not be the best thing for the future protection and management of the sacred groves.

It is also observed that the supreme court in one of its recent judgements relating to the forests has opined that the forest conservation act applies to all the forests areas irrespective of the fact whether they are recorded in government documents as forests or not. It has asked the state authorities to take a review and report as to how much additional area can be considered as forest. It is possible that such additional areas may include the present sacred groves. In such an eventuality the forest conservation act would also apply to the sacred groves automatically. This could complicate the issue unless the existing non forest practices followed is cleared by government of India at the initial stage itself. Institutional strengthening of the forest department as well as giving it adequate discretionary powers to handle locality specific issues effectively would also help the overall cause of environmental conservation.

Wild Life (Protection) Act 1972

This act is also implemented by the forest department and extends to all forest as well as non forest areas. Wild life had been defined here as 'All uncultivated flora and fauna'. In theory this act is applicable to all flora fauna within the sacred groves even today, but has not been effectively enforced on account of perhaps administrative and logistic limitations of the forest department. Since the sacred groves have enormous natural biodiversity this act is very relevant to its protection. There is no need here to declare sacred groves as reserved forests nor to interfere with the traditional practices presently in vogue in its management. The existing rules under the wild life act can be suitably augmented by the state government to specifically cover and give importance to the sacred groves. Thus while the management of the sacred groves can carry on in the traditional manner through the temple committee, in case of any perceived danger of its wild life heritage the forest department could intervene under the provisions of the wild life act and rules made there under, as a back up supporting agency. It can be a good example of joint participatory management enthusiastically being promoted the government of India.

Linkages with the Village Ecodevelopment and the UNESCO Biosphere Reserve Concept

It is also possible to think of situation where in the village sacred grove wherever present is integrated with the village ecodevelopment programme and linked to the concept of Biosphere management as proposed by the UNESCO. Together they could set the stage for a harmonious ecorestoration and sustainable development programme in our rural country side.

The sacred grove in the proximity of village could be an integral part of the ecodevelopment plan for that village and primarily serve the function of conservation of natural biodiversity within the village environs. It is felt that the concept of village ecodevelopment is really an extension of the Gandhian concept of self sustaining Indian villages. Ecodevelopment focusess on the restoration of the village ecosystem through land, water and vegetation interventions. A fully restored and functional ecosystem is the very foundation on which other village development activities must rest including economic development if it is to be sustainable.

The biosphere reserve concept of the UNESCO (Ishwaran, 1992) is also essentially about ecosystem restoration and sustainable development. It can very effectively blend with the village ecodevelopment effort that is envisaged. For example taking the district as a unit the sacred groves occurring in that district could collectively be designated as a cluster core area and managed accordingly. A buffer zone could be developed around each of the core areas constituting the cluster and rehabilitated with emphasis on non timber forest produce generating species i.e. fruit, myrabalans, medicinal plants etc. whose usufructs could be harvested annually without felling or damaging the parent plant. The rest of the area could be treated as a transition zone in which people's activities will go on as before but against a background of an understanding of the basic ecological principles involved and their linkages for environmental well being, and sustainable development. In villages where no sacred groves exist, efforts can be initiated to start new sacred groves. Say a one hectare plot in an existing government plantation in the village area or a part of the village community land or a piece of land donated for the purpose by an enlightened local farmer could be utilised for the purpose. This plot could be planted up voluntarily by the people or by the forest department with a variety of ecologically suitable indigenous species. In course of time, because of the protection afforded to it through fencing in early stages and social acceptance, other forms of local biodiversity will also find natural shelter there.

Many of the old sacred groves during the monsoon season have abundant natural regeneration of indigenous species uncommon elsewhere. These seedlings which are in excessive numbers could be removed and transplanted in new sacred groves elsewhere. The seed from sacred grove plants could be likewise collected annually and distributed to other areas in the state for production of seedlings and eventual planting in new locations. Interested school children could be given a polythene bag containing seedling of uncommon sacred grove plants for raising it for a year as a part of their household. In the next rainy season these bags could be collected back for planting the young saplings at suitable locations elsewhere. This will generate interest and empathy for plant species in the younger generation. It will also make the children aware of the valuable biodiversity components in the sacred groves of their own village and establish linkages between the village school and the sacred grove with benefit to both. Large scale extension and motivation activity primarily aimed at enhancing environmental literacy amongst the village children and women is also very essential. It should not be treated as something cosmetic but in the light of the fundamental contribution it can make towards attitudinal change it deserves regular and serious attention.

Thus sacred groves, old or new, can be a very valuable and integral part of the village-wise ecodevelopment effort. They can make solid contribution to conservation of biodiversity. The other activities of the ecodevelopment plan like soil and moisture conservation, waste land development, Agroforestry, etc. can not only help to recover the village ecosystem but make the village truly self sustaining in its basic needs of fuel, fodder, small timber, fruits, etc. with surplus going to the market. Such pilot scale ecodevelopment activities could be initiated in selected villages on the coast, in the western ghat and in the drought prone low rainfall zone, to develop guidance modules for others to emulate with locality specific variations. Such effort will collectively go a long way in safeguarding and promoting biodiversity conservation in areas out side the government forests as well as meeting legitimate and basic needs of the rural people in a self sustaining manner.

CONCLUSION

It is clear then that in the prevailing situation the cause of the sacred groves and biodiversity conservation can be best served by implementing village wise ecodevelopment programmes blending options 3b) and 3c) such as use of existing wild life (protection) Act 1972 and the UNESCO biosphere reserve concept. The option like notification of sacred groves as forest area though very appealing at first will entail implementation of the forest conservation act to the sacred groves. Though this act is vital for safe guarding interests of the government forests, it may not be ideal for management of sacred groves because of the potential conflict situations that may arise in its wake between the implementing agency and the people. Ironically the interest in the sacred groves and their traditional philosophy of participatory management can best be perpetuated by not designating them as 'Forests' in government records.

The UNESCO promoted concept of the Biosphere reserves is considered very relevant to the above, and can be effectively blended along with the existing or new sacred groves into the village Ecodevelopment approach. It is not considered as essential to declare such areas as Biosphere reserves. What is important is to integrate the spirit and philosophy of Biosphere reserve management into the village ecodevelopment approach. This will not only give a qualitative boost to biodiversity conservation in the villages of the state but also provide valuable supplemental support to the biodiversity conservation work being done by the forest authorities through the network of National Parks and wild life sanctuaries existing in the government reserved forests.

REFERENCES

Gadgil, M. and Vartak, V.D. 1976. Sacred groves of India: A plea for continued conservation. *Journal of Bombay Natural History Society,* 72: 314-320.

Ishwaran, N. 1992. Biodiversity, protected areas and sustainable development.*Nature and Resources,* 28: 18-25.

Ramakrishnan, P.S. 1996.Conserving the Sacred: from species to landscapes. *Nature and Resources,* 32 : 11-19.

31

Sacred Groves of Tamil Nadu

P.S. Swamy, S.M. Sundarpandian and S. Chandrasekaharan

Department of Plant Sciences, School of Biological Sciences, Madurai Kamraj University, Madurai, India

INTRODUCTION

Nature worship has been an ancient Indian tradition. Sacred groves, representing one form of nature worship, are common in Tamil Nadu, as elsewhere in other parts of India and also south Asia. Some of these groves have been studied for their floristic composition (Sethi, 1993; Visalakshi, 1995) and for the traditional culture associated with them (King, *et. al.*, 1996). However, the biodiversity potential of sacred groves and scope of this traditional institution in promoting the objectives of conservation and sustainable livelihood are little understood. This article deals with the ecological and socio-cultural attributes of some of the sacred groves in State.

DISTRIBUTION OF SACRED GROVES

In general, sacred grove is a component of landscape in all the villages. The size of the grove may vary. Some groves such as Vijayakaruppar Kovil Kadu and Paduvayal Kaliyamman Kadu in district Sivagangai, Algar Kovil Kadu in Madurai district and Nagar Kovil Kadu in Kanyakumari district are considerably large (> 10 ha). Smaller groves have local village level religious attachments while larger groves are revered by the regional community.

SOCIO-CULTURAL TRADITIONS

The residing deity in most of the cases is snake god. Ayyappan, Murugan and Amman are the forest deity in some cases. Daily worships by the individuals are performed in the morning and evening in villages. On full moon days wider communities offer prayers collectively. Priests are considered to be the links between the common man and the deity. Traditional people believe that the groves directly or indirectly provide livelihood and restrict natural catastrophes.

Many groves have temples within them. Traditionally the temples used to be in the form of a small stone bearing the symbol of the deity but now larger complexes are becoming more and more common. This change is accompanied by the change in attention to the temple rather than to the grove.

Apart from sacred groves wherein natural vegetation has been protected from human disturbances, there are sacred gardens locally called Thirunandavanas created by the man. These gardens are meant exclusively for the flowers for the deity (Desayar, 1987). One could find such gardens in Srirangam, Chidambaram, Suchiindram, Madurai and Srivilliputhur. At Srivilliputhur temple trust has established a garden of medicinal plants.

BIODIVERSITY

Biodiversity potential may tremendously vary depending upon ecological conditions and disturbance regime characterizing a given grove. Within a grove, species composition could tremendously vary. In larger groves with a temple complex, generally the area near temple complex is more disturbed compared to the area away from temple complex and habitation. Trees like *Acacia* spp., *Caesalpinia bonduc, Capparis spinosa, Catunaregam spinosa, Hemicyclia sepiaria, Strebulus asper, Miliusa eriocarpa* were observed in highly disturbed area around in the vicinity of Algar Kovil kadu. *Artocarpus integrifolius, A. heterophyllous, Ficus* spp., *Mangifera indica, Eranthemum capense* were the tree species confined to the less disturbed areas. Herbaceous species *Ageratum conyzoides* and *Blepharis repens* dominate the highly disturbed sites while *Bryophyllum pinnatum* and *Pseudoranthemum malabaricum* were indicators of less disturbed environments (Table 1).

WATERSHED FUNCTION

Most of larger groves provide ecosystem services through watershed function. Water bodies associated with them are the sacred where people take bath before offering prayers and/or for irrigation purposes.

Table 1. Important plant species present in sacred groves of Kanyakumarai district in Tamil Nadu.

Name of the species	Sacred groves 1	2	3	4	5	6	7
Trees							
Aegle marmelos Corr.				+			
Ailanthus malabarica Wall.	+	+			+	+	
Albizia amara Boiv.				+			
Alstonia scholaris R.Br.			+		+		+
Areca catechu L.				+			
Artocarpus hirsutus Lam.				+			
Artocarpus integrifolia L.		+					
Azadirachta indica A. Juss.			+				+
Bassia latifolia Roxb.			+				
Borassus flabellifer L.		+					+
Caryota urens L.					+		
Cocos nucifera L.			+	+		+	+
Erythrina indica Lam.	+				+	+	
Eugenia jambolana Lam.	+						
Ficus bengalensis L.					+		+
Ficus religiosa L.	+	+	+				+
Ficus sp.	+				+		
Mangifera indica L.	+	+	+				
Morinda tinctoria Roxb.	+	+	+		+		+
Odina wodier Roxb.		+	+			+	
Sideroxylon tomentosum Roxb.	+	+				+	
Tabernaemontana heyneana Wall.			+		+		
Tamarindus indica L.	+	+	-	+	+	-	+
Tectona grandis L.f.	+						
Terminalia belerica Roxb.			+		+	+	
Terminalia catappa L.	+						
Terminalia sp.	+						
Shrubs & Herbs							
Bryophyllum sp.	+				+	+	
Cassia tora L.				+			+
Coleus aromaticus Benth.				+			
Ervatamia sp.	+				+		
Eupatorium odoratum L.	+	+		+	+	+	
Hemidesmus indicus R.Br.	+	+	+		+	+	
Hibiscus rosa-sinensis L.			+				
Ixora nigricans Br.	+	+					
Jatropha sp.				+			
Lantana camera L.	+	+			+	+	
Leea indica L.	+						
Mimosa pudica L.			+			+	
Murrya konigii Spr.				+	+		
Ocimum sanctum L.		+	+				
Phyllanthus niruri L.	+	+	+	+	+		+
Sida cordifolia L.	+	+	+	+	+		+
Tephrosia purpurea Pers.		+			+	+	

Contd.

Table1 Contd.

Name of the species	Sacred groves						
	1	2	3	4	5	6	7
Climbers and Lianas							
Butea parviflora Roxb.		+					
Calycoptis floribunda Lam.		+			+	+	
Canthium sp.	+						
Chloris superba L.	+						
Cissus quadrangularis L.					+		
Clitoria ternatea L.	+		+		+		+
Connarus wightii J. Hk.	+					+	
Jasminnum azoricum L.			+		+		+
Toddalia asiatica Lam.	+				+		
Zizyphus rugosa Lam.	+	+			+	+	

1—Mupparam Kavu; 2—Sasthanagar Kavu; 3—Chathra Kavu; 4—Kelavettu Kavu; 5—Naduvaathamurai Kavu; 6–Alappangodu Kavu; 7—Manjathopu Kavu

MANAGEMENT

The smaller groves are generally in private ownership while the larger groves are controlled by a trust. Trusteeship generally transcends as a hereditary privilege. In many smaller groves owned by the orthodox families all types of biomass resource uses are still not permitted. Traditions have been distorted in larger groves managed by the temple trusts. In such cases biomass is often utilized to derive income for maintenance/expansion of temple complex. For example, in Vaiyakaruppar Kovil Kadu deadwood on the forest floor is used for cooking food during annual celebration. Non-timber forest products such as mangoes, *Eugenia* and tamarind are marketed. About 1/3 rd of income by sale goes to the trustees while the remaining income is used for organization of annual celebration. Yet large scale and intense disturbances such as timber extraction are rare. Hunting within the sacred groves is prohibited. However, in Variyakaruppar Kovil Kadu bat catching has become common since the last few years. These bats are local delicacy and fetch good income in the market. In some areas such as Algar Kovil Kadu small areas within the grove have been fenced to restrict entry of bat catchers.

In recent period some of the groves such as Algar Kovil Kadu in district Madurai are declared as reserve forests which means transfer of ownership to the forest department. However, people at large have opposed this policy action and have filed suits in the court.

CONCLUSION

In general, the area under sacred groves is fast depleting due to the interplay of a variety of factors. Sacred groves originally maintained in the form of pristine ecosystems dedicated to the deity are now looked as a source of income. Appropriate policy interventions are needed to promote avenues of income such as eco-tourism so that traditional sacred groves promote the objective of biodiversity conservation together with sustainable development of the local communities.

REFERENCES

Desayar, M. 1987. *Temple Administration under the Pandian (A.D. 600—A.D. 1300).* Ph.D. thesis, Madurai Kamraj University, Madurai, Tamil Nadu.

King , E.D. I.O., Narasimhan, D. and Chitra Viji, 1996. Sacred groves: traditional ecological heritage. In: *Proceedings of First Indian Ecology Congress.* National Institute of Ecology, New Delhi.

Sethi, P. 1993. *Phytosociology of a tropical dry evergreen forest patch in the Puthupet sacred grove, Coromandal Coast, Tamil Nadu.* M.Sc. Thesis, Pondicherry University, Pondicherry.

Visalakshi, N. 1995. Vegetation analysis of two tropical dry evergreen forests in Southern India. *Tropical Ecology,* 36: 117-127.

PART 3

FUNCTIONAL ATTRIBUTES, CONSERVATION, MANAGEMENT AND POLICY

32

Sacred Woods, Grasslands and Waterbodies as Self-organised Systems of Conservation

Y. Gokhale, R. Velankar*, M.D.S. Chandran** and M. Gadgil****

*RANWA 16, Swastishree Society, Ganeshnager, Pune, India
**Dr. A.V. Baliga College of Arts and Science, Vivekanager, Kumta, India
***Biodiversity Unit Jawaharlal Center for Advanced Scientific Research, Jakkur P.O., Bangalore and Centre for Ecological Sciences, Indian Institute of Science, Bangalore, India

INTRODUCTION

Humans are by far the most dominant species of living organisms on the earth today. As a result people have dramatically affected, and often greatly depleted the abundance and diversity of other organisms. But humans are also the only species endowed with foresight, with an ability to appreciate their impact on the environment, and a potential to take deliberate measures to bring under check what may be perceived as negative developments. When so motivated, people formulate rules of behaviour and organize social institutions to further their implementation. Our particular focus in this paper is on rules and institutions that promote long term persistence of communities or individual populations of living organisms, through imposition of restraints on human activities, such as harvesting. We may visualize in this context two broad objectives, namely conservation and sustainable use. These, of course, grade into each other, but conservation primarily emphasizes persistence for its own sake, while sustainable use pertains to maintaining long term harvests.

Rules and institutions focusing on conservation and sustainable use of living resources-as also non-living resources such as water-are known from societies at all stages of technological developments. They change as societies have changed, primarily in response to technological developments that have tremendously enhanced capabilities of production of food, harnessing of energy, material and informational resources, and of transforming and transporting them over the surface of the earth. Our focus in this paper is on systems of conservation of living resources, and how they have changed and continue to change in the context of these broader changes. These systems have passed through three major phases, namely (1) sacred sites, (2) hunting preserves and (3) wild life sanctuaries and national parks. The sacred sites are characteristic of hunter-gatherer and horticultural, largely tribal societies; the hunting preserves are characteristic of agrarian states; and wildlife sanctuaries and national parks are characteristic of industrial nations. It is notable that the spatial scale of these conservation systems is correlated with the spatial scale of resource catchments - or area from over which are garnered the bulk of the material and energy resources employed by a society (Gadgil, 1996) (Fig.1).

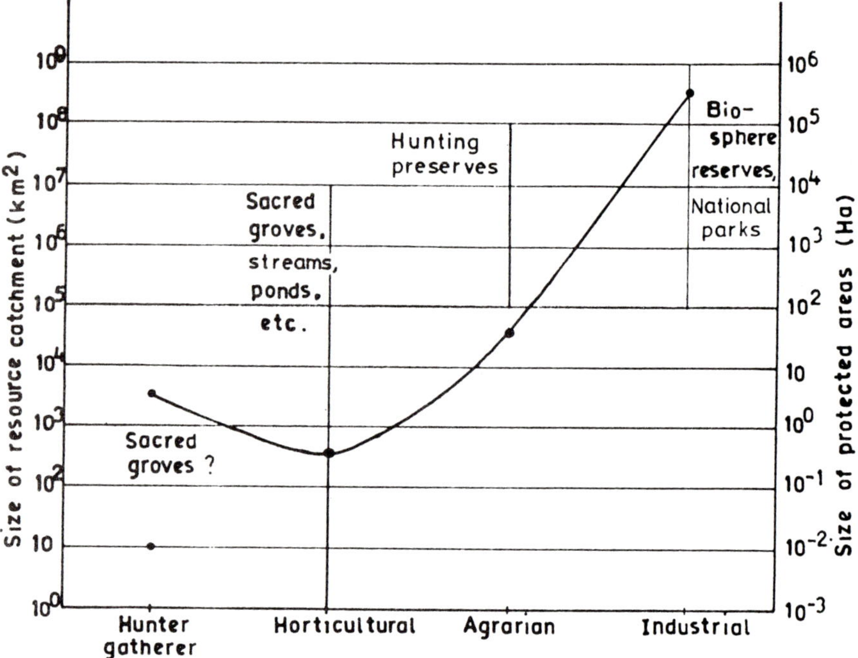

Fig. 1. Relation between sizes of resource catchments and sizes of protected areas. Pure hunter-gatherer societies have catchments of the order of 5000 km^2. It is not certain if they maintain any protected areas. Slash and burn and low input settled cultivators (i.e. Horticultural societies) have somewhat smaller resource catchments around 500 km^2 and maintain protected areas ranging in size from around 0.01 to 100 ha. More advanced agrarian societies have larger resource catchments of the order of 50,000 km^2, and they maintain hunting preserves of the order of 100 to 100,000 ha in size. The modern industrial societies have resource catchments spanning the whole biosphere and concentrate on protected areas of a hundred to a million hectares.

The first category, that of sacred sites (groves, grasslands, pond, pools, stretches of streams and rivers, lagoons, coastal areas) is typical of small scale societies largely practicing subsistence economics (Johannes, 1981; Johannes and Ruddle, 1985; Gadgil and Berkes, 1991). These may be characterized as self-organized conservation systems, as opposed to hunting preserves of the elite, or wildlife sanctuaries or national parks which are conservation systems organized by a state apparatus. Other self-organized systems of resource management include village woodlots, pastures and irrigation tanks (Ostrom, 1990). The state societies, especially the modern industrial societies have tended to dismiss the self-organized resource management systems characteristic of small-scale, subsistence societies, as primitive, unscientific, inefficient. A majority of such systems have indeed been swept aside, even deliberately destroyed by the growing centralized powers of agrarian and industrial state. However, times are changing, and there is an increasing realization of the built-in advantages, especially of cost-effectiveness and adaptability to local conditions, of the self-organized systems (Singh, 1994). This has prompted serious interest in the possibilities of persistence, revival or new emergence of such self-organized systems for managing irrigation water, woodlots and pastures. This has catalyzed a series of excellent studies, especially with reference to the irrigation systems, of the principles on which long-enduring self-organized systems of resource management have been designed, by different societies, in different parts of the world. These principles, best summarized by Ostrom (1990,1992) provide an excellent framework for viewing sacred sites as well.

We have been involved in a series of studies on sacred sites of the Indian subcontinent beginning in 1972 (Berkes, et al., 1995; Chandran and Gadgil 1993; Gadgil, 1985; Gadgil, et al., (in press); Gadgil and Vartak 1975, 1976). These earlier studies, primarily focusing on Western Ghats of Maharashtra and Karnataka and the Manipur hills, have recently been complemented by a project covering the states of Himachal Pradesh, Uttar Pradesh, Rajasthan, Bihar, Orissa, Assam, Meghalaya, Karnataka, Tamilnadu and Maharashtra. This project was undertaken in collaboration with 13 conservation oriented NGOs and NGIs. Volunteers of these NGOs initially identified a total of around 150 sacred sites and collected information relating to location, topography, vegetation, animal life, local human society, belief systems and practices pertaining to the sacred sites and impact of outside economic, political and other factors on the pattern of these practices. This was supplemented by actual field visits by Yogesh Gokhale and Raghunandan Velankar to about 50 sites. This paper draws on the results of earlier published studies, as well as the current country wide investigation (Fig. 2).

Fig. 2. Geographical locations of the full set of earlier and current study sites.

SMALL SCALE SOCIETIES

Till about 10,000 years ago humans everywhere were organized into small scale hunter-gatherer societies with bands of 50-60 individuals making up endogamous tribal groups of 1000-5000 in numbers. This organization is largely preserved in horticultural societies, dependent primarily on low input, often shifting cultivation (or nomadic herding) coupled to extensive hunting and gathering (Lenski and Lenski, 1978). Some hunter-gatherer or horticultural societies continue to this day in India, as with Sentinelese islanders of Andamans. Most others have however been gradually brought under the influence of larger scale agrarian or industrial societies, beginning first with the Indus Valley civilization five thousand years ago, and at an increasing rate in recent decades (Gadgil and Guha, 1992). The self-organized conservation systems of sacred sites originated amongst and have been characteristic of hunter-gatherer or horticultural societies largely insulated from the influence of larger scale agrarian or industrial societies.

Such societies tend to be largely self-sufficient in their requirements of biological resources, though some living resources such as honey, ivory or grain may be exported. Their own requirements of food, fodder, fuelwood, small timber, plant or animal based medicines are however almost wholly met from resources produced locally through low input cultivation or herding, or gathered from within a radius of a few, at most tens of kilometers. The catchments from over which these resources are obtained may often be under very firm control of local communities. In northeastern India, for instance, people may have had to risk their lives when venturing into territories of alien tribes in older times; and even today Jarwas of Andamans repel outsiders with bows and arrows. Elsewhere the traditional rights of local communities to prevent aliens from harvesting resources from certain well defined areas was widely recognized and respected, with other traditional rights assigned to groups like nomadic herders. With long standing familiarity with their immediate environments, such communities would have access to considerable information on how varied forms and levels of harvesting have influenced the local stocks of living resources. Such communities would have been entirely or largely self-governing with community level institutions permitting collective choice to prevail in a transparent fashion. Those violating community based norms of behaviour-including those pertaining to living resources-would be subject to sanctions, imposed flexibly, often in a graduated fashion.

DESIGN PRINCIPLES

Elinor Ostrom (1990, 1992) in an insightful analysis of self-organized systems of resource management suggests that many of these features of small scale societies tend to favour long term persistence of such systems. These are captured in the seven principles of design of such systems suggested by her (Fig. 3).

Design Principle One

The Long Term Benefits Flowing from the Restraints on Resource Use Should be Commensurate with the Costs Incurred by the Community

The benefits could take many different forms: (a) Provision of ecosystem services such as watershed conservation: The villagers of Gani (Shrivardhan taluk, Raigad district, Maharashtra) wished to save their Kalakai sacred

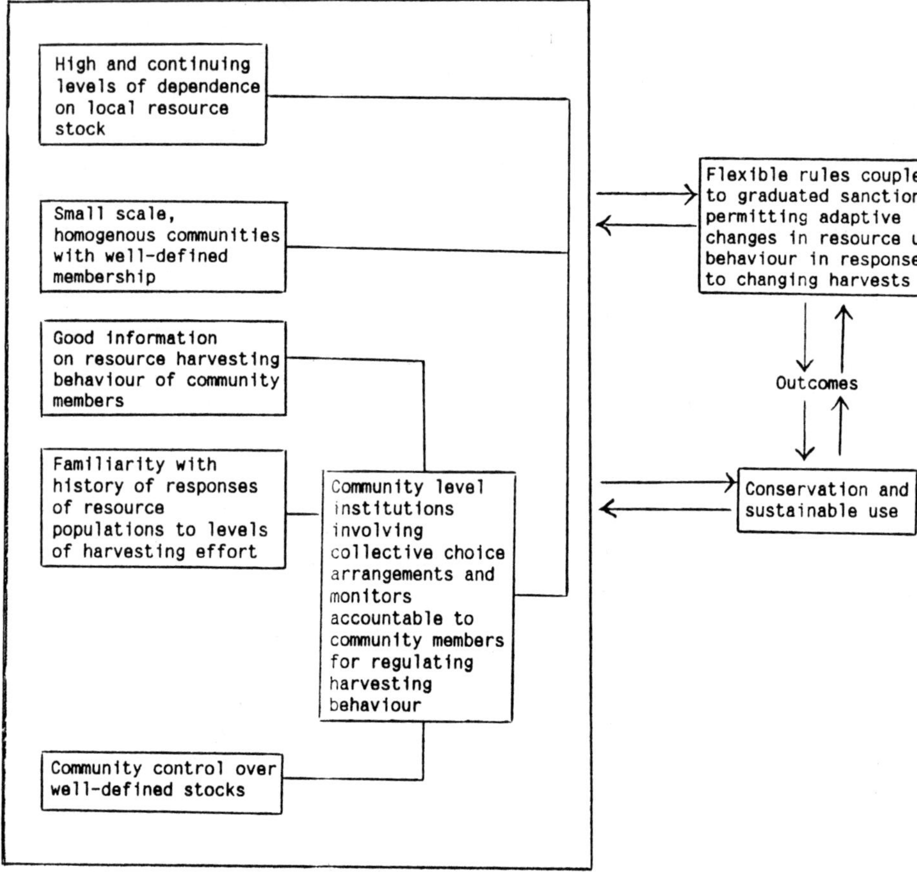

Fig. 3. The organization of resource management systems of small scale societies.

grove of 10 ha as forming the catchment of the only remaining perennial stream (Gadgil and Vartak, 1975). (Wingate, 1888) noted that the Kans or sacred groves of district Uttara Kannada in Karnataka were of "great economic and climatic importance. They favour the existence of springs, and perennial streams, and generally indicated the proximity of valuable spice gardens, which derive from them both shade and moisture." Many major sacred groves are located in the catchments near the origins of rivers, e.g. Bhimashankar (Khed taluk, Pune district, Maharashtra) harbours a large grove of over 700 ha at the origin of Bhima, a major tributary of Krishna (Borges and Rane, 1992). (b) Ecosystem services as firebreaks: These may be especially important in tracts otherwise under shifting cultivation where the slashed forest is burnt prior to cultivation. Indeed realization of benefits of sacred groves as firebreaks has led to the revival of protection to groves that had been destroyed in Churchandpur district of Manipur (Gadgil et al., in press). (c) Other ecosystem services such as provision of shade: A sacred grove of goddess Shilai located amidst paddy fields at Supegaon (Murud taluk Raigad district, Maharashtra) is preserved for the shade it provides to people and cattle (d) As a refugium that helps reduce the chances of extermination of resource populations: The chances of long term persistence of any exploited population are considerably enhanced by providing complete immunity to a part of such population in some protected habitats. Thus in the classic experiments of Gause, (1939), on an experimental prey-predator system involving two species of protozoa, it was found that the predator population always overexploited the prey population leading to its extermination, followed by a crash in the predator population. This could however be averted by creating for the smaller sized prey species a habitat into which the larger predator could not enter. A certain level of prey population could then persist in this refugium. The predator had access to the prey population outside of this special habitat. Such a set up permitted long term coexistence of the prey-predator populations. Many sacred sites serve such a function. Thus in the Dharamshala district of Himachal Pradesh certain stretches of hill streams are fully protected against fishing, providing an excellent refugium for fish populations. These fish populations are then fished beyond a distance of 100 meters on either side of the sacred pool. Similarly in village Mangaon (Velhe taluk, Pune district, Maharashtra), barking deer are not hunted inside the Janni sacred grove of 10 ha; but hunted outside (Gadgil and Vartak, 1975). The local villagers report that they have much better chances of hunting barking deer compared to neighboring villages which provided no such refugium. Apart from mobile animal populations, these refugia may also serve as seed sources that promote regeneration of plants outside the sacred sites. (e) As a biological community subject to much lower levels of exploitation than elsewhere and hence much more likely to continue to provide certain resources on a long term basis: Sacred woods or grasslands need not be totally protected from all harvests. Thus in Aajivali (Mawal taluk, Pune district, Maharashtra) villagers have revived protection to a sacred grove that harbours several *Caryota urens* palm trees that are tapped for toddy. A sacred grove near village Gani (Shrivardhan taluk, Raigad district, Maharashtra is valued as a source of *Entada pursaetha* used to treat cattle against snake bite. (f) As a resource that serves as an insurance to be exploited only in an emergency. Thus in Ghol, (Velhe taluk, Pune district, Maharashtra) the sacred grove is otherwise left undisturbed, but furnished timber for rebuilding after many huts were destroyed in fire (Gadgil and Vartak 1976).

Apart from these more tangible benefits, sacred sites may serve other aesthetic, cultural, religious functions. (g) A sacred grove of the B.R.T. hills (Yelandur taluk, Mysore district, Karnataka) harbours a *Michelia champaka* tree of enormous proportions, famous over a large region,and once regularly worshipped by the Maharaja of Mysore as well. Humans may thus protect sacred sites that inspire awe and wonder. Indeed many mountain peaks such as Nandadevi in Himalayas are also worshipped. Along with mountain peaks, the surrounding areas for approaching the peak are also considered as sacred. In Garhwal Himalayas while approaching the Nandakaht-one of the sacred peaks, various places in between are considered as sacred one of them is Devi ka Aangan. Local people have their rituals associated with all such sacred places. People maintain strict norms about human behaviour like, they go to these areas bare footed for worship, plucking of flowers before specific time is not allowed etc. (h) Other sacred sites may be set aside as a mark of respect for organisms disturbed by humans. Thus it was a tradition in Kerala to set aside a portion, sometimes reported as 1/7th of land being newly brought under cultivation and habitation as a sacred grove dedicated to snakes driven away from the remaining part of their habitat (Nayar, 1987; Unnikrishnan, 1995) (i) Humans may also protect sacred sites in the belief that this would please supernatural forces and attract bounties such as enhanced fertility of fields, or of humans themselves. This seems to be the benefit expected of sacred groves called saranas in many parts of Chhotanagpur plateau in states of Orissa, Bihar and Madhya Pradesh. Similar tradition is followed by Dimasa tribe in northeast India where the sacred grove 'Madaico' is used to please supernatural forces. (j) In case of Aasahpoorna Devi Ki Oran at Devikot, (Fatehgad taluk, Jaisalmer district, Rajasthan) the actors i.e. the old people used to predict the climate for coming season, year according to condition of Oran at the time rituals. Local level management about crops, cattle etc. was regulated on the basis of these predictions. Thus, Oran-the sacred grazing land used to play an important role for villager's economy.

The costs to be weighed against these benefits would be opportunity costs of diverting land or water to other uses, or of harvesting resources at greater rates. The balance would change depending on the values placed on different forms of benefits and costs, at the moment and in future. The balance would tend to tilt against conservation under a variety of conditions: (a) If resource scarcities mount, the opportunity costs of conservation would be considered to be higher. Thus as pressure on cultivated land has increased, sacred groves have been felled to make way for either cultivation, or habitation. (b) If values of resources being conserved go up,as with remoter areas being connected with roads, the opportunity costs of conservation would be perceived as having increased (c) If resources earlier valued are now obtained from elsewhere, or in other ways, as for instance, with use of medicinal herbs being abandoned in favour of allopathic drugs, or of leaf manure being given up in favour of chemical fertilizers the opportunity costs of conservation would be considered to have increased (d) Lastly, if future is discounted, with immediate interests being given more weight, the opportunity costs of conservation would be perceived to have gone up. In general, rapid technological progress has favoured such discounting of future value of resources, in the belief that technological solutions would always be available to deal with any resource scarcities.

Design Principle Two

The Conservation System Should Deal with a well Defined Resource Under Reasonably Secure Control of a Well Defined Social Group

In India the tribal and the subsistence rural societies probably had considerable *de facto* control over a well defined area near their settlements in pre British times, and therefore met this condition. In post independence times, the provisions for Schedule Areas also create such a situation. In north-eastern state of Meghalaya, by constitution people have rights on their natural resources. At Cherrapunji the earlier sacred forests of Khasi people are under control of local councils as per their traditional system of administration. Due to Christianity, the sacred status of forest is now no more, but the local authority-Siem of Sohra (king of Cherrapunji) has maintained the earlier utility status of specific forest patch; which is called supply reserve- 'Law Adang'. Earlier ownership of sacred groves used to be with the Siem Darbar i.e. the higher authority for administration. But after Christianisation the rituals associated with sacred groves were stopped resulting into loss of faith with the sacred grove. Hence, people started cutting trees from it. To prevent this higher authority took the decision that village Darbar i.e. lower authority will also take care of these earlier sacred groves. It is reported from far villages around Haflong that 'Madaico' of Dimasa tribes were encroached by outsiders. Outsides did not have respect for the tradition, so they cut plants from the sacred grove. The local people were not organized to resist outsiders from encroaching the sacred grove.

Design Principle Three

The Group Responsible for the Conservation System Should be Effectively Organized to Administer the System

Traditionally tribal and caste groups had such social organization;

there were also effective organizations at the multi-caste village community level in pre-British times. These community level organizations were largely dismantled during the British reign. There have been some attempts to rebuild appropriate organizations at the community level following independence. These are now gathering strength following the 73rd amendment to the constitution relating to the Panchayat Raj and the 1996 Act extending Panchayat Raj institutions to Schedule Areas. It is reported that during 1970-75 in Rajasthan, commercial extraction of *Acacia senegelensis* gum and *Commifera vitae* was done with the help of commercial contractors. The contractors were also allowed to extract from the Orans or sacred groves, now owned by the government. For extracting Acacia gum the contractors probably used some chemicals which increased the productivity but the tree used to die in a short period. This extraction badly affected both these species and vegetation in Oran was thinned. Since for many Orans, ownership of lands was with government, nobody could resist to stop this exploitation of resource.

Design Principle Four

Existence of a Monitoring Machinery Accountable to and Respected by the Actors

Effective implementation of a conservation or other resource management system requires monitoring of the behaviour of actors affecting the resource stock by some machinery which will be accountable to the actors,and respected by them. For the self organized systems of sacred sites this function in served by the local deities, nature spirits and ancestral spirits, as well as by the whole local community, especially the priests and leaders who take on the role of enforcing the will of the gods and the community. The earliest religious beliefs of humanity include belief in spirits specific to particular localities, the *sthaldevatas (sthala = locality, devata = deity)* concerned with the well being of nature and people in that locality. People tend to believe such deities to have the interests of the local environment at heart, as well as to be omniscient, so that no action, even intention can escape these monitors. These monitors are often believed to punish violators with a variety of undesirable consequences. But additionally, all members of local communities who tend to be intimately familiar with all local happenings also take on the function of monitoring. The priests and other local leaders are then expected to take appropriate action on behalf of the deities against any infringement of prescribed behaviour. The local community and leaders may also serve as monitors on their own, not necessarily on behalf of deities. Such monitoring tended to be quite efficacious in small scale societies. In case of Viratra Ki Aan or Oran at village Dhok (Chauhatan taluk, Barmer district, Rajasthan) people and priests have themselves been monitoring the area. The Oran is spread on a huge area about 13 sq. km. People can graze their cattle and with prior permission are allowed to use deadwood. In 1971, the war refugees across the border from Pakistan came to settle in the village. They violated the traditional restriction of Oran for getting wood. Local people warned the refugees and gave an idea about the wrath of the goddess. But refugees kept on violating. As a tradition people believed that goddess would punish the violators. After some years the refugee colony caught fire due to unknown reason and people considered it as wrath of goddess. In this case, one can notice the warning given by local people who were monitoring such a large area on the India-Pakistan border having very sparse population.

Design Principle Five

Collective Choice Agreements

The actors affecting the sacred sites would themselves be involved in relevant management decisions. This would be possible in relatively small scale societies where the whole group would be directly or indirectly involved in the decisions.

Design Principle Six

Flexible Rules Relating to Resource Use Patterns

Long term viability of conservation system is promoted by flexibility, accepting uses otherwise prohibited in emergencies; for instance extraction of timber for house construction in case of fire. In informal, community based systems of small scale societies such flexible arrangements are possible. In village Irani located very remote in Chamoli district of Garwhal Himalaya at about 2500 meters people have to face exacting climate. They have protected a small sacred grove, which can be useful to them for fodder to their cattle at the time of emergency in winter. Hence, in winter if snow fall is heavier, they can get the fodder from a nearer place from the village instead of going far away.

Design Principle Seven

Graduated Sanctions Against Violation of Management Rules

Community based management systems may incorporate flexible sanctions taking into account the overall record of the violator and circumstances of the violation, promoting long range endurance of such systems. In the north eastern state of Mizoram many villages have still protected safety reserves around their villages serving as fire breaks. Due to Christianity people do not have any sacred relation with the safety forests which they had in the past. But shifting cultivation is going on and hence they need fire breaks. To protect safety reserves the local village councils in villages namely Hrianmum and Teikhang which are in Sialkal area of Aizwal district are charging fine according to the nature of violation; for big trees fine is more, for climbers less fine is charged.

SOCIAL TRANSFORMATIONS

Figure 3 attempts to depict the organization of resource management systems of small scale societies. It is evident that the system conforms to the seven design principles discussed above, suggesting that the conservation systems such as sacred sites, of these societies may have been both effective and long enduring. This is not to contend that there were no instances of extermination of living resources in such societies. Indeed there were many such accompanying the colonization of new territories such as Americas, Pacific islands and Madagascar (Diamond, 1991). It appears however that as people settled down and acquired a deeper understanding of the living environment they developed much more effective conservation systems such as sacred sites (Gadgil and Berkes 1991; Gadgil et al., 1993; Gadgil, 1995; Heywood, 1995).

Such small scale societies and their conservation systems have been radically altered by a whole series of technological developments that have absorbed them, first into agrarian states and subsequently into industrial nations. This emergence of larger scale societies is grounded in access to ever increasing levels of material, energy and informational resources. Thus productivity of agriculture is increased initially by input of animal energy through bullock and horse power and of water through irrigation systems, and later by input of fossil fuel energy for a whole range of agricultural operations and for pumping water, and by inputs of synthetic fertilizers and pesticides. This permits the production of increasingly greater levels of agricultural surplus by cultivators. Technological developments also facilitate transport of this surplus over long distances with the use of bullock power, wind power and then energy of fossil fuels. These possibilities of large surpluses of food being transported over substantial distances catalyse extensive division of labour and growth of towns and cities. The emerging occupational specializations include priesthood, bureaucracy, military, trade and artisanal activities. This new social organization creates larger scale social, economic, political units in the form of chiefdoms, kingdoms, nation states (Service, 1975). These larger scale societies depend upon large outflows of natural resources from rural hinterlands; resources such as surplus grain, wild spices, timber, wood charcoal, bamboo, animal skins, ivory, coal and other minerals. These outflows are often against the interests of forest dwelling and rural populace, and require a loosening of the control of local communities over their territories. The centralized states therefore take measures to strengthen the authority of the state apparatus and their allied economic and political interests and weaken that of local communities.

The centralized machinery and its allies controlling the resources are not dependent on the resources of any particular locality. They always have other options if the natural resources of any particular locality are exhausted. Furthermore, the pace of technological developments also picks up, especially in the industrial societies; and new technologies open up the possibilities of substitution when a resource is exhausted. Thus, the West Coast Paper Mill at Dandeli (Haliyal taluk, Uttara Kannada district, Karnataka) could successively bring in bamboo from further and further away, from Andhra Pradesh, Garhwal, Assam and Nagaland as these stocks are exhausted; as also switch to using eucalyptus as bamboo supplies dried up. These possibilities have promoted a pattern of sequential exploitation and exhaustion of resources, with commercial interests concentrating at any given time on the kinds of resources, and on the localities which yield maximal levels of profits; shifting to others as these are exhausted (Fig. 4).

Such exhaustive resource use does prompt corrective responses in large scale societies, responses that are primarily guided by the recreational interests of the elite. In agrarian kingdoms they take the form of hunting preserves of aristocrats. The two thousand year old manual of statecraft, Kautilya's Arthashastra prescribes the maintenance of such preserves (Kangle 1969). The Mughal emperors also maintained huge areas as their hunting preserves. The British planters in Western Ghats set up similar game preserves. On independence, as populations of the larger wild animals and birds declined many of these hunting preserves have been converted into wild life sanctuaries and national parks. These tend to emphasize the preservation of flagship species like tigers, and concentrate on elimination of all demands of local tribals, herders and peasants through a guns and guards approach as their main concern.

IMPACT ON CONSERVATION TRADITIONS

The process of absorption of small scale societies into large scale societies has progressed to different degrees in different parts of the world. It has

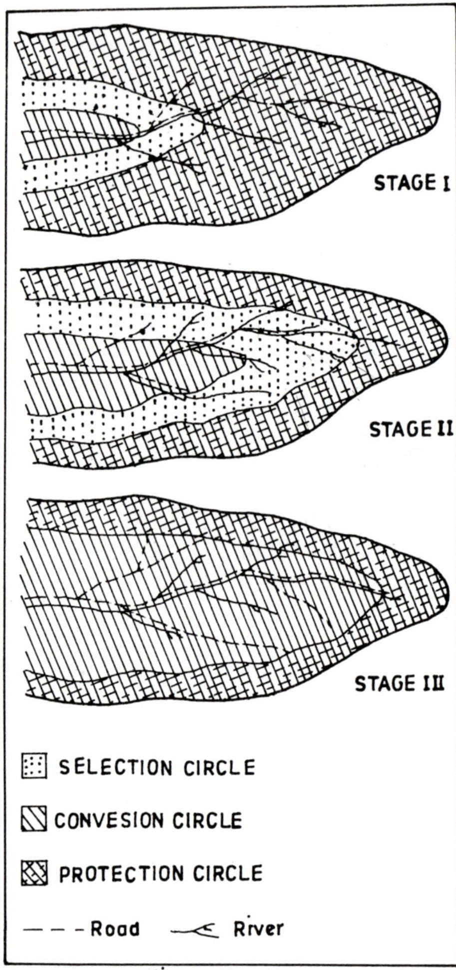

Figure (4a).

a. Working plans of the Ranni forest division of Kerala, 1950-80. This Figure shows that the protection zone (set aside for watershed conservation), selection zone (earmarked for extraction of a limited number of trees on a sustainable basis) and conversion zone (devoted to clearcutting and raising of monoculture of commercial species) have kept on shifting in response to exhaustive resources use (after FAO, 1984).

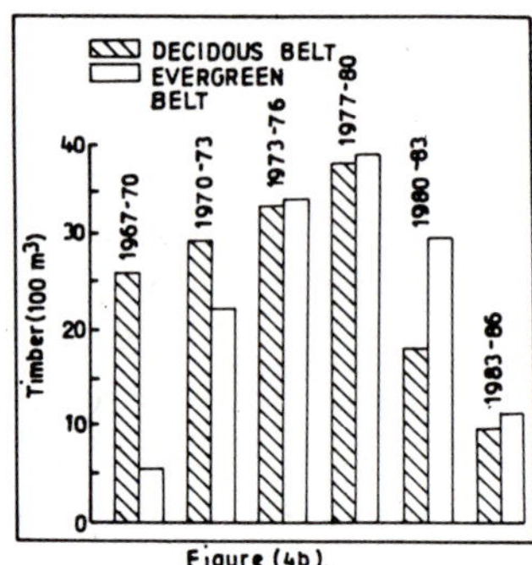

Figure (4b).

b. Timber harvests by the plywood industry from the Uttara Kannada district of Karnataka State, 1967-85. It is evident that pressures of non-sustainable use have forced this industry to shift from more accessible deciduous forest tracts to less accessible evergreen forest tracts, and that timber supplies from both zones have falled off.

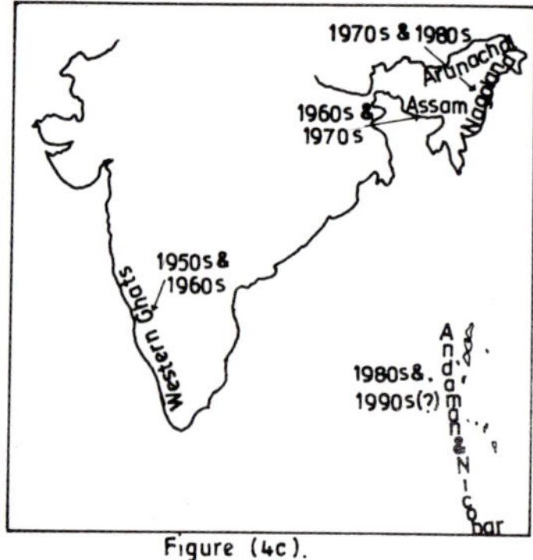

Figure (4c).

c. Geographic shifts of the plywood industry in India. The non-sustainable resource use methods of this industry have kept it shifting over time to more and more remote areas of the country as resources from more accessible areas near exhaustion.

Fig. 4. Patterns of sequential exhaustion of resource use at different spatial scales.

gone farthest in countries of western Europe where industrialization was first set in motion over two centuries ago. It is just beginning to take off in countries like New Guinea. Within a country too the process may proceed at variable rates. Thus in India, it has gone relatively far in the industrial belt of Punjab and Haryana, the least in parts of northeast such as Arunachal Pradesh. India may indeed be remarkable in harbouring within a single nation state the greatest range of variation in terms of transition from small scale to large scale societies. Thus, some 7% of the Indian population is classified as belonging to scheduled tribes. Hunting-gathering and shifting cultivation characteristic of horticultural societies, still remain significant in the economy of these tribal populations. The tribal communities also retain many features of the social organization of small scale societies, including the role of community based decision making in prescribing norms of social behaviour in relation to the members of the community. However these tribal communities are no longer self sufficient, nor are they in full control of their own resource base, barring special exceptions such as Sentinelese islanders of Andamans. Almost everywhere they are now linked to the larger economy; the forest produce they gather (or fell with their axe) are largely for sale; they also buy many commodities from the market. They have variable, and rather uncertain, level of control over the living resources in their own localities. On the peninsula, this control has largely passed to the hands of state authorities; in the northeastern states it is still to fair extent vested with local communities, or individuals or tribal chiefs. Even here however there is a lack of clarity as to exactly who controls what resources, so that there is little of firm community control. The cultures, religions of these tribal communities have also changed substantially on contact with the large scale society. These changes include conversion to Christianity or absorption into mainstream Hinduism. These tend to weaken the belief in *'sthala-devatas'*, spirits, deities, rooted in particular localities. Other, non-tribal, rural communities have changed similarly, to a greater or lesser extent, beginning, in particular, with the British rule, and at an accelerating pace in recent years. These changes have, by and large, tended to erode the self-organized systems of conservation, the systems of sacred sites, throughout the country. These processes of erosion may best be assessed in terms of their consequences for the seven principles of design considered above.

Design Principle One

Balance of Benefits and Costs

The continuance of conservation systems would evidently be affected by the relative levels of benefits and costs perceived by local communities. The benefits may be tangible; as ecosystem services (e.g., watershed conservation, firebreaks) or as resources obtained occasionally or at low levels of harvest (e.g., grass rather than wood). They may also be intangible (e.g. respecting deities or gigantic trees). The value ascribed to such benefits would vary with how dependent the local communities, as well as others with access to the locality are on such resources.

For a variety of reasons values of these benefits have substantially depreciated in the perceptions of people with access to the sacred sites and their resources. Modern technological developments have prompted changes in resource use practices of local communities, as well as brought to them access to newer kinds of resources and resources from elsewhere reducing their dependence on local resources. Thus, Gangtes, a group of Kuki tribes of Manipur have traditionally depended on shifting cultivation. During the slash and burn operations involved fires may spread, and sacred groves surrounding the habitations have traditionally served the function of firebreaks. However, in villages close to towns, such as Koshbong (Churchandpur district, Manipur State), pineapple cultivation on terraced hill sides has replaced shifting cultivation. Sacred groves are therefore no longer of significance as firebreaks. In consequences the practice of protection to sacred groves has been discontinued, in part for this reason.

Yet another reason that has contributed to the disappearance of the sacred grove of Koshbong is that it would no longer serve the function of showing respect to the local deities. This is because the tribal animist beliefs are now replaced by acceptance of Christianity. Christianity prescribes faith in a supreme god who is everywhere. As a corollary the god does not reside in any particular locale, and cannot be associated with any particular patch of forest, or spring, or tree. In effect, religions like Christianity and Islam desacralize nature and eliminate the rationale for respecting a sacred site (White, 1967).

Hinduism too does this, though to a lesser degree. It is a more eclectic faith based on absorbing the local deities of small scale societies, as incarnations of deities of Hindu pantheon, particularly the god Shiva and the goddess Durga. Unlike Christianity, Hinduism can therefore accept the continuing veneration of sacred groves, or ponds, or of trees like peepal or animals like hanuman langur as a part of legitimate religious practice. However the more formal Hinduism, also tends to emphasize the worship of idols and temples over tree and forests. Hindu priests therefore often encourage the liquidation of a sacred forest to be replaced by a temple to the presiding deity. Indeed the priests may do so on behalf of timber contractors assuring the local people that they would perform appropriate rites to placate the deities of the sacred forests, who might otherwise be offended at the cutting down of the forest (Nayar 1987). Furthermore veneration of sacred sites tends to be a function of local generally non-Brahmin, priesthood. As small scale societies come in contact with the larger society, the Brahmin priesthood is replacing this rural non-Brahmin priesthood and with that the tradition of respecting and protecting sacred sites. Thus in Mensi village (Sirsi taluk, Uttara Kannada district, Karnataka), the sons of priest officiating are not inclined to carry on the tradition, the grove has been permitted to be encroached upon by a neighbouring farmer belonging to the Brahmin caste.

Escalation of perceived costs is, of course, the obverse of depreciation of perceived benefits. Contacts with large scale societies in many ways contribute to such escalation of perceived opportunity costs of continued protection to sacred sites. These opportunity costs may involve foregoing uses of living resources, such as timber from sacred groves, or foregoing alternative uses of land or water, as for cultivation or aquaculture. Indeed the yields from arecanut orchards in Uttara Kannada depend on the availability of substantial amounts of leaf manure from forests. The clamour for leaf manure for arecanut gardens made the British allot sacred Kans as 'betta' or leaf manure forest, following forest settlement in Sirsi and Siddapur in early part of this century. Where individual bettas were not granted some of the kans were thrown open as commons for extraction of leaf manure and other biomass. In eastern Sirsi, according to Collins (1922) 769 ha of Kans were added to the minor forest. The demand for such manure has escalated with development of marketing facilities and higher prices for arecanuts. An arecanut orchard owner has therefore converted the sacred grove of Mensi (Siddapur taluk, Uttara Kannada district, Karnataka) into a patch of forest dedicated to supply leaf manure for his orchard.

It is, of course, not just the opportunity cost of protecting the living resources for local communities that is of relevance. Often what is more relevant is the opportunity cost to the outsiders, including the government agencies and commercial interests. All over India state controlled forest resources have been made available to forest based industries such as plywood manufacture at highly subsidized prices. They have then been overexploited and sequentially exhausted. The plywood resources of Uttara Kannada district of Karnataka had been subject to such overuse and depletion following 1940's (Fig. 4) . With plywood resources of reserve forests largely depleted by early 1970's, some of the larger sacred groves of the district harboured the only surviving good stocks of well grown trees suitable for plywood manufacture. Many of these sacred groves had been taken over as part of reserve forests in the forest settlements of late 19th century, but not subjected to extraction. In early 1970's the Karnataka Forest Department clearfelled several of the larger ones, such as a 21 ha sacred grove at

Menisi in Sirsi taluk and planted them with eucalyptus. The district Coorg of Karnataka also had several large sacred groves, commented upon as among the best he had seen by Dietrich Brandis, the first Inspector General of Forests of British India (Brandis 1897). Many of these were also felled by the forest department in early 1970's; as well as encroached upon for coffee plantation and habitation (Kalam 1996).

Sacred ponds, often attached to temples met the same fate when the technology of freshwater fish culture became available. The Government Fishery Departments then took over many of the larger sacred ponds, poisoned and removed the diverse aquatic communities that they supported and converted them to fish culture. The fish so cultured were auctioned to traders with revenue accruing to the Government. This for example was the way Devikere, a large sacred tank of the town of Sirsi (Sirsi taluk, Uttara Kannada district, Karnataka) was treated in 1970's.

Sacred sites may also be affected when the land or water bodies concerned have alternative uses that become more attractive with time. Population growth and commercialisation of agriculture have greatly increased the demand for land for cultivation. This has often led to encroachment on sacred groves, as for example, at Golgodu (Siddapur taluk, Uttara Kannada district, Karnataka). In fact a farmer of Golgodu who has thus encroached on the sacred grove has done so against the wishes of many other members of the community. However the community is no longer in a position to enforce its will on the encroacher. Such pressures have also affected devara-kadus of Coorg.

Pressures for alternative uses from outside the local community may also affect sacred sites. These include construction of reservoirs for hydroelectric or irrigation projects, leading to submergence of lands, including sacred groves. A large number of sacred groves have thus been submerged under the Panshet dam west of Pune city (Velhe taluk, Pune district, Maharashtra). Many other sacred groves that remained above the submersion zone were cut down by farmers whose cultivated lands were been submerged and who expected to be settled in alternative sites several kilometers away (Gadgil and Vartak, 1976). In Mayurbhanj district of Orissa, Subarnarekha Irrigation project has destroyed about 62 'Jahiras' -sacred groves recently. The main canal of the project is also going to threaten many more Jahiras.

Design Principle Two

Well Defined Resource Boundaries and Social Boundaries

The development of larger scale agrarian or industrial societies demands dissolution of resource boundaries on smaller community controlled scale to facilitate extensive resource transfer. Along with the resource fluxes come migrations of people blurring boundaries between communities. Both these open up sacred sites to influences from outside and lead to their erosion. Thus fishing in the more hilly areas was earlier a monopoly of members of local community, especially those belonging to fishing castes. This is no longer so with the opening up of transport routes, so that fishermen from outside, and others such as army people may come and fish in streams everywhere. They may also employ highly destructive methods such as dynamiting.

The outsiders do not respect practices such as sacred sites either. Thus in village Machiyal (taluk Nagarotta ? district Dharamshala, Himachal Pradesh) the army people were reported to have fished in the sacred pond. "On 26 May 1996, tragedy struck the congregation of mahseer fishes at Shishila near Sulya, a small village in Dakshina Kannada district, Karnataka. The entire lot of fish, all mahseers, zealously guarded and protected by the local villagers in the river Kapila flowing by the side of Sheshileshwara temple was poisoned by fishermen with severe toxic compounds, killed mercilessly and thus made unfit for any purpose. The reason was that being a protected place, a temple sanctuary, where fishing has been strictly prohibited from 1938, the villagers reprimanded four fishermen for fishing illegally in the night. The fishermen in turn, wrought destruction on the fishes." (Jayaram, 1997). A similar incident also took place in Sringeri taluk of Chikamagalur district of Karnataka.

We have already mentioned above several instances of Government machinery intervening to take over sacred groves or sacred ponds, as local communities have lost their control.

Design Principle Three

Social Organization Competent to Manage Sacred Sites

Through much of India's local community based social organizations competent to manage sacred sites have given way to new social organizations subservient to larger polity unable to do so. On the mainland this process has gone on for well over a century; but in some of the more remote areas such as in Manipur it is more recent. Thus amongst the Kuki tribe called Gangtes in Churchandpur district of Manipur the traditional social organization persisted till 1950's. Over the last 40 years it has gradually given way especially in localities close to the market town of Churchandpur. Thus traditionally the Chief of any village was nominally the owner of all land and property, and was given a certain portion, about 5% of the agricultural produce by all members as the rent. The Chief however was expected to provide hospitality to all visitors and discharge other social responsibilities with the help of the supplies so received. The council of elders and others made all community wide decisions, including sanctions against members violating rules governing the management of sacred sites. This social organization has totally broken down close the market town, with the Chief asserting his ownership rights over all land and demanding that others buy it from him for cultivation and pay him royalty for cutting wood from forested patches. At the same time, protection to sacred sites has also fully eroded from these localities (Hemam, 1997).

Design Principle Four

Monitors Accountable to and Respected by the User Group

Traditionally the protection to sacred sites was maintained by belief in omniscience and powers of local deities complemented by enforcement of the deity's will by priests and leaders of local community. Just as the social organization has broken down, so has the faith in omniscience and power of the deities. This is particularly evident in the Christianised tribal communities of northeastern India, where protection to old sacred sites has been largely abandoned along with the belief in local nature spirits. But elsewhere in India too the belief in these supernatural monitors, the local deities is weakening, to be replaced by deities of Hindu pantheon, who may be worshipped in man-made temples without any association with sacred trees, or groves. This weakening of protection through loss in faith in local deities is greater, greater the proximity to market towns in predominantly Hindu state of Karnataka as well. Thus a comparison of an area close to the market town of Kumta and much further away in the district of Uttara Kannada in Karnataka shows a significantly lower level of association of deities with sacred trees and sacred groves in the area more exposed to market influences.

Design Principle Five

Collective Choice Arrangements

With the dissolution of local community control over resources, and local social organization, decisions on resource use, including protection to sacred sites pass into the hands of more centralized state machinery with little or no scope for arrangements reflecting collective choice by local communities. On takeover with the Forest Settlement beginning 1880's, the British policy towards the Kans of Uttara Kannada was one of stringent protection. Yet there were many lapses. In Uttara Kannada sacred kan demarcation during the period of early forest settlement (1880-1920) was imperfectly done. Also some were converted into 'betta' (leaf manure forests) and some others into 'minor forests' open to all for exploitation which in practice was carried out in unregulated fashion during the British period itself (Collins, 1922). Kans were, during post independent period included in forest working plans for selective felling of particularly industrial timber and even fuelwood (Shanmukhappa, 1966; Thippeswami, 1963). All this has contributed to erosion of protection to sacred sites, as in the village Golgodu (Siddapur taluk, Uttara Kannada district, Karnataka) mentioned above where an individual farmer has encroached on a sacred grove against the wishes of a majority from local community. In case of Oran in Devikot village in Jaisalmer district of Rajasthan, Muslim community does not believe in the restrictions on cutting trees from Oran. At the same time they are also dependent on Oran as like other villages to graze their cattle. But other villagers cannot take any legal action against the violators as the Oran land is under control of the government and not of the community. So if some initiative starts from villagers towards conservation of the Oran, villagers cannot legally enforce it.

Design Principle Six and Seven

Flexible Rules and Graduated Sanctions

The centralized bureaucratic procedures pertaining to the management of natural resources are rigid and insensitive to particular local context. They are also unsympathetic to local traditions including those of sacred sites. This is evident in several examples of the state interventions leading to liquidation of sacred sites narrated above.

SCALE OF EROSION

All the trends in recent past thus seem to militate against the seven principles for successful functioning of self-organized conservation systems such as sacred sites. There is indeed abundant evidence of the consequent erosion of such systems. Such erosion took place much earlier in Europe as Christianity spread buttressing states with concentration of power in the hands of the aristocracy and the church. (Hughes, 1994) summarises much of the pertinent evidence. In India such erosion gathered pace as the consolidation of British rule after 1857 was followed by land settlements and forest reservations that denied the legitimacy of community control over land and water resources. Dietrich Brandis the first Inspector General of Forests (1864-1882) was a witness to this early phase of destruction of India's sacred groves (Brandis, 1897). He records: "Very little has been published regarding sacred groves in India, but they are, or rather were very numerous. I have found them in nearly all provinces. As instances I may mention the Garo and Khasia hills.. the Devara kadus or sacred groves of Coorg ... and the hill ranges of the Salem district in the Madras Presidency .... Well known are the Swami shola on the Yelagiris, the sacred forests on the Shevaroys. These are situated in the moister parts of the country. In the dry region sacred groves are particularly numerous in Rajaputana..... In the southernmost states of Rajaputana, in Partabgarh and Banswara, in a somewhat moister climate, the sacred groves consist of a variety of trees, teak among the number. These sacred forests, as a rule, are never touched by the axe except when wood is wanted for the repair of religious buildings, or in special cases for other purposes".

This attrition continued through the British regime. Its pace picked up again after independence as rapid development of transport and communication networks and of forest based industries led to another spurt of deforestation in areas that were earlier much more difficult to access such as northeast India. In these areas the destruction of sacred groves took place primarily in 1950's and 1960's (Gadgil et al., in press).

Can we then estimate the extent of such erosion that has occurred? Chandran and (Gadgil, 1993) attempted to do so for a limited area of 25 km^2 in Siddapur taluk (Uttara Kannada district, Karnataka). They constructed a landscape map labelled with local names of the landscape elements. These names provide clues to the historical pattern of land cover and land use. On this basis they surmise that 5.85% of the land was earlier under a system of sacred groves; this had come down to 0.3% at the time of their field study in 1991. Gadgil et al., (in press) conducted extensive interviews with Gangtes, a group of shifting cultivators of Churchandpur district of Manipur in 1997. Many of those interviewed were personally familiar with the system of sacred sites that prevailed prior to their destruction beginning in 1950's. Their accounts suggest that somewhere between 10% to 30% of land and waters were earlier treated as sacred sites. In Kerala there is a tradition that 1/7 of the land must be set aside as a sacred grove dedicated to snakes when bringing new forest land under cultivation. These findings suggest that it is possible that 10% or more of land and waters may have been covered under sacred sites in pre-British times. What may the total area of sacred groves be in present day India? The level of 1/3% noted above for Siddapur is an unusually high level today. Such a level may hold for only a small proportion of the better forested tracts of India which cover parts of Western Ghats, Chota Nagpur plateau and Northeastern India. These amount to less than 10% of the total country. In rest of the country the levels must be far lower. An estimate of the latter may be obtained from an admittedly incomplete inventory of sacred groves of Maharashtra prepared in 1973-74 (Gadgil and Vartak, 1981). The total area of sacred groves in this inventory amounts to 3570 ha which is a mere

(1.16 /100)% of the state of Maharashtra. The total area of the sacred groves of India as a whole is however more likely to correspond to this level; it would then come to 33000 ha. If the original area of sacred sites amounted to 10% of the land and waters, then this amounts to a decline by a factor of 1000.

REASONS FOR PROTECTION OF SACRED GROVES

Nevertheless sacred sites continue to be protected, albeit to a much lower degree in present day India. This may be because of (a) persistence i.e. continuation of traditional protection, (b) revival, i.e., resumption of earlier practices of protection that had lapsed, or (c) emergence, i.e., institution of protection on a site without an earlier tradition.

The seven design principles remain relevant in these cases of persistence, revival as well as emergence. An appropriate balance of benefits and costs is particularly relevant in cases of revival and emergence; persistence may be found in certain conditions, even in face of an adverse benefit-cost balance.

Persistence

The village Mathigar (Siddapur tatulk, Uttara Kannada district, Karnataka) herbours a patch of pristine evergreen forest of 1 ha in size, rich in species like *Vateria indica* right next to paddy fields on level ground. There are fairly extensive stretches of *Acacia auriculiformes* and *Casuarina equisetifolia* plantations as well as degraded evergreen and moist deciduous forests and tree savannas in the neighbourhood, so that the sacred grove is by no means the only source of woody vegetation of the locality. In fact there is a taboo on the removal of any woody matter, including dead and fallen wood from the sacred grove. Neither is the grove important in providing any ecosystem services such as watershed conservation. There are thus no tangible benefits of the grove; retaining it does incur to the community substantial opportunity costs. These opportunity costs are in terms of income which may be accrued by felling the wood, income which is very large relative to their earnings from small scale agriculture. The opportunity costs are also substantial in terms of the level land on which the grove stands and which may be converted to paddy cultivation.

The villagers of Mathigar are then maintaining protection to their sacred grove despite adverse balance of tangible benefits and costs. It is notable that this is happening in a single caste Karivokkaliga village. Karivokkaliga is a small community of erstwhile shifting cultivators who are still very poorly linked to the market economy. They retain strong community level organization for deciding on disputes amongst their own community members. They retain strong faith in the deities believed to be resident in the sacred grove and have community level ability to regulate the behaviour of their own members so as to ensure that they do not violate the taboos against interference with the sacred grove. The situation of Mathigar then does conform to several of the other six design principles for successful functioning of self-organized conservation systems.

The story of village Gani (Shrivardhan taluk, district Raigad, Maharashtra) narrated by Gadgil and Vartak (1975) provides an illustration of a sacred grove maintained by the local community of Kunbis because of an attractive benefit/cost balance. These authors came across the villagers of Gani during the course of a survey of sacred groves of the Maharashtra Western Ghats. This village had a sacred grove of 10 ha dedicated to goddess Kalkai in the catchment of the only remaining perennial stream of the village, all the other forest in the neighbourhood having been liquidated earlier. The forest was also important as a source of shade to the cattle during the afternoons. The Maharashtra Forest Department had marked the trees in the grove for clearfelling. This felling would not have benefited the local villagers in any way, except through some temporary employment as wage labour. They were therefore very much interested in saving the forest, but had no way of influencing the remote central authorities of the forest department in any way. However a sympathetic local Range Forest Officer informed them of the authors' ongoing survey of sacred groves, suggesting that they may be able to help them out. They wrote to the authors who visited the village and the sacred grove and carried their request to the Chief Conservator of Forests of the state. This official agreed to the request on grounds of personal friendship, but remarked that he saw no merit in saving these "stands of overmature timber". It is evident that local ecosystem services, which the villagers valued were of no concern to the centralized authority focussed on wood as a commercial commodity.

Apart from an appropriate balance of benefits and costs for the villagers of Gani, several other principles of design were satisfied as with Mathigar. The local Kunbi community is socially homogenous and not strongly linked to market forces. It has community level decision making machinery to regulate the behaviour of members of their own community. They believed in the monitoring power of their local deity. Importantly of course, the rules governing how to treat the sacred grove are no longer in control of this local community, but framed by a rigid, centralized authority with no accountability to local community. That is of course why the grove was about to be felled.

Revival

The tradition of sacred groves is not only being maintained in some cases, in other cases it is being revived. Such cases of revival seem dependent on local communities perceiving tangible losses of benefits on liquidation of the sacred grove of a level sufficient to offset the costs of revival of the tradition. Some of the most interesting cases of such revival come from northeastern states like Manipur and Mizoram where the once extensive network of sacred groves was largely destroyed in 1950's on development of a transport network and a lucrative market for timber coupled to conversion to Christianity. But in this tract where shifting cultivation prevailed, some of the sacred groves encircling the settlements served as firebreaks during the slash and burn operations. In several villages of inhabited by Gangte tribals of Churchandpur district of Manipur such revival of sacred groves encircling habitation has taken place. Since now the community has embraced Christianity, the groves are no longer being viewed as abodes of deities. In communication with outsiders they are called "forest reserves", or as Malhotra (1990) reports for Mizoram "safety forests". The term used for the grove in their own language however remains as before 'Gamkhal'. Protection to these gamkhals continues to be organized through monitoring by local community members and implemented through sanctions for violation imposed by the community leaders. This is possible because in the more remote Gangte villages traditional community level organization is still functional.

There is however a clear pattern in the spatial distribution of villages where such revival has occurred. In villages close to the market town of Churchandpur the whole landuse pattern has changed with all land now belonging to individuals; terraced and bought under permanent cultivation of commercial crops like pineapple. In these villages the function of sacred groves as a firebreak is irrelevant, nor do these villages retain the community level organization capable of monitoring and enforcing protection. In such villages there has been no revival. The cases of revival become more frequent as one moves to the interior, away from the market town and roads, to settlements which continue the practice of shifting cultivation and retain more traditional forms of community organization.

Emergence

Finally, there are interesting lessons to be learnt from new emergence of sacred groves in places where none existed earlier. One can see three types of contexts in which such emergence may occur. (a) It may employ religious beliefs, but serve a tangible function (b) It may relate to the traditional religious beliefs without serving any tangible function (c) It may relate to the state machinery attempting to ensure protection through the medium of traditional religious beliefs. The sacred forests newly established by people in the areas of Almora and Pithoragarh districts in Uttar Pradesh are instances of case (a), serving tangible functions. In Dharamgarh area which is on the border of the two above mentioned districts, about 25 villagers are protecting extensive forest areas dedicated to the local goddess. The area is at around 2400 ft. from mean sea level in very remote hills of Kumaon Himalaya. All the economy is mainly dependent on agriculture and people are directly dependent on forests to meet daily requirements like firewood, fodder, timber etc. Forests were also under threat because of market demand for timber. Though the local panchayats were having ownership over the area, they were unable to protect the forest. Finally villagers decided to handover the forests to local goddess—Kokilamata who is supposed to be the goddess of justice. They prepared an agenda for protecting the forest and offered it to goddess. Then in a ritualistic fashion they also marked the boundaries of forests offered to goddess. People were allowed to cut the twigs of trees, collect firewood and deadwood. Nobody was allowed to cut the live plant; otherwise goddess will punish the violator. The process was started around 1982 from the village Jakhani in Almora district and got spread in more than 25 villages around. In villages like Madigaon, Phanku, Dharamgarh people are getting tangible benefits to meet complete their requirements of forest produce.

The initiative for the establishment of this sacred grove probably came from local people, although today panchayat members as well workers from a Gandhian Ashram at Dharamgarh claim credit for this experiment. Today people are permitted to harvest dead trees and green twigs as well as graze their cattle in the sacred grove. There is no commitment to protect animals with the goddess. People hunt various animals for various purposes e.g. medicinal treatment, meat etc. This low level regulated harvest of living resources from the sacred grove is clearly providing sufficient tangible benefits to offset costs of protection.

Two other examples of recent emergence of sacred groves in villages Bada Bhilwada and Shyampura (Zadol taluk, Udaipur district, Rajasthan) also relate to adequate tangible benefits being available. Boda Bhilwada and Shyampura are situated in foothills of Aravalli mountains in southern Rajasthan. A new tradition has emerged in these villages for the conservation of forest; which they call as "Kesar Chirkav", The story of this emergence was started in a literacy programme by a worker of NGO Sevamandir, based at Udaipur, Rajasthan. The worker told the villagers about such a "Kesar Chirkav" reported from Sagwada village, Udaipur district because of the influence of the famous deity "Kesariyaji" situated about 80 kms. south of Udaipur. People also came to know about the tangible benefits because of the new tradition and they decided to do the same in their villages namely Bada Bhilwara and Shyampura coming under Panchayat-Bichiwada.

In both the villages Joint Forest Management was initiated with the help of Sevamandir organization. Villagers brought sacred 'kesar' i.e. saffron from the "Kesariyaji" temple. They prepared saffron water using that saffron,and in ritualistic fashion they sprinkled the saffron water along the boundary of the area to be protected under "kesar chirkav". They also included JFM land in "kesar chirkav". Since that day in 1994 there was total ban on cutting trees from the area, harvesting of grass is allowed. With the help of forest department, plantation was done on "kesar chirkav" area. When the benefits started coming out, then there were different experiences in village Shyampura and village Bada Bhilwada.

Bada Bhilwada villagers experienced a lot many cases of violation of the restrictions. A few times the Forest Protection Committee also asked for the help of forest department for legal action against the offenders. People have faith for "Kesariyaji" but the belief of wrath due to violation of restriction is not deeply rooted.

Unlike Bada Bhilwada, Shyampura villagers deeply believe in wrath of "Kesariyaji" due to violation of restriction of not cutting trees. The forest protection committee has not used its powers since 1994. Today, grass harvest is the cause of tangible benefits available to people immediately. In Shyampura, villagers harvest the grass as per the rules of the forest protection committee. The committee charges Rs.5/- per sickle. Villagers harvest the grass and can easily get around Rs.30/- for a headload in the local market.

The villagers of Shyampura are happy with present scenario, but their neighbouring villages are encroaching the land of "Kesar Chirkav" as some of the land was part of the resource catchment of those other villages. Also sometimes because of vested interest they try to violate the restrictions.

We also have on record one case where a sacred grove has been established, by an individual farmer on his own land, apparently without any motivation relating to tangible benefits. This is a small grove of 750 m^2 dedicated to a local female deity, Yakshi: in village Mathigar (Siddapur taluk, Uttara Kannada district, Karnataka). But such cases are very likely truly exceptional at the present time.

Karnataka Forest Department, which in early 1970's, clearfelled large sacred groves, such as one at Mensi of 21 ha. to supply softwood to plywood industry, has also now programs of establishment of sacred groves called 'Pavitravanas', first initiated in 1988. As mentioned above the cream of Karnataka's forest resources was largely exhausted through a process of sequential overexploitation by early 1970's. This led to a gradual phasing out of commercial exploitation of natural forests with the imposition of a ban on green fellings in the evergreen forest belt by 1985. As the stocks of commercially more attractive forest resources depleted, the Forest Department's motivation of maintaining a tight control over the resource base weakened. Simultaneously, democracy, first instituted in 1947 has been growing stronger with more and more power being devolved to people. It was in this setting that the program of decentralized administration with considerable powers vested with village cluster (or Panchayat) and district (or Zilla) level elected bodies was introduced in the Karnataka state in 1986. This weakening of the hold of centralized bureaucracy and political set up and simultaneous strengthening of decentralized Institutes is clearly favourable to the functioning of self-organized systems of resource management. A number of initiatives have therefore emerged over the last decade, including those of Joint Forest Planning and Management and Water User's Associations.

The experiment of Pavitravana,an initiative of the state forest department to create new protected sites with the co-operation of local communities and specifically designated as pavitra or sacred is an outcome of these forces. In 1988, Karnataka forest Department established a Pavitravana called as 'Sridhar van' in village Salkani of Sirsi Tahsil of district Uttara Kannada. Youth organization in the village took initiative in the process. Plantation of species used for ritual performances like yajnas is done on 2 acre land on hill top.

Now in the same village people are demanding to afforest the barren hill tops with useful species for NTFP, fuel, fodder etc. The later proposal is getting support from villagers because the plantation will serve as wind break and the Areca and Coconut gardens will be saved from heavy winds.

In another experience in the village Bakkal in Sirsi taluka of Uttara Kannada district with the initiative from people, state forest department has established a Pavitra vana spread over 28 ha. Earlier forest department was planning to have Acacia plantation on that area. But the people opposed it; with the support from the then Conservator of forests, Shri Yellapa Reddy, the Pavitra vana was constructed. People also consulted the local priest, Shri Nagendra Bhat and provided a religious and mythological touch to the 'Pavitra vana'.

The Pavitravana is based on Hindu scriptures which state about various plant species as favourite of gods. Planting of these species confers some merit to the persons who do it and also obviously some ecological functions these species perform.

Each auspicious star of Hindu mythology has its own favourite plant species-mostly a tree. Thus 'Ashwini' has *Strychnos nux vomica*, 'Bharani' *Emblica officinalis*, 'Kritika'-*Ficus glomerata* and so on. Similarly each

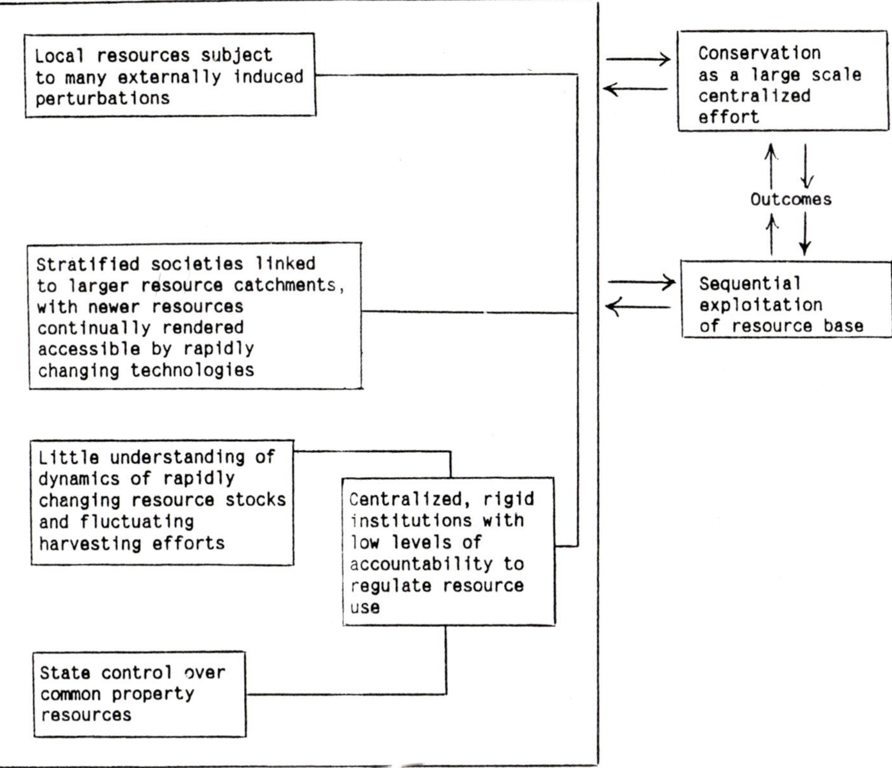

Fig. 5. System of management of natural living resources characteristic of industrial societies

planet and each zodiac sign has its own associated plant species. The Hindu gods also have their own favourite plant species. For example, *Aegle marmelos* and *Calotropis gigantica* are favourites of Shiva and *Ocimum sanctum* for Vishnu. The Bakkal experiment shows that planting together of these various species mentioned in the scriptures in one place-organized under star, planet, zodiac gardens and plants favourite of gods in separate clusters do not however make up for the alarming degradation of sacred groves of natural vegetation which have been selectively felled, or clear-felled and converted into plantations or allocated as leaf manure forests (bettas) or added to the minor forests. At best then Pavitravanas uphold the supremacy of the more formal, Sanskritised traditions. They can also function as excellent arborata. However, planting together of species from diverse ecosystems will not give rise to a natural ecosystem.

For instance in the star garden we find grown together *Acacia catechu* (Mrigashira star) and *Calotropis gigantea* (Shravana star) from dry open on rocky habitats grown in the company of evergreens *Artocarpus heterophylla* (Uttarashada star) and *Mesua ferrea* (Ashlesha star) and *Pinus longifolia* (Jyestha star). They are intermingled with various deciduous tree species like *Butea monosperma* (Hubba star) and *Spondias mangifera* (Hastha star). Nevertheless these groves can be of educative value as well as re-emphasize man's bonds with plants.

PROSPECTS

What then is in store for the sacred sites, sacred groves, ponds, grasslands in coming years ? There are obviously signs of strengthening of community level institutions which would favour the persistence, revival or emergence of such self-organized systems. But there are also continuing trends of weakening of traditional religious beliefs, especially of small scale societies, of 'little traditions' of faith in sthaladevatas-deities rooted in particular localities. That means that intangible benefits flowing from respecting the sites where such deities reside are likely to continue to be devalued. Nor would such deities be relied upon to help local communities monitor adherence to prescribed norms of behaviour. If the system of self-organized conservation sites is to continue, then such sites must provide to local communities tangible benefits, be they as firebreaks or in terms of dead and fallen wood. But such benefits would often turn out to be as inadequate as gathering market forces lead to an escalation of the opportunity costs of maintenance of sacred sites. The larger society will then have to invest in augmenting tangible benefits through special financial or other rewards (Gadgil and Rao 1994, 1995).

Simultaneously with availability of appropriate tangible benefits, the prospects for such self-organized systems of conservation would depend on effective community organization and devolution of authority to communities. As discussed above, there are many encouraging trends in this direction. It is then entirely reasonable to expect that these self-organized systems of conservation would continue to play a role worthy of their history in years to come.

ACKNOWLEDGEMENTS

The authors are thankful to a very large number of individuals for their manifold help. Unfortunately, it is not posssibe to mention all of them by name. Amongst the more notable of them are: Rajeev Aahal, G.K. Bhat, Omprakash Bhat, Jethusingh Bhati, Ketaki Das, Soumyadeep Datta, Prahalad Dubey, Ajay Dolke, N.R. Hegde, K.Kamaraj, Devashish Kar, Iqbal Singh

Kaundal, K.C. Malhotra, S. M. Monhot, K.Murugan, Shubu Patwa, Ishwar Prakash, D.S. Srivatsava, Sarvajeet Thousain, B.K. Tiwari, R.S. Tripathi, V.D. Vartak, PRAVA, Orissa, Sevamandir, Rajastan, Vasundhara, Orissa, Young Mizo Association (Central), Mizoram.

We are also indebted to the residents of all the villages surveyed for patiently answering our questions. Thanks are due to the State Forest Departments of the states of Himachal Pradesh, Rajasthan, Uttar Pradesh, Maharashtra, Karnataka, Tamilnadu, Kerala, Orissa, Bihar, west Bengal, Assam, Meghalaya and Mizoram. We thank WWF-India for making their funds available for this work.

REFERENCES

Berkes, F., Folke, C. and Gadgil, M. 1995. Traditional ecological knowledge, biodiversity, resilience and sustainability. In: C.A. Perrings, K.G.Maler, C.Folke, C.S. Holling and B.O. Jansson (Eds.) *Biodiversity Conservation: Problems and Policies.* Kluwer Academic Publishers Group.

Borges, R.M and Rane, U. 1992. *A Study of Wildlife/People Conflicts. The Bhimashankar Wildlife Sanctuary as a Case Study. Final Report.* WWF India and Bombay Natural History Society.

Brandis, D. 1897. *Indian Forestry,* Oriental Institute, Woking

Chandran, M.D.S. and Gadgil, M. 1993. Kans-safety forests of Uttara Kannada. *Proceedings at IUFRO Forest History Group meeting on Peasant Forestry.* Frieburg, Germany. Nr.40, pp. 49-57

Collins, G.F.S.1922. *A Report into General Conditions of Forest Administration in Siddapur Taluk,* 15 June. Forest Settlement Office, Karwar.

Diamond, J. 1991. *The Rise and Fall of the Third Chimpanzee.* Vintage, London.

Gadgil M. 1985. Social restraints on resource utilization: the Indian experience In: J.A. McNeely and D. Pitt (Eds.) *Culture and Conservation: The Human Dimension in Environmental Planning,* Croom Helm, pp.135-154

Gadgil, M. 1995. Traditional conservation practice In: William A.Nierenberg (Ed.) *Encyclopedia of Environmental Biology,* Volume 3, Academic Press. pp. 423-425.

Gadgil, M. 1996. Managing biodiversity. In: K.J.Gaston (Ed.) *Biodiversity: A Biology of Numbers and Difference,* Blackwell Science Ltd., Oxford. pp. 339-360.

Gadgil, M. and Berkes, F. 1991. Traditional Resource Management Systems *Resource Management and Optimization,* 18: 127-141

Gadgil, M. and Berkes, F. and Folke, C. 1993. Indigenous knowledge of biodiversity conservation, *Ambio,* 22: 151-156.

Gadgil, M. and Guha, R. 1992. *The Fissured Land. An Ecological History of India,* Oxford University Press.

Gadgil, M. and Rao, P.R.S. 1994. A system of positive incentives to conserve biodiversity, *Economic and Political Weekly, Vol. 29, No.32,* August 6.

Gadgil, M. and Rao, P.R.S. 1995. Designing incentives to conserve India's biodiversity, In: S. Hanna and M. Munasinghe (Eds.) *Property Rights in a social and Ecological context.* The Beijer International Institute of Ecological Economics and the World Bank, Washington D.C. pp. 53-62.

Gadgil, M. and Vartak, V.D. 1975. Sacred groves of India: A plea for continued conservation *Journal of Bombay Natural History Society,* 72: 314-320.

Gadgil, M. and Vartak, V.D. 1976. Sacred groves of Western Ghats of India, *Economic Botany,* 30: 152-160.

Gadgil, M. and Vartak, V.D. 1981. Sacred groves of Maharashtra: an inventory. In: S.K. Jain (Ed.) *Glimpses of Ethnobotany,* Oxford University Press, Bombay, pp. 279-294

Gause, G.S. 1939. *The Struggle for Existence.,* Williams and Williams, Baltimore.

Hemam, N.S. 1997. *The Changing Patterns of Resource use and its Biosocial Implications: An Ecological Study Among Gangtes of Manipur,* PhD. thesis, Indian Statistical Institute, Calcutta.

Heywood V.H.(ed.) 1995. *Global Biodiversity Assessment.* Published for the United Nations Environment Program, Cambridge University Press, Cambridge.

Hughes, J.D. 1994. *Pan's Travail—Environmental Problems of the Ancient Greeks and Romans.* The John Hopkins University Press, Baltimore and London.

Jayaram, K.C. 1997. Massacre of mahaseer fish. *Current Science,* 72: 901

Johannes, R.E. 1981. *Words of the Lagoon,* University of California Press, Berkeley.

Ruddle, K. and Johannes, R.E. (eds.) 1985. *The Traditional Knowledge and Management of Coastal systems in Asia and the Pacific,* UNESCO, Jakarta

Kalam, M.A. 1996. *Sacred Groves in Kodagu District of Karnataka.* Pondy paper on Social Sciences, French Institute, Pondicherry.

Kangle, R.P. 1969. *Arthashastra.* University of Bombay.

Lenski G. and Lenski J. 1978. *Human Societies: An Introduction to Macrosociology.* McGraw-Hill, New York.

Malhotra, K.C. 1990. Village supply and safety forest in Mizoram: a traditional practice of protecting ecosystems. In: *Abstracts of the 5th International Congress of Ecology,* Yokohama, Japan.

Nayar, P.K.B. 1987. *Religion, Mythology and Ecosystem.* Report submitted to the Ministry of Environment and Forests, New Delhi.

Ostrom, E. 1990. *Governing the Commons: The Evolution of Institutions for Collective Action.* Cambridge University Press, New York.

Ostrom, E. 1992. *Crafting Institutions for Self-Governing Irrigation Systems,* Institute for Contemporary Studies, San Francisco, California.

Service, E.R. 1975. *Origins of State and Civilization: The process of Cultural Evolution,* W.W.Norton and Co., New York.

Shanmukhappa, G. 1966. *Working plan for the unorganized forests of Sirsi and Siddapur.* Mysore Forest Department, Bangalore.

Singh, K. 1994. *Managing Common Pool Resources,* Oxford University Press.

Thippeswami, S.C. 1963. *Sirsi Town Firewood Supply Plan.* Mysore Forest Department, Bangalore.

Unnikrishnan, E. 1995. *Sacred Groves of North Kerala an Eco-folklore study* (in Malayalam), Jeevarekha, Thrissur.

White, L. 1967. The historical roots of our ecological crisis. *Science,* 5: 155.

Wingate, R.T. 1888. *Settlement proposals for 16 villages of Kumta taluk.* No.210 of Dec. 1888, Forest Settlement Office, Karwar.

33

Policy and Institutional Aspects of Sacred Groves: Tending the Spirit, Sustaining the Sacred

A.K. Gupta

Indian Institute of Management, Ahmedabad, India

INTRODUCTION

There was a Chola king by the name Parivallal in ninth century in south India who was very famous for his generosity and concern for nature. He was passing through a forest once when he felt thirsty. He stopped his chariot near a stream and went to drink fresh water and also relax a bit. After a while when he came back he noticed something that stopped him from boarding the chariot. He decided to walk back to his palace on foot. A tendril of a creeper (locally called as nilotpal) had grown and twined around the spoke of the wheel. If he moved his chariot, the tendril would have broken. Whether the king recognized the 'rights' of a creeper or his 'responsibility' towards it, is difficult to say but the irrelevance of such a story in a discourse on modern institutions of conservation may certainly appear out of place of many (Honey Bee, 1994).

And yet it is such a discourse that merits our attention in the current popular upswing of consumerist and utilitarian world. When we seem to price everything and decide to allocate conservation priority only on what is considered useful or important by the most, the case for not so important and not so relevant needs to be made.

In this paper I attempt such a case. I argue that the concept of sacredness is at the very root of our civic consciousness and whenever any boundary of sacredness is violated, we are reduced to that extent in our civil consciousness. Our humane urge to relate to all that we adore. respect and some times even fear (of losing) provides some basis of what we consider sacred. But memories of some one we love and respect seems to ascribe sanctity and sacredness to anything and every thing that we attach that memory with. Is sacredness then only a function of individual ability to recall, rever and relate to one's memories. Obviously, sacredness goes beyond individual criterion and concern for adorable past. Communities and cultures define sacred symbols that grow with time and also erode with time. Collective memories, myths, legends and spiritual code of conduct evolve over a period of time sanctified through sanctions, materials as well as moral.

The traditional ecological knowledge systems continually evolve, intermesh with contemporary influences and some times become part of Flokloric memoryscape. Memories as assume the forms of landscape and a metaphor becomes a reality; an absence becomes the presence. Cultural reserve of ecological knowledge also creates and/or dissolves institutions for conservation. Ecologists are often confronted with situations in which the social perceptions of nature and their interactions do not parallel with biological explanations of the same. And yet the ecological ethics stems from such social perceptions. Does knowledge and knowability of nature demand restraint, if only to nurture these perceptions. Or is it that a new language of discourse has to be invented so that what we know but do not accommodate in our rational tool box does continue to influence our frame of reference. How else do we explain so many traditional institutions for conserving common property resources and other natural sub sets on grounds other than purely utilitarian logic. If technology is treated as words, institutions may be like grammar (Gupta, 1992). The paper deals with the elements of several traditional institutions of conservation particularly scared Groves which achieve what modern consciousness and institutions may seldom even attempt to achieve. There are songs and stories like the one which reminds people about their responsibility towards birds in a drought year when there is not enough to eat for human beings, or another one in which a lovelorn woman negotiates with a lake to tell her when her lover would come back if she removes the layer of algae blocking her (the lake's) breathing.

Metaphors have low entropy and thus have been used by societies from time immemorial to code certain crucial rules for respecting nature in cultural terms. But there are also functional elements in traditional knowledge. There are no less important and have in fact become a major precursor of modern search for leads for drugs and other recipes. In this search, we have become so obsessed with the results in the form of patented drugs and obvious profits that we seldom reflect on the long term future of this knowledge systems itself. We do not realize that erosion of knowledge is sometimes taking place at faster rate than the erosion of resource itself. Devaluation of traditional knowledge and contemporary innovations (which have received much less attention) takes place partly out of ignorance but partly on account of a short sighted calculus. This calculus neither takes into account historical experimentation that has been gone into by societies nor recognizes ethical responsibility of researchers towards providers of knowledge as well as resources situations.

This paper provides understanding of how traditional ecological knowledge systems around sacred consciousness have evolved and how could we draw upon these in grafting modern institutions for conservation. The blending of modern and informal science so achieved may even enrich the scientific pursuits but not just that. It may nurture human urge to cross frontiers of consciousness that bridge sacred with profane, holism with reductionism, and ethereal with material.

How else the nature with in and with out may communicate and create enduring landscapes of memories, meanings and metaphors of conservation.

In part one of the paper I review debate on sacredness in the context of the sacred groves as well as species. Idea is to broad base our search for organizing principles and building blocks which can help us in building upon sacred institutions and in fact further help in grafting modern institutions upon those foundations.

In part two I review global experiences with sacred groves as well as sacred species particularly insects. I isolate principles and options for developing future, strategies in the third part. In this part I thus discuss, the way we can graft modern institutions on these sacred principles drawn from cultural, spiritual and social streams of consciousness. It is argued that sacred and profane (and secular) and reductionism and holism intertwine like a double helical structure of DNA, one without the other is not sustainable.

PART ONE: SACRED GROVES-DEBATE, DISCOURSE AND DIVERSITY

The divergence between the perspective of Kalam (1996) and Gadgil and Vartak (1976, 1981) provides one end of the spectrum of discussion on roots of sacredness. The review of research on sacred insects, trees, lakes and other sites and species providing not just utilitarian but more importantly symbolic and cultural basis for conservation provides another end of the spectrum. Both these perspectives require institutional understanding to analyze continuance as well as discontinuance of the conservation ethic. Kalam argues that standard narrative of fear or respect for God in the earlier times being the reason for evolution and conservation of sacred groves may not be tenable. If such were the case (and people and innate faith in such deities), he asks as to why should then sanctions be needed-a view not entirely consistent with the theories of institutions. He also asks that if one went in past several centuries ago, the functional need for conserving certain species would not hold because at that time pressures on resources were negligible or non existent. He believes that the sacredness of land scape or a site was the point of origin and deities were located at such sites much later. This too happened because people had a desire to propitiate the fearful god who brought floods, rains, droughts etc., and these gods were supposed to reside in the sacred or unknown places.

The argument by Gadgil and Vartak that sacred groves originated out of more secular causes such as utility of some plants of medicinal or other uses and not out of any sacred concern per se is also disputed by Kalam. (1996) disagrees with the notion that the sacred groves might have been the sites of human and other sacrifices or that these have evolved during transition from hunter gatherers to cultivating cultures.

There are several other aspects to this debate. Gerden and Mtallo (1990) while describing sacred forest reserves in Tanzania take a slightly different angle to this issue. This views the existence of Scared groves as part of taboos evolved historically over several generations to provide a site for culturally crucial social interactions. For instance, circumcision of boys, discussions on village and tribal matters, creation of cultural conformity, restrictions on tree cutting but mature trees allowed to be cut for building houses with collective sanction etc. It may be recalled that the empirical studies about absence of many mature trees in sacred groves may be partly explained by such rules in different parts of the world.

The important issue however is the need for institutions sanctified by sanctions. It is here that Kalam commits to mistake. The need for sanctions whether internal (moral, guilt, curse, blessing etc.,) or external (punishment, fines, social censure, shame etc.,) have been used as a necessary feature by every social institution for managing boundaries of the institution. I have argued that there are always at least three kinds of rules in any institution (Gupta, 1990, 1996) which partly parallel the eight fold classification of rules by Ostrom. Two types of rules are primary viz. (a) rules for boundary management and (b) rules for resource allocation. The interface of the two would generate conflicts and thus would arise the third type of rules to resolve conflicts. These rules could be coercive or persuasive and sanctified by moral or material sanctions.

Thus, Gadgil and Vartak are not wrong in suggesting that sacred groves as an institutions might have evolved for performing sanctions-a very important function of any communication to keep itself together. Whether these sanctions included human sacrifices or not at any time is something on which I will not speculate. But suffice it may to state that sanctions are an inevitable part of any arrangement which involves evolution of rules that have to be enforced.

Where I disagree with Gadgil and Vartak is on their suggestion that sacred groves need to be only utilitarian in nature. Utility may emerge as a consequence because of conservation but it need not be always a primary precursor. If such was the case then we are taking our present day values and projecting the same into past. More importantly, we will lost the potential of sacred groves as an social institution of contemporary relevance by restricting our analysis to only functional analysis—important that is.

In fact, as Gerden and Mtallo (1990) have shown that may rules in vogue in such sacred groves may be not acceptable in contemporary sense and their context has to be understood before any change is attempted. For instance they show that rules include (a) no iron tool or axe to be taken inside the grove to cut any tree and only mature trees are allowed rarely to be cut for house construction, (b) women are not allowed to enter the grove at all, (c) no non-member is allowed to enter when the sacred ceremonies are going on in the grove, (d) no bees are allowed to be kept inside, (e) any one can pass through the sacred grove when rituals or any meeting is not going on the grove.

The sanctions include psychological pressures put on the offender, collective prayers to shame and curse offender, and ultimate sanction is exile.

Beliefs, Myths, Norms, Values and Ideologies

The continuum if beliefs and ideologies has to be understood carefully to fully appreciate the role of sacredness.

Beliefs are meanings attached to human experiences that does not need to be explained every time or everyday. some argue that beliefs deal with those human experiences which can not be explain such as taboos. That is not true. Taboos are only one kind of beliefs. Beliefs are like a mechanism of calibrating of our compass in the basic directions we follow in life but not entirely.

Those beliefs which can not be retained in their explicit form but which are crucial to explain and reinforce identity of a community become myths. Myths have always an element of metaphor i.e. partial maps of reality. The fusion of imaginary and real is inherent in a metaphor and that is what makes it so powerful.

The beliefs and myths which have normative ascendance in human order of preference become norms. Such beliefs are allowed lesser degree of freedom.

The relationship between beliefs, myths, and norms is a matter of degree to which meanings are explicit or implicit, shared or held individually and sanctity attached by the community. when certain norms are widely shared and accepted, these become part of values. The cultural values i.e. beliefs widely shared and considered desirable as well as crucial for the core identity of the community can be distinguished from the spiritual and other kinds of values.

When values are internalized so much that these provide not only a way of doing things but also mobilizing collective will for purposive action, these acquire the form of ideology.

The sacred beliefs, myths, norms, and values have not become ideologies in the context of the forest in many cases though have the potential to become so. For instance, when some states in India encouraged people to plant trees in memory of their near and dear ones, who had departed, the memory forest (Smruti Vana) was a value being shaped to become an ideology for action. If it has not become so, it could be because of our inability to understand the sensitivity with which sacred beliefs are created and, therefore, moulded.

Sacredness has another root and that is our reciprocity with non-human sentient beings such as insects, animals, trees, and other aspects of nature. Many indigenous communities seek permission from the deity of the forests before harvesting any produce or hunting any animals. The notion is that while destruction of some life is necessary for reproduction and continuity of other lives, there is no moral superiority of one over other. Chinese, and Hindu, Jain, Buddhists, and many other faiths do not recognize inherent superiority of human needs and wants over that of other living beings. The dilemma was identified by K.M. Munshi, a man of literature and former union minister of food and agriculture when he recognized the destiny of communities and cultures that rever nature so entirely that they become themselves extinct. In contrast there were other cultures which dominated nature so much that natural resources and several environmental properties got extinct or modified adversely. The challenge obviously is to define the zone of total reverence vis-a-vis the zone of total dominance. Sacred groves could thus become and have been one end of the spectrum while cultivated fields could be the other.

What has to be appreciated is that even the cultivated fields may have elements of sacredness as sacred plants, stones and other species. Similarly sacred groves may be used for harvesting medicinal plants, seeds and other products within some constraints. The issue is not whether we need to identify the scope of sustainability only from the point of view of one extreme or the other. Instead, to what extent internally regulated behaviour through beliefs, norms and values can supercede the need for external regulation and sanctions while generating institutions for conservation. It is this ability across species and ecological contexts which we need to appreciate before we can understand and design the institutions for conserving not just the forest but also underlying sacred consciousness.

PART TWO: REVIEW OF SACRED CONSCIOUSNESS IN NATURE

Sacred Groves

Just as Indian communities have evolved sacred beliefs around certain groves, lakes, mountains, etc., similarly local communities around the world have evolved similar institutions. Godbole (1996) describes the nature of sacred groves in western ghats and demonstrates the role of fear, economic utility, indirect benefits, and participation in governance of these groves, one needs to also augment these through supplementary plantation. Native Americans have struggled for a long time to preserve their sacred sites which have come under stress more and more due to expanding urban boundaries, prospecting for various mineral and other natural resources, and location of recreational, scientific and other facilities (Bodine, 1973)

Jain sutra in Acaranga-sutra, states:

> And just as it is the nature of a man to be born and grow old, so is it the nature of a plant to be born and grow old... One is endowed with reason, and so is the other; one is sick, if injured, and so is the other; one grows larger and so does the other; one changes with time, and so does the other... He who understands the nature of sin against plants is called a true sage who understands karma...

In the European context, for instance, alder is supposed to have protective and oracular powers, Birch has fertility and healing powers, Fir indicates vision of long term future and what is beyond and yet to come, Hazel is supposed to help in gaining knowledge, etc. Just as Basil, banyan, and many other trees have religious and spiritual significance in our own country. The point is not to establish a causality between functional use of these plants and their symbolic and cultural value.

In Borneo, the relation between sacred people and sacred land. The spiritual connection between self, spirit, nature and the God is seen through a continuity of consciousness and experiences. There is an unique example of a tree shrine that became a church around the oak of Allon-ville-Bellefosse. This is perhaps the most famous living tree in France estimated to be over thousand year old with a trunk measuring about 45 ft. in circumference. This has been a church since 1696 when it was consecrated to the Virgin Marry.

In U.K. an association of Nature Reserve Burial Grounds networks 25 or so woodlands burial grounds where a tree is planted in place of having a headstone. The initial cost of purchasing a land was met through sales of cemetery plots at well below the going rate for the burial industry but sufficient for the purpose. The initial participants included those whose family could not afford high cost of conventional burial lots, baby boom environmentalists, and wilderness buffs. This is an effort to reestablish the value of sacredness implicit in cycles of life in the western society by bringing mystery of unknown as well unknowable around our present day life.

An ancient sacred tree called as Eboga discovered thousands of years ago has provided a very important medicine called as Ibogaine used for de-addiction among people as well as for curing AIDS. Brothershine, representing African Descendants' Awareness Movement (ADAM), cautioned people that this drug was not a magic bullet. It worked only when a knowledgeable counsellor was available and patients were willing to change their lifestyle and also they were re-socialized. The relationship between psychic, social and biological healing may become more and more manifest as e bring sacredness back into our life.

Chapman (1995) describes in considerable detail the historical account of sacred groves of Britain most of which have disappeared since. He quotes Ellis to suggest that sacred groves (Sarna) were left for local deities while clearing the forest for pastor and crop cultivation. Sarnath was one such grove in India where Buddha was born. He speculates that ritual enclosures and burial mounds of the bronze age and neolithic landscapes in Europe may have evolved in the similar way.

In ancient Egypt, there were sacred lakes around temples similar to Mansarovar lake in Tibet and other such lakes in Africa. The origin of rivers and streams also have been given sacred status in most cultures.

A large number of studies on sacred sites and their relationship with human rights in different parts of the world are available. It is interesting to see such a connection because the cultures which have kept sacred groves alive often (though not always) may be in minority of one kind or the other and thus may come under pressure from outsiders intending to violate the sanctity of the sacred groves for commercial reasons.

In Japan, there are traditions where people go to a particular site for communicating with the trees. The Tamaki shrine is located in a forest of ancient sacred cedars, one of which is claimed to be three thousand year old. In Kumano, there are people who are reputed to be able to communicate with trees. Just as there are also people who practice 'Takigyo' under freezing cold water falls to purify both mind and body. The sacred forest and sacred water falls provide opportunities to seek communion of human spirit with nature (Nara Web) particularly their burial grounds.

The sacred Kaya forest have been studieds not only in their historical context but also for their contemporary relevance. In coastal Kenya, there is a network of sacred Kaya forest, which are the sites of cultural discourse, rituals and other ceremonies. The Kaya has remained the spiritual centre of several communities in the region. Despite being fragmented and patchy in nature, these forests provide may small but imprint watersheds located in the coastal areas. There are conflicts emerging on account of outsiders interested in using these locations for tourism and local communities trying to conserve their sacred traditions. In some cases, the elders were willing to make adjustments and allow tourism to visit so long as the sacred forest was preserved.

A recent WWF study listed on WWF web site notes that more than half of Kenya's rare plants occurred at the coast, most being found within these sacred forests.

Induchoodan (1988) describes ecological characteristics of a sacred grove near Perumbavoor. Like in many other temples, the origin of temple in this place is also linked with discovery of miraculous property of the local rocks leading to establishment of temple. In one-and-half hectare area, more than 2310 plants of over 10 cm gbh could be recorded. The most interesting thing about the sacred grove is that it has expanded between 1973 and 1987 to at least some extent. This is contrary to the trend witnessed by Kalam (1996) who noticed that the sacred grove he has studied had expanded from 10865 acres to 15506 acres during 1873 to 1905. However, between 1905 and 1985, it lost 9558.77 acres. Even what is left is not under uniformly thick forest. Paradoxically the attempts are being made, as he reports, to sanskritise the deities to make them vegetarian and in conformity with orthodox Hinduism.

Nipunage et al. (1988) describe Sagdara grove in Pune district which is surrounded by parent land. Joint teak trees preserved through religious sanctions seem to be a unique provenance with considerable potential for breeding in future. The authors refer to almost a 100 year old study by Talbot (1909-11) to suggest that teak seldom attained a large size in that ecological region. Sacred grove, authors suggest, could be an important source for high quality seedlings.

Hughes (1984) traces the origin of the concept of sacred grove in the European and Greeko Roman context. He quotes Pliny the Elder (23-79 A.D.) that these sacred groves or sanctuaries were the first temples of God. Also was as Greek term used for grove consecrated to a deity called Temenos, a 'cut off' or de-marcated place. The Latin equivalent were Nemus (grove) and Templum i.e. a marked out space. The latter was the origin of the word Templum. The examples of leasing out private groves for commercial purposes in First Century A.D. and Fouth Century B.C.

imply that pressures for weakening moral or religious motivations had started emerging in urban fringe areas even at that time. Another interesting dimension of sacred groves recalled from the historical documents in Roman era is about their being experimental forest to demonstrate the potential of sustained yield through proper management. The positions of forest guards or forest custodian (hyloroi) are recalled by Aristotle (4th Century BC) signifying the need for protection and, therefore, external regulation. It seemed that priests protected the sacred groves and forest guards protected the secular forests for state use.

It is a different matter that the sustainability of the former institution seemed much higher than the latter. In several of the contemporary sacred groves, the norms of usage have evolved according to the local needs. For instance, the grazing is allowed but no green trees are allowed to be cut. The dry wood, however, can be collected for fuel purposes with the permission of the village and, of course, the god. The ownership is with the revenue department but people have informal control. In another case, the priests of the temple in the sacred grove claimed to have the right to take legal action against anyone who cut down the trees. The forest department and the temple trust have planted about 8000 plants of *acacia* and *eucalyptus* in the sacred grove. In the third case, the sacred grove was supposed to have originated more than 2000 years ago and no tree was allowed to be cut. Only those trees which fell down due to storms or some disease could be used. The grazing was practiced round the year. The farmers had planted only bamboo in the sacred grove and rest of the species were naturally grown.

US President Clinton's Executive Order on Indian Sacred Sites (May 27, 1996) states provides new identity to the rights of indigenous people in USA on sacred sites:

Section 1. Accommodation of Sacred Sites.

(a) In managing Federal lands, each executive branch agency with statutory or administrative responsibility for the management of Federal lands shall, to the extent practicable, permitted by law, and not clearly inconsistent with essential agency functions, (1) accommodate access to and ceremonial use of Indian sacred sites by Indian religious practitioners and (2) avoid adversely affecting the physical integrity of such sacred sites. Where appropriate, agencies shall maintain the confidentiality of sacred sites.

(b) "Sacred site" means any specific, discrete, narrowly delineated location on Federal land that is identified by an Indian tribe, or Indian individual determined to be an appropriately authoritative representative of an Indian religion, as sacred by virtue of its established religious significance to, or ceremonial use by, an

Indian religion; provided that the tribe or appropriately authoritative representative of an Indian religion has informed the agency of the existence of such a site.

Section 2. Procedures.

(a) Each executive branch agency with statutory or administrative responsibility for the management of Federal lands shall, as appropriate, promptly implement procedures for the purposes of carrying out the provisions of section 1 of this order, including, where practicable and appropriate, procedures to ensure reasonable notice is provided of proposed actions or land management policies that may restrict future access to or ceremonial use of, or adversely affect the physical integrity of, sacred sites.

(b) In all actions pursuant to this section, agencies shall comply with the Executive memorandum of April 29, 1994, "Government-to-Government Relations with Native American Tribal Governments".

The executive order does not fulfill all the aspirations ot native American communities but at least is a step in the right direction.

Sacred Species

There is considerable discussion on higher animals which have been treated as sacred by various religious and cultural communities. But the role of insects has been rather neglected. Since sacred groves, lakes and marshy patches provide a very rich sanctuary or different kinds of insects, it may help us to look at the roots of sacredness differently.

Hogue (1987) explains the evolution of cultural entomology as a discipline and observes, 'Human spend their intellectual energies in three basic areas of activity: surviving, using practical learning (the application of technology); seeking pure knowledge through inductive mental processes (science); and pursuing enlightenment to taste a pleasure by aesthetic exercises that may be referred to as the "humanities". Entomology has long been concerned with survival (economic or applied entomology) and scientific study (academic entomology), but the branch of investigation that addresses the influence of insects (and other terrestrial Arthropoda, including arachnids, myriapods, etc.) in literature, language, music, the arts, interpretive history, religion, and recreation has only recently been recognized as a distinct field, This is referred to as "cultural entomology".'

He adds, 'Insect products have also helped to determine the direction of civilization's march. It could be said that the Chinese Empire was largely founded on the silk trade. Commerce in dyestuffs derived from the bodies of the cochineal insect reached global proportions by the 18th century, and proved so lucrative that the insect and its cactus host were introduced to various parts of the world from their native America. In the adopted countries the plant spread and became a noxious weed that rendered vast tracts of land unusable. Trade in other insect products such as honey and shellac has had similar economic significance. The Israelite band that founded the Jewish nation survived on "manna" during its extended trek through the Sinai Desert. This nutritious substance is thought to have been extruded by scale insects on the tamarisk plant.

Several important personages were aided in difficult times and inspired to lofty deeds by insects and spiders. The Chinese inventor of paper, Ts'ai Lun (89-106 AD), according to legend, was shown the process by wasps making their nests by chewing tree bark and mixing it with their saliva.

Insects have generally influenced human history, principally by forcing shifts in pivotal events. Battles have been lost, expeditions foiled, and population decimated through the direct involvement of insects, usually as carriers of disease.

Many creation myths involve insects: The Hops explained the origin of the world by the actions of the Spider Grandmother; According to the Yagua Indians of Peru the Amazon River was created by the wood-eating insects; and fire came from a mythical campfire ignited by fireflies, according to the Jicarilla Apaches of New Mexico.

A basic topology of public attitudes towards invertebrates and presents various potential reasons for prevalent societal anxiety, aversion, and antipathy.

Aesthetics

Primary interest in the physical attractiveness and symbolic appeal of invertebrates.

Humanistic

Primary orientation one of strong emotional affection for invertebrate animals.

Moralistic

Primary concern for the right and wrong treatment of invertebrates, with strong ethical opposition to presumed cruelty towards invertebrate animals.

Naturalistic

Primary interest in direct outdoor recreational contact and enjoyment of invertebrates.

Dominionistic

Primary interest in the mastery and control of invertebrates.

Ecologistic

Primary concern for interrelationships among invertebrates and other species, as well as between invertebrates and natural habitats.

Negativistic

Primary orientation a fear, dislike or indifference towards invertebrates.

Utilitarian

Primary interest in the practical value of invertebrates or the subordination of invertebrates for the material benefit of humans.

Scientific

Primary interest in the physical attributes, taxonomic classification and biological functioning of invertebrates.

All these attitudes have to be assimilated while developing a new strategy for conserving sacred groves and also mobilizing civil society. One will have to do a kind of market segmentation to as to develop strategies which will suit people with different kinds of attitude mix.

PART THREE: INSTITUTIONAL AND MANAGEMENT ISSUES IN CONSERVING SACRED GROVES

1) Blending Secular with Sacred

Recognizing moral boundaries of restraint in using natural resources is an imperative in the current context of increasing alienation of modern mind from some of the traditional cultural consciousness. The paradox is that at the same time when moral order is declining, the fundamentalist forces are becoming a stronger all around the world. Apparently, the ideology of sacred relationship with nature is not able to fill the void created by collapse of other ideologies.

2) Fusing Reductionism with Holistic Thinking

Repairing an ecological system disturbed by various activities ideally may require changing all the parts at the same time. This is a task very difficult to achieve by the NGOs as well as other civil society institutions, public bureaucracies, etc., because of their own fragmented capacity and understanding. In any case, resources being limited, one couldn't perhaps repair all faults at the same time. It is in this context that a prudent fusion of reductionist approach with holistic perspective may be required. The technological intervention may be reductionist but the context for prioritizing, monitoring and evaluating the impact may be as holistic as possible. The sacred groves would need external inputs in many cases. The sequence, source, and style of these inputs will have to be determined keeping the spirit of conservation into mind. The example of US executive order illustrates why intervention at the highest level become necessary when conflicts among local communities and state bureaucracies around sacred sites become a rule rather than an exception.

3) Augmenting Sacred Groves by Moulding Markets

The pressure for managing livelihoods in marginal environments has been increasing. In the regions of high biodiversity, people are often found to be poor. Obviously, we cannot conserve diversity by keeping people poor. One of the opportunities that may arise for generating incentives for conserving diversity as well as knowledge around it is through value addition in local knowledge of medicinal uses and germ plasm characteristics of diversity. In many cases, the boundaries of the people having rights in a sacred grove may be much larger than the boundaries of the local village in which the gove is located. The assignment of entitlements will have to contend with the fuzzy boundaries.

4) Process Theory of Institutions: Participative Evolution of Institutional Arrangements for Enforcing Sanctions

Does it matter whether a rule for managing common properties is evolved through different processes? Can a dish have the same flavour, no matter who cooks it and how? This aspect gets neglected compared to the structure base theory in our anxiety to develop generalized frameworks for analyzing institutions. Traditionally, different cultures did stress the importance of various pathways to reach the same goal. However, despite increasing appreciation for cultural pluralism in design and implementation of institutions, the literature is not a eloquent on the process of rule making as on rules per se. A database of seventy one indigenous institutions on common property resources has been developed at SRISTI in collaboration with Indian Institute of Management (IIM) and other members of Honey Bee Network. Drawing upon this database, we can show how institutional pathways for sustainability could vary significantly depending upon the process through which different rules are evolved. The need for building upon local ecological knowledge in the study of institutions has been emphasized by several researchers (Gupta, 1992, 1995). Similarly the importance of blending sacred and profane in the design of institutions has been noted though not to the same extent. But what has been left under-explored is the process by which rules for boundary and resource allocation evolve and accordingly the rules for conflict resolution evolve. We contend that this neglect partly also explains the lack of patience among practitioners as well as theory builders with more participative discourse in which local communities and individual assert not just rights of physical participation but also intellectual participation-an issue persistently raised by Honey Bee network of grassroots innovations for last eight years. How else one explains the irony that people are considered good enough for building CPR institutions which work for centuries but not good enough for critiquing theories about their institutions. They could not do so unless discourse taken place in vernacular language. The connection between language and culture is similar to the one between words and grammar and technologies and institutions. The technologies have been considered like words and the institutions like grammar (Gupta, 1992).

A: Transaction Costs Framework

Three propositions can be made about the way TC theory applies to evolution of rule making process, (a) Some of the *ex ante* transaction costs may be inversely related to *ex ante* transaction costs (Gupta and Prakash, 1992). For instance, if time and resource spent on negotiation are higher, the costs of enforcement may come down because most of the members of the community may have internalized the logic of why rules have been framed in the first place; (b) if arena of negotiation is more open and inclusive, the cost of monitoring may come down. Many times due to penchant for multi tier institutional structures, the arena of negotiation is severely restricted. The result is that lots of people involved in compliance of rules do not know the whole process that went into evolving the rules. Once the arena is wider and wider and more inclusive, the logic through which rules were evolved is known to larger number of people and thus monitoring costs may go down; (c) if negotiations have been particularly segmented with special emphasis on incorporating naughty boys or girls, the cost of redrawing the contract or rules may come down. The old saying of making the naughty boy (the girl) the monitor may have a ring of truth around it. Not every one has the same propensity to violate the rules and create chaos. If the potential violators are allowed responsibility during the process of negotiation, as and when new rules have to be drawn, the process involves less road blocks because the blockers may become the facilitators.

B: Socio-Ecological Paradigm

The eco-institutional framework (Gupta, 1992, 1995,1996) suggests that interface between access, assurance, ability and attitudes of members of an institution with ecological endowments, technology, institutions and culture provide a way of looking at the design possibilities more comprehensively. However, the limitation was that one could not incorporate the principle of 'sequential synergisms' (Gupta, 1989). That is which factor and rule is incorporated in what sequence could make substantial difference to the outcome of an institutional dilemma. Those who practice cooking would vouschafe that same spices and ingredients added in different sequences may generate different flavours. Why then have we neglected this issue so much in the discussion of rule making process?

C: Playful Portfolios of Multi-level, multi-market and multi-actor choices

The households in particularly marginal environments derive their survival need from various resource markets over space and time is well known. What is less well recognized that one can analyze the validity or otherwise of any choice of various actors in one resource market, or at one level, or in one institutional regime (private, markets or public) without reference to similar choices in other resource will seem highly complex to an analyst. But in day to day life, such an analysis kinship and ethnic relationships with all kind of reciprocities and obligations, there is not another way of looking at the world except through the prism of portfolio of interests, involvement and incentives. Playful aspect of this portfolio of from an insight (Gupta, 1990) that people deal with stress or conflict not just by manifesting, enduring, managing and transforming it but also by laughing about it. Much of the earlier work of the author (Gupta,1989) suffered from this inadequacy i.e., it was full of pathos while describing and thus analyzing the survival options of poor people in high risk environments such as dry regions. but then the light dawned and it was realized that creativity of knowledge rich and economically poor people lies in not just struggling with stress but also laughing about it, singing about it and some times making it into riddles.

The process of rule making thus we submit is consistent with what I call Godel's theorem of institutions. Godel was a famous mathematician who demonstrated that at least one assumption had to be made outside the framework of any set of propositions to be proved. One could not formulate a framework and prove all the assumptions by making reference to that framework only. The Godel theorem of institutions could thus be formulated by suggesting that rule making process and structure of rules make sense only in the living world and context of chaos (natural, spontaneous, or deliberately designed and 'institutionalized') and playful manner of oscillation between order and chaos across fuzzy boundaries.

5. Rediscovering Roots of Sacredness

Inventing new institutions for restoring the place of sacredness in our day-to-day life is important. Far too many concrete monuments have been created in the memory of those who sometimes cared and sometimes not for the non-human sentient beings. Time has come when a large global initiative needs to be taken for encouraging civil society initiatives for creating sacred sites, groves and lakes all around us. The motivations need not be based on fears or heavenly retribution for our carnal mistakes. Instead the respect for perfect strangers i.e., unknown and unknowable may be underlined together with the cultural investment of memory space in physical landscape.

We can resurrect the spirit of sustainability which requires faith in future as well as in the past heritage that has made it possible for us to experience the serenity only visible in sacred groves. The challenge is to combine utilitarian logic with spiritual concern for conservation. A new ideology to fill an old gap is what we are talking about.

REFERENCES

Bodine. J.J., D.J. Hughes, J.D., Jim, M. Keegan, J.M., Lubik . G., Martin, E. and Milne, C. 1973. Sacred Lands Graves Web Sites. In: *Native American and the Environment Web Site,* pp. 1-2.

Chapman, C. 1995. *Sacred Groves of Britain.*

Elkin, A.P. 1964. *The Australian Cycle of Life :Aborigin.* New York: Doubleday and Co, New York

Gadgil, M., and V.D. Vartak 1976. Groves of India—A plea for continued. *Journal of Bombay Natural History Society,* 72: 314-320.

Gadgil, M., and V.D. Vartak 1981. Groves of Maharashtra: An Inventory. In: S.K. Jain (Ed.) *Glimpses of Indian Ethnobotany.* Oxford University Press2, Bombay, pp. 279-294.

Gerden, C.A. and S. Mtallo 1990. *Traditional Forest Reserves in Babati District, Tanzania.* Swedish University of Agricultural Sciences International Rural Development Centre, Uppasala.

Godbole, A. 1996. Role of Tribals in Preservation of Sacred Forests. *Ethnobiology in Human Welfare.* pp. 345-348.

Gupta, A.K. 1989. *Managing Ecological Diversity, Simultaneity, Complexity and change: An Ecological Perspective.* W.P.No. 825. IIM Ahmedabad. P115, 1989, Third Survey on Public Administration, Indian Council of Social Science Research, New Delhi.

Gupta, A.K. 1990. Eco-Sociology of Household Risk Adjustment and Commons: Performance in an Uncertain World. Presented at *International Conference on Designing Sustainability on the Commons, International Associations for the study of the Common Property. Duke University, USA September 26-30, 1990.*

Gupta, A.K. 1991. *Household Survival Through Commons: Performance in an Uncertain World.* IIMA Working Paper No. 940. Indian Institute of Management in Administration, Ahmedabad.

Gupta, A.K. 1992. Nature within and without: Understanding environment by being part of it. Presented at University of Aarhus, Denmark on November 11, 1992.

Gupta, A.K. 1995. Sustainable Institutions for Natural Resource Management: How do we participate in people's plans? In:. Syed Abdus Samad, Tatsuya Watanabe and Seung-Jin Kim (Eds.) *People's Initiatives for Sustainable Development: lessons of Experience.* APDC, pp. 341-373.

Gupta, A.K. 1996. Social and Ethical Dimensions of Ecological Economics. In: Robert Constanza, Oleman Segura and Juan Martinez-Alier (Eds.), *Getting Down To Earth: Practical Applications of Ecological Economics,* Island Press, Washington DC, pp. 91-116.

Gupta A.K. and Prakash, A. 1992. *Choosing the Right Mix: Market, State and Institutions for Environmentally Sustainable Industrial Growth,* IIMA W.P.No. 1066, Indian Institute of Management in Adm

34

Sacred Groves and Biological Diversity: Providing New Dimensions to Conservation Issues

S. Deshmukh, M.G. Gogate** and A.K. Gupta****

*Principal Scientist, Bombay Natural History Society, Mumbai 400023, India

**Chief Conservator of Forests (WL), Maharashtra State Forest Department, Nagpur, India

**Professor, Indian Institute of Management, Ahmedabad 380 015, India

INTRODUCTION

Since the ancient days, setting aside pockets of forest lands has been the practice for centuries in India. To protect these biological resources more meaningfully, a religious tag was attached which served over the years as a key factor in genetic conservation through the mechanism of sacred groves. Sacred groves thus represent a tradition of conservation by the people much prior to the modern concepts of "Wildlife reserves".

Studies on sacred groves of Maharashtra were pioneered as early as in 1970s (Gadgil and Vartak, 1976). While studying the sacred groves of Maharashtra (Deshmukh and Gogate, 1997), it was observed that traditionally, sacred groves were established with a view to preserve, share and save water resources from the region where they existed. The concept of sacred groves grew over time when some of the important ecological and economic species of plants or of animals were conserved (or protected) in a grove. Sacred grove was one way of expressing the gratitude of man towards the vegetation which sustained and supported life under respective agro-ecological condition. However, studies on socio-economic aspects as well a complete inventory of sacred groves representing different agro-ecological regions of Maharashtra are still lacking.

This paper poses several questions to address issues relating to conservation and development of sacred groves and also provides criteria for genetic conservation based on primary information on each of the sacred groves that are exiting today. We hope that the study will eventually lead to defining the critical role of scientific community as well as Forest Department in the legal context which will not only help it organise peoples' movement in recognising the importance of sacred groves, but also achieve their conservation and development for posterity.

WARNING SIGNALS

A detailed summary of available knowledge on the number and size of sacred groves in Maharashtra is given in Table 1. Before one works out the future strategies for conservation and sustainable utilisation of sacred groves, it would be useful to look at the problems which were encountered during the exploration of various forest areas in Maharashtra for inventorying sacred groves. Some of the warning signals are:

1. strained status to the groves due to depletion of old, indigenous flora as well as land-use alienation;
2. depletion in the natural resources, such as drying of streams;
3. calamities such as intensifying droughts as well as cyclones and floods;
4. loss of grazing lands vis a vis decrease in the common property resources;
5. increasing biotic and abiotic stresses on areas surrounding sacred groves thereby affecting agriculture;
6. heterogeneity in size of sacred groves and absence of systematic, scientific efforts to establish values to a reasonable extent;
7. alteration in the land use patterns in villages as well as in the urban areas due to explosive growth of rural and urban population as well as unemployment.

Nature protection can not be done without people's cooperation and understanding. Therefore it would be desirable to derive simple and reliable methodology to determine relevance of a sacred grove to the micro-level biodiversity conservation and thus to livelihood and well-being of the persons concerned. Following paragraphs highlight some of the unanswered questions, the answers of which could lead to defining strategies for conservation and utilisation of sacred groves.

UNDERSTAND SACRED GROVES: SOME UNANSWERED QUESTIONS

Ecological History

a. *Age of Sacred Groves*

If the sacred grove is one generation old (i.e.,contemporary) then the question is what events took place that the current sacred grove was evolved? What are the reasons for identification of this particular area for establishment of a sacred grove? Which trees were planted first?

One of the interesting observations in terms of religious practices of communities is in the form of establishment and/utilisation of burial grounds. These practices have also lead to protection of a few forest patches.

b. *Changing Proportion of Species*

What is the changing proportion of species over time? Whether we can highlight this aspect, and if yes, then we will have to look at the issues relating to the institutional aspects. Existence of the sacred grove could be a long term impact of some events which have helped sacred grove maintain its identity so far, such as (a) catastrophic events; (b) epidemic; (c) deforestation due to crisis, such as drought, etc.; (d) unusual events such as visits of pilgrims of particular religion, to the area for some time; and (e) propaganda of particular God/religion among the local people.

c. *Occurrence of Uncommon Species*

What are the attributes for occurrence of one or two species which are not native of that religion?

d. *Decrease in Plant Cover*

What are the reasons (known and unknown) for decrease in the vegetation cover of the sacred groves? Formation of barriers such as creation of roads (which also means bunds) could lead to decrease in water table to the land under sacred groves. In other words, looking at recent/ancient infrastructural events could be one such reason for understanding the role of deforestation in sacred groves.

e. *Regeneration in Sacred Groves*

Regeneration in the sacred grove is one of the most important aspect to the understanding of extent of anthropogenic pressures. It is important to evolve a few more categories in addition to the natural regeneration, which facilitate regeneration process. They are:

i. Natural regeneration which itself is a very complex phenomena; and

ii. Artificial regeneration: this need not always be linked with efforts of individuals in physically planting seedlings in the sacred grove, but could be in terms of (a) preventing loss of species, (b) sustenance of the vegetation cover, (c) preventing grazing in the area, and (d) dispersal of seeds (by birds and other fauna).

f. Extraction of Wood from Sacred Groves

It will be interesting to note the major events/calamities-such as epidemic or drought-have forced people to utilise sacred groves for meeting their livelihood. Thus, it is clear that partial or total extraction of wood as well as non-forest produce by indigenous communities has led to decrease in the vegetation cover of sacred groves. Such attributes of documented in relation to time, could prove crucial in understanding the role of people in conservation as well as sustainable utilisation of sacred groves.

Deity in the Sacred Groves

During our studies in the State of Maharashtra, where more than 950 sacred groves were listed (are known to exist) as seen from Table 1, it was observed that the number of deity per sacred grove was found to be more than one in quite a few cases (Deshmukh, 1997). In such cases, it would be important to know details about (a) their history; (b) establishment/ commissioning of the Deity; and (c) reasons for introducing the second deity and/or may be the third deity would be necessary. Related questions to the above points could be:

i. Which specific flowers and fruits are liked by the Deity? Can the occurrence/abundance of these species be attributed to the occurrence of specific flowering as well as fruiting trees? Whether these trees were planted with specific intention?
ii. Traditional beliefs relating to various deities in the sacred groves, e.g. in some of the rural areas in Maharashtra, it is believed that use of iron equipment/tool is disliked by the deity in the sacred grove which signifies that no trees are cut from that region.

Indigenous Communities/Tribals

It will be important to know the time/period of settlement of the communities in the respective areas where sacred groves are currently located. Influence of the traditional as well as cultural practices of such communities has been observed to have its own bearing on the conservation ethics of biologically rich areas such as sacred groves.

CRITERIA FOR GENETIC CONSERVATION OF SACRED GROVES

In view of the above questions as well as data provided in Table 1, it is clear that the patches of biological wealth occurring in near-natural state are required to be preserved. It would therefore be useful if certain criteria for their conservation are defined. For example, out of 953 sacred groves that are known to exist in Maharashtra, about 455 are those which are less than one hectare in size. It may not be useful to preserve all of these as they may not possess required amount of intra-specific diversity for their prolonged sustenance. Instead a substantially large (in size) sacred grove with required amount of genetic diversity may qualify for protection for posterity.

Swaminathan et al. (1995) proposed ten broadly classified criteria for identification of suitable sites for conservation and sustainable utilisation of mangrove forest resource. They are:

1. *Genetic aspects:* Potential for *ex situ* preservation, visual polymorphism.
2. *Ecological aspects:* Species richness, nutritional level, environmental degradation.
3. *Utilisation level:* Human use and exploitation.
4. *Neighbouring flora and fauna:* Associated ecosystem.
5. *State of the forest:* Levels of degradation.
6. *Accessibility:* distance from main population centres, transportation facilities.
7. *Personnel:* Training, turn-over etc.

Table 1. Available knowledge on total number and area-wise categories of sacred groves in Maharashtra (Source: Deshmukh, 1997).

District	Area-wise (ha) categories							*Total
	1	2	3	4	5	6	7	
AHMEDNAGAR	2	3	-	-	-	-	-	5
BEED	4	-	-	-	-	-	-	4 + 10*
BHANDARA	3	-	-	-	-	-	-	3
CHANDRAPUR	1	1	-	-	-	-	-	2 + 2*
JALGAON	-	4	-	-	-	-	-	4
KOLHAPUR	62	52	7	2	-	-	-	123 + 6*
NASHIK	4	-	-	-	-	-	-	4
PUNE	98	42	10	6	-	1	1	158 + 21*
RAIGAD	1	1	3	9	3	-	1	18 + 3*
RATNAGIRI	113	60	11	5	2	-	-	191
SATARA	13	6	-	-	-	-	-	19 + 2*
SINDHUDURG	152	124	17	16	5	6	3	323 + 24*
THANE	1	13	3	3	5	1	-	26 + 1*
YEOTMAL	1	-	-	1	-	1	-	3 + 1*
Sub-Total	455	306	51	42	15	9	5	883 + 70*
						Grand Total : 953		

* Area-wise categories : 1 = 1-5 ha; 2 = > 1 upto 5 ha; 3 = >5 upto 10 ha; 4 = >10 up to 25 ha; 5 = >25 upto 50 ha; 6 = >50 up to 100 ha, and 7 = >100 ha.

8. *Anthropogenic factors:* Threats, level of dependence, etc.
9. *Socio-political factors:* National commitment, land tenure and public awareness.
10. *Land size:* Area available at the proposed site.

The above criteria with some modifications could be applied in case of sacred groves to achieve their genetic conservation.

AGENDA FOR FUTURE

It would be useful if the scientific community takes a coordinated approach to find solutions to the questions raised in this paper. The Scientists will have to adopt during explorations of sacred groves, strategies for appropriate education, strengthening of local communities, sustainable management and protection of the sacred groves without alienating local/indigenous population. This alone could help conservation of biological diversity for posterity besides satisfying aesthetic, cultural, recreational and utilisation needs of mankind. Participants of the National Workshop on "Conservation and Development of sacred groves", organised by the Regional Centre (Western Zone) of the National Afforestation and Eco-development Board of the Ministry of Environment and Forests, Government of India (Rajkot, November 18-20, 1997), realised this need and recommended that agenda for achieving the conservation and development of sacred groves could be:

a. identify locations of sacred groves and their association with communities in specific areas,
b. conduct a detailed inventory of existing flora and fauna in sacred groves and their relevance to the surrounding ecology; also list rare and endangered species in the groves,
c. quantify the social impact on the sacred groves which may then lead to anthropogenic impact assessment,
d. identify external threats, such as cyclones, forest fires (natural threats), and/or submergence (due to human activities) to the groves,
e. find means for conservation of sacred groves with the help of forest department, local communities, interested organisation (both government and non-government) as well as with the help of individuals,
f. develop a projectise approach for creation, protection, conservation, and development of sacred groves providing technical, financial and management inputs to strengthen the symbiotic relationship between the sacred groves and people, for their sustenance, and
g. preserve existing sacred groves but also try and create new (if possible) with preference on endemic as well as indigenous species

which could serve as genetic warehouses of natural forests, for posterity.

ACKNOWLEDGEMENTS

S. Deshmukh would like to express his gratitude towards the Forest Department, Govt. of Maharashtra for providing financial assistance to Bombay Natural History Society (BNHS), Mumbai to undertake this important project. He would also like to express his thanks to his research colleagues, particularly, Mr. Umesh Mundlye and Mr. Balakrishna Bhatkhande for providing some insights to this conceptual paper. Thanks are also due to the Director, BNHS, for this constant encouragement and support.

REFERENCES

Deshmukh, S. 1997. *Conservation and development of sacred groves in Maharashtra.* Quarterly Report on World Bank aided Maharashtra Forestry Project submitted to the Forest Department, government of Maharashtra. Bombay Natural History Society, Mumbai.

Deshmukh, S. and Gogate, M.G. 1997. Conservation and development of sacred groves in Maharashtra: a review. *Proceedings of the National Workshop on "Conservation and development of sacred groves" (November 18-20, 1997), Rajkot. RC-NAEB (Mumbai)* Ministry of Environment and Forests, Govt. of India (in press).

Gadgil, M. and Vartak, V.D. 1976. The sacred groves of Western Ghats in India. *Economic Botany,* 30 : 152-160.

Swaminathan, M.S., Deshmukh, S. and Balaji , V. (Eds.) 1995. *Establishment of an international network for the conservation and sustainable utilisation of mangrove forest genetic resources.* Mimeograph submitted to the International Tropical Timber Organisation, Yokohama, Japan. M.S. Swaminathan Research Foundation. Chennai.

35

Anthropological Dimensions of Sacred Groves in India: An Overview

K.C. Malhotra

Indian Statistical Institute, Calcutta, West Bengal, India

INTRODUCTION

In present day Indian society, there are several endogamous populations that continue to practice many forms of nature worship. One such significant tradition of nature worship is that of dedicating and according protection of patches of forests to some ancestral spirits/deities. These vegetation patches have been designated as "Sacred Groves". Although there is variation in the way these groves have been described/defined, most of the scholars emphasize two aspects, namely, the climax nature of the vegetation and the preservation of the vegetation on religious beliefs.

The institution of sacred groves has been studied by anthropologists from the socio-cultural and religious point of view, and the taxonomists and ecologists have studied the biodiversity and conservation aspects of the institution. There is now a vast body of information available on various aspects of the sacred groves.

In this paper, however, an attempt has been made to provide an overview of the various anthropological dimensions of the sacred groves. The materials described in this paper are based on (i) review of anthropological literature on the sacred groves, and (ii) field studies conducted by us in Orissa and West Bengal. We would like to emphasise that it is a Herculean task to review the numerous anthropological studies that are available on sacred groves. Within the constraints of time, we have done our best to include many as studies we could lay hand on, and hope we have not missed many important studies. Thus, what is presented here is not at all a comprehensive overview. The strategy followed in this overview is to highlight different anthropological aspects, that have been studied and to identify gaps in the knowledge and to suggest areas of future research in this field.

The anthropological dimensions of sacred groves (SGs) included in this overview are :

(i) The antiquity of the SGs
(ii) Spatial distribution of SGs in India
(iii) Number and size of the SGs
(iv) Legal status and management of SGs
(v) Ethnicity and SGs
(vi) Gender and SGs
(vii) Interface between people and SGs
–Sacred
–Socio-cultural, economic and political
(viii) Areas for future research.

THE ANTIQUITY OF SGs

Several scholars have noted that SGs are a very ancient and a widespread phenomenon in the old world cultures. According to Kosambi (1962) the institution of SGs in India is very ancient and dates back to pre-agrarian hunting stage, before humans had settled down to raise the livestock or till the land (also see Vartak and Gadgil, 1981; Patil, 1982). These inferences have been drawn on the basis of the nature of religious cults associated with the SGs.

If the above inference is correct then we expect that this institution should be present among the present day hunter-gatherer societies in the subcontinent. There are only a few such populations on the mainland of India-like Lodhas in West Bengal and Orissa, Birhor in Bihar, Phase Pardhis in Maharashtra (and groups having ethnic similarity with the Pardhis found in several south Indian states). The other nearly true hunter-gathers are Onge (till 1972), Jarawas and sentenelese in the Andamans.

Interestingly, Deb and Malhotra (1997) have reported SGs among the Lodhas of West Bengal. We have not found evidence of SGs among other groups. This, however, needs confirmation.

SPATIAL DISTRIBUTION OF SGs IN INDIA

The sacred groves are widely present in different parts of the country. They have been reported from Rajasthan, Western Ghats of the states of Maharashtra, Karnataka, Kerala, Tamilnadu, West Bengal, Bihar, Madhya Pradesh, Andhra Pradesh, Orissa, Meghalaya, Sikkim and other north-eastern states (see among others Brandis, 1897; Hughes and Chandran, in this volume).

The groves, as would be expected, are known by different names. For example, in central India they are known as jankor, sarana, in West Bengal jaher, in Rajasthan orans, in Kerala kavus, in Maharashtra Deorai, in Kodagu Devarakadus, in Meghalaya Lakyntang, in Karnataka kans etc.

However, sacred groves have not been reported from Jammu and Kashmir, Gujarat, Punjab, Haryana, UP, and Andamans and Nicobar islands. It is likely that studies have not been carried out in these states in respect of sacred groves, or may be groves are absent in these states. It is important to note that except for the states of Kerala, Meghalaya and some regions of Karnataka (Kodagu district), Maharashtra (Western Ghats) and Orissa (Koraput district), our knowledge regarding the presence or absence of sacred groves in other parts of the country is scanty and rather inadequate.

There is, therefore, an urgent need to systematically inventorise the sacred groves in various parts of the country.

NUMBER AND SIZE OF THE SGs

As noted above, although sacred groves have been reported from many parts of the country, however, we as yet do not have even an estimate of the number of such groves in India. In Table 1, available data in terms of numbers of groves in India is given. Somewhat detailed investories are only known for certain regions of the country. Gadgil and Vartak (1981) report 483 groves in Maharashtra. Actually they reported only 233 groves, but the total area of 3570 ha included area from Ratnagiri district also. Godbole et al. (in this volume) report the existence of over 250 groves in Ratnagari district. Thus adding these 250 to 233 makes the total to 483. Kalam (1996) reports 1214 groves in the Kodagu district of Karnataka. In Meghalaya 79 groves have been listed by Tiwari et al. (this volume). Rajendraprasad (1995) reports the existence of over 2000 well preserved sacred groves in Kerala. In one block Semiliguda of Koraput district in Orissa Malhotra et al. (1997) surveyed all the 220 villages and found in 169 villages 322 sacred groves. Pooling together these groves and other listed in Table 1, it appears that at least 4125 groves have been reported so far.

It may be pointed out here that this figure of 4125 is only an indication of the extent and magnitude of the presence of the institution of sacred groves in the country. It is certainly difficult to make a guess regarding the total number of sacred groves in the country, but the known presence and pattern of distribution of sacred groves in Madhya Pradesh, Bihar, Andhra

Table 1. Reported area of the sacred groves in India

Location	No. of Sacred Groves	Area (ha)	References
Maharashtra[1]	483	3570	Gadgil and Vartak, 1981
Kodagu	1,214	5947	Kalam, 1996
Meghalaya[2]	79	26326	Tiwari et al., this vol.
Kerala[3]	2,000	500	Rajendraprasad, 1995
Tamil Nadu	10	127	Swamy et. al., this volume
Rajasthan	1	83	Singh and Saxena, this volume
Rajasthan	8	158	Jha et al., this volume
UP	1	5500	Sinha and Maikhuri, this volume
Orissa	322	50	Malhotra et al., 1997
West Bengal	7	2	Malhotra et al., 1997
Total	4,125	39,063	

1. includes districfs of Thane , Kolaba, Jalgaon, Pune, Satara, Kolhapur, Yeotmal, Bhandara, Chandrapur and Ratnagiri; Gadghil and Vartak, however, estimate that the total area in Maharashtra under sacred groves is likely to be 5,000-10,000 ha. Godbole et al. (in this volume) report existence of over 250 groves in Ratnagiri district, these have been added to 233 reported by Gadgil and Vartak.
2. Khiewtam and Ramakrishnan (1989) reported 21 sacred groves in Cherrapunji and adjoining areas in Meghalaya. The total are under protection was estimated as about 5000 ha (50 Km^2). It is not clear whether in the 79 groves reported by Tiwari et al (this volume) includes these 21 groves.
3. Rajendraprasad (1995) estimated about 2000 well preserved sacred groves in Kerala. However the area is based on only 360 groves which obviously is an underestimate of the area under sacred groves in Kerala.

Pradesh and several north-eastern states for which inventories are not available, we strongly feel that the number of SGs in India are likely to be between 100,000 to 1,50,000.

The limited data that are available suggest the existence of inter-district (region) variation in the extent of presence of sacred groves. Thus, an examination data of Gadgil and Vartak (1981) and Godbole (in this volume) pertaining to Maharashtra reveal : (i) that the number of groves in the Western Ghat districts of Maharashtra (see Table 2) are much higher compared to Khandesh (Jalgoan district) and Vidharbha in eastern Maharashtra, (ii) even the 6 districts of Western Ghats shows considerable variation, the observed range is from a minimum of 15 in Kolaba to a maximum of 250 in Ratnagiri district; and (iii) there seems to be a gradient of increase in number of groves from north to south, thus while in Thane and Kolaba districts the number of groves are 21 and 15 respectively, further south in Pune and Kolhapur district, the number is 70 and 97 respectively and the southern most Ratnagiri district has over 250 groves. We certainly cannot take this inter-district variation at the face value, since we do not know whether the density of groves is a function of the size of geographical area and number of villages and their ethnic composition

Table 2.The size distribution of sacred groves in Maharashtra, inter-district variation.[1]

District	No. of Sacred groves	Size of sacred groves in ha				
		upto 2	2-5	5-10	10-15	> 15
Thane	21	2 (9.5)	7 (33.4)	3 (14.3)	1 (4.7)	8 (38.1)
Kolaba	15	1 (6.7)	0 (0.0)	3 (20.0)	3 (20.0)	8 (53.3)
Pune[2]	66	54 (81.8)	6 (9.1)	2 (3.0)	3 (4.6)	1 (1.5)
Satara	16	15 (93.7)	0 (0.0)	1 (6.3)	0 (0.0)	0 (0.0)
Kolhapur	97	80 (82.5)	11 (11.3)	4 (4.2)	1 (1.0)	1 (1.0)
Jalgaon	4	4 (100.0)	0 (0.0)	0 (0.0)	0 (0.0)	0 (0.0)
Vidharba[3]	7	6 (85.7)	0 (0.0)	0 (0.0)	0 (0.0)	1 (14.3)
Total	226	162 (71.7)	24 (10.6)	13 (5.7)	8 (3.6)	19 (8.4)

1. Data given by Gadgil and Vaitak (1981) in Appendix I has been reanalysed to examine inter-district variation in number and the size distribution of sacred groves in Maharashtra.
2. Out of 70 sacred groves, area for four was not available thus reducing the total to 66.
3. Sacred groves from Yeotmal (n = 3), Bhandara (n = 3) and Chandrapur (n = 4) have been pooled together; however, area for only 7 groves was available.

in different districts, or is a function of the intensity of the institution in itself. Only future well planned studies in the region will clarify the situation.

Literature review and our own studies in Orissa and West Bengal suggest that the size of the groves varies considerably from a cluster of a few trees to several hectares. Unfortunately, not many scholars have given sizes of individual sacred groves. However, pooled total area are available for some states/regions/localities. In Table 1 we have documented the available data in terms of the area under sacred groves in different parts of the country. The data reveal that there exists vast inter-state variation in the extent of area under sacred groves. It appears that the sacred groves in Western Ghats, Kodagu, Meghalaya, Tamilnadu and Rajasthan are relatively larger compared to groves in Orissa and West Bengal. Data is terms of area are available for 4,125 sacred groves. These groves cover an estimated area of 39,063 ha.

According to Gokhale et al. (in this volume) the total area of sacred groves in India as a whole would be about 33000 ha or 0.01% of the total area of India. There seems to be some error in this estimate as just 4125 sacred groves reported so far cover over 39000 ha. Although based on the rather incomplete available data it is not possible to come up with a reasonable estimate, however, it can safely be said that the area under sacred groves will be many-fold more than Gokhale et al. have estimated.

As noted above, there also exists inter-state/region variation in the number as well as size of the sacred groves. It is likely that there also exists intra state variation in size of the groves. The inventory prepared by Gadgil and Vartak (1981) provides excellent opportunity to examine this aspect. These authors documented sacred groves in 9 districts of Maharashtra. This data is reanlysed by the author to capture inter-district variation. The results are presented in Table 2. It is evident that the size of the groves depict considerable variation between the districts. Thane and Colaba districts have much larger groves compared to the adjacent southern districts Pune, Satara and Kolhapur. Over 80% of the groves in these later districts are up to only 2 ha, while in the former districts such groves are less than 10%. Likewise there are a significant number of groves measuring above 10 ha in Thane and Colaba (20 out of 36, i.e. 55.6%) in southern districts they are only 6 out of 179, i.e., 3.4%.

Further, there seems to be inter-regional variation in the size of the groves in Maharashtra; compared to groves on Western Ghats, the groves in eastern Maharashtra are much smaller in size.

A gradient of increasing number of sacred groves from north to south in the Western Ghats was observed. In terms of size of the groves, interestingly the reverse pattern is seen i.e., the number of groves with larger size show an increasing trend from south to north.

The nature of data available on the number and size of individual groves are rather scanty and often in many studies ethnicity of the localities are not described (Kalam, 1996; Godbole et al., this volume; Gadgil and Vartak, 1981; Pushpangadan et al., this volume). From the available data the following tentative observations can be made :

(i) The distribution of over 2000 sacred groves in Kerala are almost entirely found in districts having very little tribal population. Therefore, it can be inferred that sacred groves in Kerala seem to be stronly associated with different Hindu castes. However, from the data it is not clear with which castes the groves are associated;

(ii) In Kodagu (Kalam, 1996) also it seems that the groves are primarily associated with non-tribal Hindus, but the caste-affiliation can not be inferred;

(iii) On the Western Ghats region of Maharashtra the groves are associated with both tribals (Mahadeo Kolis) and non tribals (Kunbis and several other castes) (Vartak and Gadgil, 1981; Roy Burman, 1996);

(iv) In south West Bengal the groves are associated predominantly with the tribals;

(v) In Semiliguda Block of Koraput district of Orissa the groves are predominantly associated with the tribals;

(vi) In Meghalaya the groves are associated with the Khasi tribals;

(vii) In Rajasthan the groves have been reported among the Bishnois and Rajputs;

(viii) The only grove reported from Garhwal (Sinha and Maikhuri, in this volume) is associated with Hindus; and

(ix) The sacred groves reported from Tamilnadu are associated with non-tribals.

LEGAL STATUS AND MANAGEMENT OF SACRED GROVES

Literature on this aspect though sparse and scanty suggest the existence of vast variation in the legal status and management of the sacred groves in the country. An understanding of these aspects i.e., who owns a particular grove is a prerequisite and crucial in not only intrepreting the present cultural and biological status of the groves but also in initiating long-term conservation strategies of these sacred sites.

It appears that in terms of the legal tenurial rights, the sacred groves fall under three categories :

(i) under the control of state forest departments;

(ii) under the control of revenue and other government departments; and

(iii) privately owned.

A large number of sacred groves in Maharashtra are under the control of Forest Department. Gadgil and Vartak (1981) have documented 233 such groves. Legally all sacred groves in Meghalaya are under the control of District Councils (Tiwari et al., in this volume).

Kalam (1996) reports that deverakadus in Kodagu district of Karnataka are under the control of revenue department. Godbole et al. (in this volume) and Roy Burman (1996) mention that many sacred groves in western Maharashtra are under the control of revenue department. Roy Burman further mentions that a few thousand temples and their groves in western Maharashtra were brought under the scrutiny of the government by forming the Paschim Maharashtra Deosthan Prabodhan Samiti in 1960s.

There are also several sacred groves owned privately. There are examples of groves being owned by a family, a group of families, a clan (s), a village, a cluster of villages etc. Chandrashekara and Sankar (in this volume) give examples of groves in Kerala like Ollur Kavu owned by a single family, like S.N. Puram kavu owned and managed by several families and a kavu like Iringole managed by the Temple Trust.

In terms of management of the sacred groves i.e., upkeep, protection, performance of rituals and festivals and harvesting of biomass, etc., also there exists vast variations. To cite a few examples: Orans in Rajasthan are usually managed by the Gram Panchayat (Jha et al., in this volume); the Haryali grove in Garhwal is managed by a temple committee made up from members of three villages (Sinha and Maikhuri, in this volume); Roy Burman (1996) mentions that among the Mahadeo Kolis of Pune district, the management of the sacred groves is usually vested with the clan elders, whereas among the Kunbis of Kolhapur district the groves are managed by village elders; the Kantabanshini Thakuramma sacred grove in Koraput district is managed by two clans of Proja tribe (Hemam et al., 1997); clan based management appears to be a wide spread practice among the Santhals, Oroan, Munda, Kharia and other tribes of Central and eastern and north-east India.

ETHNICITY AND SACRED GROVES

The processes of peopling of India stretching over indeed a long period brought in not only different human biological traits, but also a variety of cultural, religious and technological characteristics. Contemporary India is an agglomeration of over 40,000 endogamous groups (Malhotra, 1984). An estimated 37,000 groups are structured in the Hindu caste system. Each group (caste) belongs to one of five varnas: Brahmin, Kshatriya, Vaishya, Sudra and Puncham. The remaining 3000 groups constitute different tribes, religious communities and other historical migrant populations like Parsis and Siddis. In other words, there is a bewildering heterogeneity in the Indian society in terms of religious beliefs, culture, language and pattern of livelihoods.

In this section an attempt is made to examine if there exists any pattern in terms of association of the institution of sacred groves ethnicity.

A few tentative inferences in terms of association of sacred groves with different ethnic groups that can be drawn from the materials described earlier are: (i) that sacred groves are found both among tribals and non-tribals; (ii) there is regional association in terms of ethnic association; (iii) the association with castes of different varnas is not clear; (iv) in states where we have both tribals and non-tribals like Bihar, MP, Orissa, West Bengal etc., the presence or absence of groves in non-tribal areas is not clear, i.e., whether in districts with no tribals or districts away from the tribal areas in these states depict presence of sacred groves comparable with districts with higher tribal population; and (v) whether the institution is found in regions/villages inhabited by Christians, Muslims, and other religious communities.

GENDER AND SGs

The role of gender in sacred groves can be analysed at atleast four levels. Firstly, the gender of the diety associated with the sacred groves, secondly, the gender of priesthood serving the groves, thirdly, the nature and extent of access to men and women in various rituals, festivals and ceremonies that take place in the groves, and harvest of biomass from the groves, and forthly who has say in the management of the sacred groves (this is very important in terms of any possible management interventions that may be initiated in future).

A random literature search reveals that by and large a majority of the sacred groves are associated with female deities. Gadgil and Vartak (1976) among 21 sacred groves in Maharashtra found 15 associated with godesses and 5 with male deity (phallic worship) and one with ancestor worship. Chandran (1995) reports occurrence of both male (Bhutappa, Jatakappa) and female (Choudamma) deities associated with sacred groves in Karnataka. In south-west Bengal and in Koraput district of Orissa, the deities are mostly female (Hemam et. al., 1997; Malhotra et al., 1997).

Regarding the gender of the priest, it appears that without an exception the priesthood rests with males (see among others; Vartak and Gadgil, 1981; Roy Burman, 1996; Godbole et al., in this volume; Sinha, in this volume). However, this aspect needs to be further studied, as many studies do not provide explicit details on the gender of priests. Of great interest will be the situation among the matrilineal populations such as Nairs in Kerala, Khasis and Garos in north-east.

The data in terms of access to sacred groves by women are also very scanty. It appears, as could be inferred from some studies, that women are not permitted in entering into the groves after attaining puberty. Roy (1912) while describing sacred groves, known as sarana or Jaher, among the Oroan of Chottanagpur mentioned that the main festival associated with sarna is sahrul. However, women are not allowed in the sarana, but take part in dance at the akhara which is located close to the grove (also see Fernandes (1993) for more details). Roy Burman (1996) has reported similar phenomenon among Mahadeo Kolis of Pune district and among Kunbis of Kolhapur district in Maharashtra. Malhotra et al., 1997) observed taboo against entry of women in sacred groves among the tribes of south-west Bengal and among the tribes of Koraput district in Orissa.

Although many studies have mentioned about harvesting of certain type of biomass from the sacred groves, it is not clear whether women are allowed to gather the same (see among others Unnikrishnan, 1990; Mitra and Pal, 1994; Viji, 1995).

Lastly, nothing is known at all, the kind of role women play in decision making regarding management of the sacred groves. It will be of immense value to examine whether women are represented in numerous trusts that are managing sacred groves, in particular in Maharashtra, Karnataka and Kerala. The limited information that is available from our own field observations in West Bengal and Orissa suggest practically no role of women in the management of the sacred groves.

INTERFACE BETWEEN PEOPLE AND SACRED GROVES

In this section we shall try to examine the role of sacred groves in the lives of the people from two aspects: (i) sacred (religious); (ii) socio-cultural, economic and political.

(i) Sacred (Religious)

There is a category of sacred groves among many communities that are associated with a certain deity(ies). In such groves annual rituals and ceremonies are performed to propitiate the deity. During these rituals sacrifices of animals (such as fowl, goat, pig, buffalo) are made. In other 'sanskritized' groves offerings of vegetarian items are given. These rituals are performed for the well-being of the inhabitants, animals, crops, etc. Anthropological literature in full of such detailed accounts (see among others Roy, 1912; Gurdon, 1914; Biswas, 1956; Nath, 1960; Kosambi, 1962; Vidyarthi and Rai, 1977; Patil, 1982; Hembram, 1983; Danda and Ekka 1984; Sinha, 1989; Hemam et al., 1997).

The presiding deities not only look after the well-being of the people, but also protect the groves by punishing (mostly death) the offenders. The practice of oath/vow taking in the sacred groves is fairly wide spread in the country (among others see; Roy, 1912; Sisodia and Malhotra, 1963; Kalam,1996). People take vow for wish fulfillment when there is a crisis, particularly bearing on health, and offerings mostly of terracota of animals, birds, humans etc are made. We observed in some of the sacred groves of WEST BENGAL HEAPS of such terra-cotta offerings of elephants and horses (Malhotra and Ketaki, unpublished).

There seems to be a hierarchy of sacred groves in terms of their geographical influence. At least five such hierarchical levels are discernible (Table 3) : village/hamlet level where only inhabitants of the village/ hamlet or even different ethnic groups/different castes in multiethnic situations have their own groves. Roy Burman (1996) reports existence of such groves for different castes in villages of Kolhapur and Pune districts. Malhotra et al., (1997) observed in Kendua village of Jamboni taluk of

Midnapore district where tribals Kora and Santhal have their separate sacred groves. Such a pattern seems fairly wide spread among the tribes of central India, Bihar and Orissa. Such groves are mostly managed by the local community(ies)-a family, group of families, clan, or entire village.

Table 3. Hierarchical levels of sacred groves in India, a tentative frame work.

Hierarchical levels of sacred groves	Management of sacred groves
V PAN INDIA	by Trust
IV REGIONAL	by Trust
III LOCAL	by whole village / community / local committees
II VILLAGE	by whole village
I INTRA-VILLAGE	by separate communities

The local level is in which people from somewhat larger geographical area usually couple of districts come to worship a particular sacred grove. Example of such groves are Iringole in Kerala and Kantabanshini Thakumaa in Orissa (Hemam et al., 1997). Such groves are usually managed by local community and/or some Committees.

The regional level sacred groves are in which people from several districts/states participate. Such an example is of the Sabarimala sacred grove in Kerala. Such groves are usually managed by temple trusts.

The next level of sacred groves are of Pan Indian character, in which people from many parts of the country participate. The example of such a grove could be Hariyali sacred grove in Garhwal Himalayas (Sinha and Maikhuri, in this volume). Such groves tend to be larger and managed by temple trusts.

Another category of sacred groves are those that are considered abodes of ancestral spirits. Often these groves are in fact burial grounds. Such groves have been reported from a number of places. A few illustrative examples are: Vidyarthi (1963) reported such groves called Masani among the Maler of Bihar; Godbole et al., (in this volume) report several such groves (e.g., in the villages of Muradpur, Hativ, etc.) in Sangameshwar tehsil of Ratnagiri district in Maharashtra; Unnikrishnan (1990) mentions that theyyam ritual in north Kerala is ancestor worship; Fernandes (1993) mentions about Sasan sacred groves as burial grounds among Chottanagpur tribes; Nath (1960) reports such groves among the Bhils of Ratanmal. It may be mentioned that sometimes a grove may serve both the functions, i.e., deity worship and ancestor worship. Unlike the groves associated with deity (ies), the groves associated with ancestor worship, in particular burial grounds, do not seem to have a hierarchical pattern.

ii) Socio-cultural, Economic and Political

Several scholars have reported that besides the sacred functions associated with the sacred groves, in these places several festivals are also performed. The literature on this subject in indeed very vast and detailed accounts of socio-cultural functions performed are described in several earlier ethnographic studies. Just to mentioned a few as illustration: Nath (1960) mentions that once a year in Deepawali, offerings of food and liquor are made in groves among Bhils of Ratanmal; Deb and Malhotra (1997) among the tribals of south west Bengal report that gatherings take place in these groves on the occasion of Salui and Karam festivals, as well as wedding ceremonies; Vidyarthi and Rai (1977) report that different tribes of Bihar celebrate their major festivals at the sacred groves; sacred groves among the Mahadeo Kolis of Pune district play a significant role in their social life, such as marriage ceremonies; Fernandes has stressed the role of sacred groves in the socialization of youth among the tribes of Chottanagpur; Godbole et al. (this volume) report that important village festivals like Holi, Navratri, Devdiwali are performed in village sacred groves in Ratnagiri district of Maharashtra; Kalam (1996) mentions offering of miniature cows and buffaloes to deity when cattle take ill in Iyappa Devarakadu in Kodagu; Troisi (1978) mentions that the association of a village with Jaherthan expresses the unity of the group. No two village share the same Jaher, and this serves as important criterion to ascertain village membership and geographical boundary; Paranjpye (1989) highlighted that the function of the sacred grove is to maintain a hierarchical order within the community; Roy Burman (1996) observed in western Maharashtra that while among the Mahadeo Koli tribes the sacred groves play the role of maintaining the clan control over the resources, among the Kunbi caste they assist in maintaining social hierarchy based on caste and social status.

There is a general belief that biomass resources from the sacred groves are not harvested. This is certainly true in many sacred groves found across the country. Gadgil and Vartak (1976), Roy Burman (1995) and Godbole et al. (in this volume) report several such groves in Western Ghats of Maharashtra. Malhotra et al. (1997) report such groves in south-west Bengal and in Koraput district in Orissa. Pushpangadan et al. (in this volume) report existence of numerous groves in Kerala from which plants and animals are not harvested while Swamy et al. (in this volume) reports such groves in Tamilnadu.

However, there are many groves in the country from where certain biomass is extracted, and thus the local communities derived certain direct economic benefits from the groves. A few illustrative examples are given here: Singh and Saxena (in this volume), and Jha et. al. (in this volume) report that in many orans people graze their animals; Godbole et al. (in this volume), report collection of dead wood and dried leaf litter and harvesting of certain species of trees (*Caryota urens* and *Mangifera indica*) from groves in Ratnagiri district of Maharashtra; Malhotra et al. (1997) report several groves from Koraput district from which dead wood and several NTFP are gathered; Unnikrishnan (1990) in the context of certain sacred groves in Kerala observed that certain plant species provide livelihood to many artisans; Gadgil and Vartak (1975) report that villagers of Tunbad in Colaba districts use bark of *Entada phaseoloides* Merr. for the treatment of cattle of against snakebite; from many groves dedicated to ancestor worship wood is extracted for cremation (Mitra and Pal, 1994).

Roy Burman in a series of articles (1992, 1994, 1995, 1996) has demonstrated the political dimensions of sacred groves in the local and regional context. This section draws heavily on the material synthesised by Roy Burman.

Sontheimer (1989) showed linkages of forest deities of western Ghats with the pastoral nomads as means of drawing their territorial affinity. Kosambi (1962) observed that sacred groves are usually found along the pre-agrarian trade routes and cross roads. Sawant (1990) wrote that the sacred groves at Phondaghat in Sindhdurg district of Maharashtra was a resting place for traders and that troops of Shivaji passed through it while depre dating the coastal townships. Roy Burman (1994) described that at Jhnji Mahal sacred grove in Kolhapur district, Shivaji had taken shelter before attacking Shayastaa Khan in Pune. This grove has been a hiding place for the troops and also their training center. Sacred groves have often been supported by the local rulers. Sahu Chatrapati, the king of Kolhapur use to support a sacred grove dedicated at Amba Devi (Roy Burman 1992). Kalam (1996) has mentioned about the state patronage of sacred groves in Kodagudistrict of Karnataka. Roy Burman (1993) mentions about the strategic location of sacred groves along the trade routes in Meghalaya where the moral authority of the Priests- chiefs guided the free flow of commodity. As noted earlier, among the Santhals the sacred groves serve as an important criterion to asscertain village membership and geographical boundary (Troisi, 1978). Hembram (1983) states that through the Sarna Dharma (religion) discrete ethnic groups in Chottanagpur were brought to a common platform for asserting their rights to self determination. The Sarna Dharma, in fact helped them in consolidating their common identity and solidarity between Christian and non-Christian tribes of the region. Mitra and Pal (1994) observerd that Sarna was one of the basic factors that stalled the Koel-Karo dam project in Bihar a decade ago. Roy Burman (1994) has also highlighted this aspect of self ascertain among the Gonds of Gadchiroli district of Maharashtra. The Gonds have revived the Danteswari sacred grove to assert their identity and right of self determination. Self assertion of the tribes through sacred groves is not always strong. D N (1990) interprets the construction of temples in the groves or replacing the local deities by the idol of Hanuman reflects subjugation and marginalisation of the tribal communities by 'the main stream Hindu civilization'.

CONCLUSION

From the foregoing description and analysis it is evident that the institution of sacred grove has many socio-cultural, economic, religious and political dimensions. The sacred graves, have played and continue to play a very important function at various levels-village, local, regional and pan Indian.

The review has provided glimpses on the nature and extent of literature available on this ancient institution and has also identified gaps in our knowledge regarding various anthropological dimensions of the sacred groves. A few areas of future research are listed below:

(i) There is an urgent need to prepare exhaustive inventories of sacred groves in various parts of the country to know the nature and extent of the spread of this ancient cultural and biological institution.

(ii) It will be extremely helpful to prepare a check-list of the kind of minimum information that need to be gathered in the field. This procedure will allow cross-cultural (regional) comparisons.

(iii) Field investigations should be carried out in areas without tribal population and/or forest lands to see if sacred groves are present or absent.

(iv) The role of women in various aspects of the sacred groves need to examined.

(v) A comprehensive study should be undertaken to review the legal status of the sacred groves in different parts of the country.

(vi) A comprehensive study is required to examine various management systems being followed and related these with the level of degradation.

(vii) A multi-location action oriented project using different approaches and involving scientists, local communities and NGOs should be initiated in groves which are under threat or being destroyed to see which approach(es) work better.

REFERENCES

Biswas, P.C. 1956. *Santhals of the Santhal Parganas*. Bharatiya Adimjati Sevak Sangh, Delhi.

Brandis, D. 1897. Indigenous Indian forestry. Sacred groves In: *Indian Forestry Working*. Oriental Institute. pp. 12-13.

Chandrashekara, U.M. and Sankar, S. Structure and functions of sacred groves: Case studies in Kerala. In this volume.

Danda, A.K. and Ekka, W. 1984. *The Nagesia of Chattisgarh*. Anthropological Survey of India, Calcutta.

Deb, D. and Malhotra, K.C. 1997. Interface between biodiversity and tribal cultural heritage: an exploratory study. *Journal of Human Ecology,* 8: 157-163.

D.N. 1990. Woman and forests. Economic and Political Weekly, April 14.

Fernandes, W. 1993. Forests and tribals, informal economy: Dependence and management decisions. In: M. Miri (Ed.), *Continuity and Change in Tribal Society*. Indian Institute of Advanced Study, Shimla.

Gadgil, M. and Vartak, V.D. 1975. Sacred groves in India: a plea for continued conservation. *Journal of Bombay Natural History Society,* 72: 314-320

Gadgil, M. and Vartak, V.D. 1976. Sacred groves in Western Ghats in India. *Ecoonomic Botany,* 30: 152-160

Gadgil. M. and Vartak. V.D. 1981. Sacred groves in Maharashtra—An inventory. In: S.K. Jain (Ed.) *Glimpses of Indian Ethnobotany*. Oxford & IBH Publishers, New Delhi. pp 279-294

Godbole, A., Watve, A., Prabhu, S. and Sarnik, J. Role of sacred groves in biodiversity conservation with local people's participation: A case study from Ratnagiri district, Maharashtra. (In this volume).

Gokhale, Y., Velankar, R., Subhash Chandran, M.D. and Gadgil, M. Sacred woods, grasslands and waterbodies as self-organized systems of conservation. In this volume.

Gurdon, P.P.T. 1914. *The Khasis*. MACMillan and Co., London.

Hemam, S.N,, Malhotra, K.C. and Stanley, S. 1997. The bamboo sacred grove, Kantabaushini Thakurmaa, of Sindipad village, Koraput district, Orissa,. Unpublished.

Hembram, P.C, 1983. Return to the sacred groves. In: K.S. Singh (Ed.) *Tribal Movements in India*. Manohar Publications, New Delhi, Vol. 2, pp. 87-91.

Jha, M., Vardhan, H., Chatterjee, S., Kumar, K. and Sastry, A.R.K. Status and role of Orans (Sacred Groves) in Peepasar village (district Naguar) and Khejarli in (district Jodhpur), Rajasthan. In this volume.

Hughes, J.D. and Subash Chandran, M.D. Sacred groves around the earth: an overview. In this volume.

Kalam, M.A. 1996. *Sacred Groves in Kodagu District of Karnataka (South India): A Socio-Historical Study*. Institut Francais de Pondichery, Pondichery.

Khiewtam, R.S. and Ramakrishnan, P.S. 1989. Socio-cultural studies of the sacred groves at Cherrapunji and adjoining areas in north-eastern India. *Man in India,* 69 : 64-71.

Kosambi, D.D. 1962. *Myth and Reality: Studies in the Formation of Indian Culture*. Popular Press, Bombay.

Malhotra, K.C. 1984. Population structure among the Dhangar caste cluster of Maharashtra, India. In : J.R.Lukaes (Ed.) *The People of South Asia*. Plenum Press, New York. pp. 295-324.

Malhotra, K.C., Das, K (unpublished) Interface between faunal biodiversity and local communities in south-west Bengal.

Malhotra, K.C., Stanley, S., Hemam, N.S. and Das, K. 1997. Biodiversity conservation and ethics : sacred groves and pools. Paper presented at UNESCO Asian Bioethics Conferences, Kobe, November 1997, Japan.

Mitra, A and Pal, S. 1994. The spirit of the sanctuary. *Down to Earth,* January 31, 1994.

Nath, Y.V.S. 1960. *Bhils of Ratanmal. Maharaja* Sayajirao University, Baroda.

Paranjpye, V. 1989. Deorai (sacred grove): An Indian concept of common property resources. In: N.D. Jayal (Ed.) *Deforestation and Drought*. INTACH, New Delhi.

Patil, S. 1982. *Dasa-Sudra Slavery*. Allied Publishers, New Delhi.

Pushpangadan, P., Rajendraprasad, M., and Krishnan, P.N. Sacred groves of Kerala—A synthesis on the state of art of knowledge. In this volume.

Rajendraprasad, M. 1995. *The Floristic, Structural and Functional Analysis of Sacred Groves of Kerala*. Ph.D. thesis, University of Kerala, Trivandrum.

Roy S.C. 1912. *Munda and their country*. City Book Soc. Calcutta.

Roy Burman, B.K. 1993. Indigenous systems of environmental management: Interface of Ontology and political economy. In: A. Nahring (Ed.) *Ecology a Theological Response*. The Gurukul Summer Institute, Madras.

Roy Burman, J.J. 1992. The institution of sacred grove. *Journal of Indian Anthropological Society*, 27: 219-238

Roy Burman, J.J. 1995. The dynamics of sacred groves. *Journal of Human Ecology*, 6 : 245-254.

Roy Burman, J.J. 1996. A comparison of sacred groves among the Mahadeo Kolis and Kunbis of Maharashtra. *Indian Anthropologist*, 26: 37-46.

Sawant, V, 1990. Phondaghat: Ughwai Parisar. Maharashtra times, 22 December.

Singh, G.S. and Saxena, K.G. Sacred groves in the rural landscape: A case study of the Shekhala village in Jodhpur district of desert Rajasthan. In this volume.

Sinha, A.P. 1989. *Religious Life of Tribal India*. Classical Publishers, New Delhi.

Sinha, B. and Maikhuri, R.K. Conservation through 'socio-cultural'—religious practices in Garhwal Himalayas: A case study of Hariyali sacred forest. In this volume.

Sontheimer, G.D. 1989. *Pastoral Deitiies in Western India*. Oxford University Press, New Delhi.

Sisodia, V. and Malhotra, K.C. 1963. Terra-cotta votives among the Warlis of Dharampur. *East and West*, 14: 97-105.

Subhash Chandran, M.D. 1995. Peasant perception of bhutas, Uttar Kannada. In: B. Saraswati (Ed.) *Integral vision: Primal Element of Oral tradition*. D.K. Publishers, New Delhi.

Swamy, P.S., Sundarapandian, S.M. and Chandrasekara. Structure and function of sacred groves: case studies in Tamil Nadu. In this volume.

Tiwari.B.K., Barik, S.K., and Tripathi, R.S. Sacred groves of Meghalaya. In this volume.

Troisi, J. 1978. *Tribal Religion: Religion and Practice among the Santhals*. Manohar Publishers, New Delhi.

Unnikrishnan, E. 1990. Part played by Sacred Groves in Local Environments. Centre for Science and Environment, New Delhi.

Vartak, V.D. and Gadgil, M. 1981. Studies on sacred groves along the Western Ghats from Maharashtra and Goa : Role of beliefs and folklores. In S.K. Jain (Ed.) *Glimpses of Indian Ethnoibotany*. Oxford & IBH Publishers, New Delhi. pp. 272-278.

Vidyarthi, L.P. 1963. *Maler: A study in Nature-Man-Spirit complex of a Hill Tribe of Bihar*. Bookland Pvt. Ltd., Calcutta.

Vidyarthi, L.P. and Rai, B.K. 1977. *Tribal Culture of India*. Concept Publishers, Delhi.

Viji, C. 1995. The sacred groves. Hindu, 5 September.

36

Conserving the Sacred: Where Do We Stand ?

P.S. Ramakrishnan

School of Environmental Sciences, Jawaharlal Nehru University,
New Delhi 110067, India

INTRODUCTION

An aspect of increasing interest in traditional resource management is a growing recognition of the extent and importance of religious sanctuaries and sacred places which house the Gods of village communities and are therefore taboo to human interference. These community-based living repositories provide an important contribution to the conservation of biological diversity, complementing the more recent approaches to protected area management, based on scientific knowledge, and promoted by conservation groups and government agencies. Examples of what can be called local, or vernacular conservation, can be observed at different scales and levels, including the protection of sacred species, sacred groves and sacred landscapes (Ramakrishnan, 1996; papers in this Volume). During the workshop discussions, it was increasingly felt that there are enough data on conservation biology but not enough on those aspects linking biological conservation with cultural integrity. Filling this gap was considered crucial, as it will provide an additional tool for biodiversity conservation and management.

While considering natural ecosystems in a purely biological sense, addition or deletion of species, functional or structural groups, and indeed ecosystem components, are sometimes shown to be of no/limited consequence in altering a given ecosystem level process. However, what is often missed out is that different functional groups control/govern different processes in an ecosystem, with a certain degree of redundancy attached with them, in a manner that others could take over similar function though to a limited extent when the system has to cope up with uncertainties in the environment. Biodiversity becomes crucial for the system as a whole and indeed at the landscape level, as a buffer against perturbations, and for maintaining overall ecological integrity.

However, in this reckoning, humans are often kept out of the definition of the ecosystem boundary. If humans are brought in as an integral component of ecosystem function, its implications for the role of biodiversity would change drastically (Ramakrishnan, 1995a). There are many examples where humans, through species inputs that are intentional or otherwise, have contributed to biodiversity and ecosystem function alterations. The parallelism that has often been recognised by this author between ecologically significant keystone species and socially selected key species, and its value for enhanced/altered ecosystem level functions including their value to the humans is a case in point. No where else is this linkage between ecological and social processes more obvious than in agroecosystem and traditional village ecosystem functions, and indeed at the landscape level where these human-managed ecosystems interact with natural ecosystems (Ramakrishnan, 1992a).

Structural attributes of sacred groves from different part of the globe are well documented (many papers in this Volume). What is less investigated upon is the functional attributes of these groves and their value for conservation of biodiversity and the policy issues that needs to be addressed. It is in this context we discuss the role of biodiversity in ecosystem function, in the context of conserving the sacred, considering the humans as part of the ecosystem/landscape functions. This discussion, therefore, revolves around biodiversity conservation and its management through an understanding of the linkages between ecological and social processes.

SACRED SPECIES AND ECOSYSTEM FUNCTION

What are Keystone Species of Value for Biodiversity in an Ecosystem?

While working in north-east India (Ramakrishnan, 1992a), we soon realised that species which are ecologically valuable keystone species performing key functions in the ecosystem and thereby contributing to support/enhance biodiversity are also species that are socially valued by the local community, for cultural or religious reasons. Thus the Nepalese alder (*Alnus nepalensis*) coming up as an early successional tree species in the shifting agriculture (jhum) fallow plots are protected by the local communities from slashing and burning during the 'jhum' operations. Based on intuitive experience, the shifting agriculture farmer realises that this species does good to his cropping system (Ramakrishnan, 1992a).

I do not wish to go here into the details regarding this fallow management system, but look into the value of this culturally valued species for the highly depleted nitrogen budget under a shortened jhum cycle of 5 years. We have done detailed analysis of nutrient budgets under different jhum cycles operating in the north-eastern region. Without going into the details of this, I give here a brief account of the net changes that could occur under varied cycle lengths. A comparative analysis of the nitrogen budget of the jhum system under 15-, 10- and 5-yr cycles of the Khasi tribe at Shillong (1500 m) in Meghalaya is indicative of the issues involved in nitrogen fertility maintenance under jhum (Mishra and Ramakrishnan, 1984).

The net change in the nitrogen pool (Table 1) suggests that the shortening of the jhum cycle results in a lower nutrient capital at the pre-burn stage as well as at the end of the cropping period. The loss of nitrogen at the end of one year of cropping under the 5-yr jhum cycle is higher only if the values are put on a time scale of 15 years: these then would lose three times more than that from a plot under a 15-yr cycle and twice that from a plot under a 10-year cycle.

Table 1. Net change of nitrogen (10^3 kg ha^{-1} yr^{-1}) in the soil under jhum at Shillong in Meghalaya (from Mishra and Ramakrishnan, 1984)

	15-yr fallow cycle	10-yr fallow cycle	5-yr fallow cycle	
			1st yr crop	2nd yr crop
Soil pool before burning	7.68	7.74	6.40	5.98
Soil pool at the end of cropping	7.04	7.15	5.98	5.60
Net difference	0.64	0.59	0.42	0.18

If the plots under a 5-year jhum cycle now, has longer fallow cycles preceding the 20-year period, for which landuse history at the Shillong site was available (Mishra and Ramakrishnan, 1984), then the system seems to have lost, in terms of initial capital, about 1.28 x 10^3 kg of nitrogen (7.68-6.40 x 10^3 kg) during the last 20 years. Thus, while 15-or 10-year jhum cycles are long enough to restore the original nitrogen status in the soil before the next cropping, it seems unlikely that about 600 kg of nitrogen lost (the difference between the nitrogen capital before and after two croppings) could be restored under a 5-year cycle. One of the important disadvantages of the 5-year jhum cycle lies in the reduced nitrogen capital with which the agroecosystem has to operate due to increased frequency of fire and cropping with too short a fallow phase. Similar conclusions also arise for the jhum systems elsewhere in the region (Swamy and Ramakrishnan, 1988).

It is in this context the value of Nepalese Alder becomes significant. This early successional tree species in the north-eastern hill region above an altitude of 500 m going up to 1900 m, as part of the jhum fallows has nodulated roots, colonised by *Frankia* occurring as an endophyte, and is effective in biological nitrogen fixation (Sharma and Ambasht, 1988). Under short jhum cycles of 5 to 6 years, we have already seen that about 600 kg ha^{-1} of soil nitrogen is lost during one cropping season and it takes a minimum of 10 years of natural fallow regrowth to recover all this nitrogen back into the system (Mishra and Ramakrishnan, 1984). It is in this context of putting back the 600 kg ha^{-1} nitrogen over a period of 4 to 5 years of the fallow phase that an early successional species such as the Nepalese alder becomes important (Ramakrishnan, 1992a). Apart from nitrogen fixation, the production of nitrogen-rich litter and mineralisation also contributes to biological build-up of soil fertility. Indeed, the Nepalese alder coming up during early succession is critical for replenishing nitrogen in the jhum system under a shortened jhum cycle.

Thus, this species could be used for fallow management with community participation, since the people can identify themselves with a value system that they understand and appreciate. One of the short-term (5-10 year) pathways considered for improving shifting agriculture in north-east India was to strengthen the agroforestry component of the shifting agriculture system using locally identified and acceptable species, such as the Nepalese alder (*Alnus nepalensis*). Indeed, the Nepalese alder technology is being promoted through the Village Development Boards of Nagaland (Gokhale, et al., 1985).

Keystone Species at the Ecological/Socio-cultural Interphase: A Less Understood Area of Study

Whilst the Nepalese alder and its role in agroecosystem and forest ecosystem function has been studied in detail, our knowledge in this inter-phase area of keystone species is still very limited. Bamboo species, *Dendrocalamus hamiltonii, Bambusa tulda and B. khasiana* which are important early successional species and which conserve nitrogen, phosphorus or potassium in the system are also highly valued by all communities in north-east India; they are often grown along the margin of agricultural plots and as part of the home garden system, or used for a variety of traditional activities, such as, house construction, making household utensils, tubing for water transportation, thatching material, etc. These and other ecologically important keystone species are also socially selected keystone species, and these through their functional attributes help in enhancing biodiversity both in space and time.

A broader survey done by this author in the Indian context also suggests such a close parallelism existing between ecological and social processes (e.g., the locally valued *Prosopis cinerarea* in the arid state of Rajasthan and *Azadirachta indica* throughout the sub-continent, and many others discussed in this paper). We have already considered the ecological value of socially selected *Quercus* species of the Kumaon and Garhwal region of the western Himalayas as an important component of the mountain forest ecosystem function (Ramakrishnan et al., 1994a). They help in improving soil fertility through efficient nutrient cycling, conserving soil moisture partly through humus build-up in the soil and partly through a deeply placed root system which has root biomass uniformly distributed throughout the soil profile. Consequently, they support a rich biodiversity in the ecosystem.

The implication is that linking up ecological and social processes is significant for enhancing biodiversity in the ecosystem, building up of biodiversity based on ecological integrity of the system (because keystone species identified from a given ecosystem type would promote biodiversity that is compatible within that system), and indeed for ensuring rural peoples' participation as an integral part of the ecosystem function, rather than being a manipulator from outside (Ramakrishnan, 1993; Ramakrishnan, 1995b). This last point is extremely important for rehabilitation of degraded ecosystems and for rural development (Ramakrishnan et al., 1994b). Being able to identify themselves with a value system that they cherish through the socially selected keystone species, the local communities would be able to participate in the rehabilitation programme in a more meaningful manner.

FUNCTIONAL ATTRIBUTES OF SACRED GROVES

As already discussed, the sacred grove in the fragile environment of Cherrapunji is one of the well worked out example of the role of sacred grove in ecosystem function. Operating under highly stressed environmental conditions, this delicately balanced sacred grove forest ecosystem is highly fragile. It is not surprising then that the landscape of Cherrapunji, except for the relict sacred grove, is largely balded with arrested seral grassland types (Ramakrishnan and Ram, 1988).

As already discussed (Ramakrishnan, this Volume), the relic climax sacred grove forest at Cherrapunji which has developed on a nutrient deficient soil is dependent upon tight recycling of nutrients. The fine root system developed in the surface layer of the soil, and that particularly located over the mineral soil should facilitate rapid uptake of nutrients released by the decomposing litter; in addition, nutrient wash-out by rain through direct fall, throughfall and stem flow would also be intercepted by the roots (Khiewtam and Ramakrishnan, 1993).

The presence of a fine root biomass up to a maximum of about 14,000 kg ha^{-1} in the soil to a depth of 30 cm is, therefore, helpful in mopping up the nutrients (Table 2). About 40% of the roots are located in the surface

10 cm of the soil, of which the fine root biomass above the mineral soil was about 6%. About 58% of the fine root biomass comprised roots of less than 6 mm in diameter. Further, the fine root biomass fluctuated at different seasons, with higher values during the monsoon. This fluctuation was more pronounced in the surface 10 cm layer of soil than in deeper layers. Further, the increased root biomass during the monsoon season corresponded to the period of rapid nutrient release (Table 2). This is particularly significant when one realises that roots at this time are growing at their fastest rates and that the concentration of nutrients becomes diluted by production of new root material (Khiewtam and Ramakrishnan, 1993).

Table 2. Seasonal variation in fine root biomass (kg ha^{-1} ± SE) for five different size classes of roots from different soil depths in the sacred grove at Cherrapunji i north-eastern India

	Biomass (kg ha^{-1})			LSD
	Summer	Monsoon	Winter	(P = 0.05)
Soil depth (cm)				
0	745±62	882±58	517±32	
0-5	1376±89	2276±102	1184±65	555 (Soil depth)
5-10	2275±162	2912±102	2174±108	342 (Season)
10-20	3774±190	3645±232	3319±212	
20-30	3627±165	4237±222	3558±256	
Root size class (mm)				
< 1	2490±127	3166±188	2058±112	
1-3	1941±90	2431±168	1725±90	
3-6	2215±112	2419±105	1977±88	430 (Soil depth)
6-12	2490±108	2859±222	2406±141	531 (Season)
> 12	2661±118	3077±242	2586±155	

In the ultimate analysis, once disturbed, this delicately balanced ecosystem degrades to extensively desertified grassland (Ramakrishnan and Ram, 1988; Ram and Ramakrishnan, 1992). Litter and fine root dynamics play a key role in nutrient conservation and are, therefore, important for the maintenance of these fragile systems. From a management point of view, this sacred grove system illustrates that the dynamics of fine roots on the forest floor in relation to litter dynamics and nutrient release should receive more attention from ecologists and forest managers.

It is in this context that an understanding of the sacred ecosystem function (Ramakrishnan, 1992a; Khiewtam and Ramakrishnan, 1993) is significant for designing a rehabilitation strategy of the degraded landscape, with peoples' participation (Ramakrishnan, et al., 1994b). The plant biodiversity contained therein could be used as a resource for ecosystem rehabilitation based on forest successional concepts (Ramakrishnan, 1992a). The role of keystone species contained therein becomes significant in this context.

Keystone species, as already discussed, are shown to play a crucial role in biodiversity conservation in an ecosystem, through key functions that they perform. Therefore, they could be used for not only restoring pristine ecosystems under threat, but also for building up biodiversity in degraded ecosystems such as those at Cherrapunji, through appropriately conceived rehabilitation strategies (Ramakrishnan et al., 1994b).

The four dominant species in the sacred grove at Cherrapunji, namely *Englehardtia spicata, Echinocarpus dasycarpus, Syzygium cuminii* and *Drimycarpus racemosus* are keystone species in the sense that their leaf litter contain high levels of nitrogen, phosphorus and potassium compared to other species in this ecosystem (Khiewtam and Ramakrishnan, 1993); they perform key functions of nutrient conservation in the ecosystem. This role of species is crucial for rehabilitating degraded ecosystems especially when soil fertility is a constraining factor. Indeed, these four species and others needed for rehabilitation at Cherrapunji are found in the sacred grove. These and other keystone species are also socially selected keystone species, with implications for rehabilitation and biodiversity conservation with peoples' participation, as already discussed earlier.

An aspect that is often ignored is that aboveground biodiversity often influence belowground biodiversity, which in turn determine soil biological processes that contribute towards sustainable soil fertility management (Woomer and Swift, 1994), particularly while dealing with fragile environment. Earthworms are often identified as an indicator of soil fertility status. It is in this context, the qualitative differences noted between the earthworm populations in highly disturbed successional environments as compared to a bench-mark sacred grove site in Mawphlong site in Meghalaya in north-east India becomes significant (Bhadauria and Ramakrishnan, 1989, 1990). Earthworms with their important role in nutrient conservation and distribution in the soil system may influence associated soil biodiversity in a variety of ways. This is an area which offers opportunities for exploration, particularly while considering fragile sacred grove ecosystems as illustrated by the north-east Indian example.

SACRED LANDSCAPE AND FUNCTIONS

Sacred landscape is a set of inter-connected ecosystems, which influence one another in a variety of ways. Large-scale perturbations cause impacts not only upon individual ecosystems but also alter other ecosystems of the landscape in a variety of ways (Box 1). Thus large-scale deforestation that have occurred in the higher alpine zone mountain ranges in the sacred landscape of West Sikkim Yoksum region has created a whole variety of problems for ecosystem types located in the lower ranges.

According to the data collated by the Himalayan Nature and Adventure Foundation, Siliguri, the Ouglthang and Rathong glaciers are retreating rapidly, with size reduction. Retreating glaciers create several moraine dams, containing sizable quantities of water. Increased snow melt and exceptionally high rainfall in a given year could result in dam bursts and flash floods in the lower regions. The 1988 dam burst is an example that could recur, if adequate catchment treatment measures are not initiated immediately.

The vegetation in the landscape is varied ranging from alpine scrub vegetation at higher reaches to sub-tropical moist evergreen forests down below. Afforestation is essential over an area of 4290 ha. catchment area of the Rathong Chu. This task is crucial for conservation of the sacred landscape as a whole and for its ecological integrity. Implementing the above mentioned activity is crucial for controlling erosion and flash floods.

On the basis of an elaborate ecological/anthropological analysis of the ancient terraced landscape of the Ifagu tribe in the upland regions of Phillippines, UNESCO has recognised this landscape as a World Heritage Site, with a view to study, conserve and indeed develop these multi-species complex traditional agricultural systems in a sustainable manner.

Here in Sikkim (Box 1) is a much complex landscape with a whole variety of agroecosystem types, closely linked to many traditional societies of varied socio-cultural background on the one hand, and to the forest ecosystem types of high biological diversity on the other. The sacred river Rathong Chu connects all the natural and human-managed ecosystem types into a unified whole right from the snow clad alpine meadow zone to the humid sub-tropical rain forest zone down below.

Sustainable Land Use Management

In all traditional societies, particularly of the Asian and the African tropics, agriculture, animal husbandry and fisheries are very closely inter-linked with forest ecosystem function. With ecological landscape being closely interwoven into the cultural landscape and with a number of traditional societies often coexisting with one another, understanding a variety of processes linked to population dynamics (Ramakrishnan, 1994a), soil fertility dynamics (Ramakrishnan, 1994b; Ramakrishnan et al., 1994c), biodiversity issues (Ramakrishnan, 1996), forest management concerns (Ramakrishnan, 1992b), socio-economic and cultural dimensions of local communities indeed

Box 1. Demojong in West Sikkim district-An unique example of a sacred landscape

The region is richly endowed with biological resources spread over a a variety of ecosystem types over a range of altitude, from the alpine *Rhododendron* dominated scrub forest through conifer forests with *Abies densa* and *Tsuga demosa* getting down to mixed evergreen forests dominated by species such as *Castanopsis* spp., *Quercus lamellosa, Lithocarpus spicatus, Elaeocarpus lanceaefolius, Michilus edulis, Michelia* spp., etc. The region under consideration here has all these types over a very short transect running down from the alpine to the sub-tropical zone. Orchids are abundant. The rich wildlife is represented by Himalayan Black Bear, Musk Deer, Fishing Cat, Leopard Cat, Black Capped Langur and a rich bird life in this unique landscape.

The region is rich in medicinal plants of value to traditional Tibetan pharmacopoeia, nurtured in Sikkim by the Buddhist monasteries. Conserving these plants and their cultivation would ensure the survival of one of the oldest system of medicine stretching back to more than 2500 years. With recent attempts to revive traditional medicine, possibilities of providing sustainable livelihood to local communities through cultivation of medicinal plants are immense.

With a whole variety of traditional societies living in the area, they do smaller perturbations to the system through a whole variety of ancient terraced multi-species complex agricultural systems, complex valley land wet rice cultivation systems, and highly complex multi-species home garden systems. Small-scale perturbations are also caused through extraction of forest-based items such as such as fodder, fuelwood and timber by local communities, in a sustainable manner. There is an urgent need not only to understand the linkages between these different land use systems, but to develop sthem sustainably for improving the quality of life of the local communities.

The air, soil, water and the biota are all sacred to the people, because of the inter-connections that they perceive to exist. Any human-induced major perturbations are considered by the Sikkimese Buddhist's to spell disaster to Sikkim as a whole, because of the disturbance caused to the rulinging deities and the treasures ('ters') placed in the landscape. Sustainable rural activities, apparently, are small-scale perturbations only. It is in this context the proposed hydro-electric project was vs viewed by local people, and which now stands abandoned. Does this then imply that the traditional societies made a distinction between small-scale versus large-scale perturbations in ecosystem function, which we now talk about in Ecology!

Yoksum is an area which what Sikkimese perceive as the very basis for the presentent Sikkimese culture. The entire region right from the Kangchendzonga to the Yoksum lowlands (the sacred region of 'Demojong') is most appropriate to be declared a 'National Heritage Site', with all its people ecological and cultural heritage - the land and the landuse systems, all the water bodies (the obvious and notional lakes), the Rathong Chu, the monasteries, the historical sites, and the rich biodiversity-for conservation and sustainable development in a truly holistic spirit.

become important (Ramakrishnan, 1992a). While dealing with traditional societies in the north-eastern hill areas of India (Box 2), which is one of the most thoroughly worked out landscape analysis of its kind which is relevant to our discussion, though not part of a sacred landscape, these linkages in a landscape analysis was shown to be crucial for designing strategies for sustainable livelihood of traditional societies in the short-term and for sustainable development of the region from a long-term perspective (Box 3). What we are looking for, then is to look at sacred landscapes from a similar perspective and sustainably develop the natural resources within them for the welfare of the local communities.

Box 2. Shifting agriculture (Jhum) in north-eastern India and social disruption (from Ramakrishnan, 1992a).

North-eastern India has over 100 different tribals, linguistically and culturally distinct from one another; the tribes often change over very short distances, a few kilometers in some cases. Shifting agriculture or jhum as the tribals call it, is the major economic activity. This highly organized agroecosystem with up to 40-45 crop species and much more associated biodiversity, is based on empirical knowledge accumulated through centuries and was in harmony with the environment as long as the jhum cycle (the fallow length intervening between two successive croppings) was long enough to allow the forest and the soil fertility lost during the cropping phase to recover.

Supplementing the jhum system is the valley system of wet rice cultivation and home gardens. The valley system with many rice cultivars is sustainable on a regular basis year after year because the wash-out from the hill slopes provides the needed soil fertility for rice cropping without any external inputs. Home gardens extensively found in the region have economically valuable trees, shrubs, herbs and vines and form a compact multi-storied system of fruit crops, vegetables, medicinal plants and many cash crops; the system in its structure and function imitate a natural forest ecosystem, and is self-sustaining. The number of species in a small area of less than a hectare may be 30-40 going up to about 100; it therefore represents a highly intensive system of farming in harmony with the environment.

Linked on the this land-use are the animal husbandry systems centred traditionally around pigs and poultry. The advantage here is primarily that they are detritus-based or based on the recycling of food from the agroecosystem unfit for human consumption.

Increased human population pressure and decline in land area resulting from extensive deforestation for timber for use by industrial man have brought down the jhum cycle to 4-5 year or less. Where population densities are high, as around urban centres, burning of slash is dispensed with, leading to rotational bush-fallow/sedentary systems of agriculture. These are often below subsistence level, though the attempt is to maximize output under rapidly deplete soil fertility. Inappropriate animal husbandry practices introduced into the area, such as goat or cattle husbandry, could lead to rapid site deterioration through indiscriminate grazing/browsing and fodder removal, as has happened elsewhere in the Himalayas. The serious social disruption caused demands an integrated approach to managing the land use/ human interface.

Thus, conserving 'sacred landscape' becomes more complex because of the inter-connected ecosystem types wherein the humans are integrated. We need to have a sustainable development strategy, highly complex by its very definition and indeed at an enormous cost. Sustainable development involves a series of compromises (Ramakrishnan, 1995b; Ramakrishnan et al., 1994a). Such a compromise should not only be based on biophysical considerations Ramakrishnan, 1995b), but also be based on integrating the human dimension, viewing ECOLOGY in an all encompassing holistic sense.

The Sikkimese sacred landscape, like many others such as Adam's Peak in Sri Lanka or Sabarimalai in the Western Ghat region of Kerala, for example, is a unique case where ecological considerations cannot be separated from historical, social, cultural and religious dimensions of the problem. Here is a sacred landscape where the people living in the region

Box 3. Shifting agriculture (jhum) and sustainable development for north-eastern India (from Ramakrishnan, 1992a).

For improving the system of land use and resource management in north-eastern India, the following strategies suggested by Ramakrishnan and his coworkers are based on a multidisciplinary analysis. Many of these proposals have already been put into practice.

- With wide variations in cropping and yield patterns under jhum practised by over a hundred tribes under diverse ecological situations, where transfer of technology from one tribe/area to another alone could improve the jhum, valley land and home garden ecosystems. Thus, for example, emphasis on potato at higher elevations compared to rice at lower elevations has led to a manifold increase in economic yield despite low fertility of the more acid soils at higher elevations.
- Maintain a jhum cycle of minimum 10 years (this cycle length was found critical for sustainability), when jhum was evaluated using money, energy, soil fertility biomass productivity, biodiversity, and water quality, as currencies) by greater emphasis on other landuse systems such as the traditional valley cultivation or home gardens.
- Where Jhum cycle length cannot be increased beyond the five- year period that is prevalent in the region, redesign and strengthen this agroforestry system incorporating ecological insights on tree architecture (e.g., the canopy form of tree should be compatible with crop species at ground level so as to permit sufficient light penetration and provide fast recycling of nutrients through fast leaf turnover rates). Local perceptions are extremely important in tree selection, for introduction into the cropping and fallow phases of Jhum, as is being done in a major initiative in the state of Nagaland in north-east India.
- Improve the nitrogen economy of jhum at the cropping and fallow phases by introduction of nitrogen-fixing legumes and non-legumes. A species such as the Nepalese alder (*Alnus nepalensis*) is readily taken in because it is based on the principle of adaptation of traditional knowledge to meet modern needs. Another such example is the lesser known food crop legume *Flemingia vestita*, traditionally used by tribals as an important species when jhum cycles decline below 5 years.
- Some of the important bamboo species, highly valued by the tribals, can concentrate and conserve important nutrient elements such as N, P and K. They could also be used as windbreaks to check wind-blow loss of ash and nutrient losses in water.
- Speed up fallow regeneration after jhum by introducing fast growing native shrubs and trees.
- Condense the time-span of forest succession and accelerate restoration of degraded lands based on an understanding of tree growth strategies and architecture, by adjusting the species mix in time and space.
- Improve animal husbandry through improved breeds of swine and poultry.
- Redevelop village ecosystems through the introduction of appropriate technology to relieve drudgery and improve energy efficiency (cooking stoves, agricultural implements, biogas generation, small hydroelectric projects, etc.). Promote crafts such as smithying and products based on leather, bamboo and other woods.
- Strengthen conservation measures based upon the traditional knowledge and value system with which the tribal communities could identify, e.g., the revival of the sacred grove concept based on cultural tradition which enabled each village to have a protected forest once upon a time although few are now left.
- In the ultimate analysis, have an integrated approach for land use development in a given ecological/cultural landscape; base short-term sustainable livelihood strategy on building upon traditional knowledge and technology; long-term sustainable development plan should based on larger ecologic/economic considerations should gradually built up to avoid social disruptions.

are truly integrated within the landscape unit itself, in a socio-economic sense. Therefore, one has to consider sustainable development of the region as an integrated issue with vernacular conservation. Declaring the sacred landscapes as 'National Heritage sites' and their eventual recognition as 'World Heritage sites' of UNESCO, would be a step in the right direction not only for conservation of biodiversity, both natural and human-generated, but also for evolving and implementing a meaningful sustainable development action plan, with peoples' participation.

MANAGEMENT IMPLICATIONS

At one level, the sacred species occurring in a variety of ecosystem types in the sub-continent, with all the sub-specific variations that they exhibit should be conserved for human welfare, as discussed above.

During the workshop discussions, an issue that came up again and again was related to studies centred around management strategies for sacred groves and landscapes. Economic incentives, developing and managing buffer zones around, mobilization of public opinion, strengthening traditional management practices while constantly reviewing and analysing management plans and incorporating them from time to time, arriving at appropriate legal framework for management, networking interested individuals and agencies for exchange of information were some of the issues that were discussed at length.

Biodiversity management in situations where traditional societies live has to deal with managed biodiversity in their landuse systems and with biodiversity in natural ecosystems. Traditional societies as part of the ecosystem function deal with biodiversity in a variety of ways (Ramakrishnan, 1992a). Thus, The manner in which the tribals manipulate crop and associated biodiversity at species and sub-specific levels-under varied Jhum cycles of shifting agriculture, transformation of Jhum to sedentary systems, valley rice cultivation systems, and home gardens-give a clue about their response to uncertainties in the environment that they have had to cope up with in the past for sustainable land use management (Box 4). Similarly, the way traditional societies perceive species within natural ecosystems (e.g., culturally valued key stone species) and the way they perceive ecosystem types at the landscape level, provide important clues for biodiversity conservation/management at ecosystem and landscape levels. All these have implications for future response mechanisms that need to be integrated into managing ecosystems under threat from 'global change'. In the context of the concept of sacred groves/landscapes or outside of it, with a variety of natural and traditionally managed interacting ecosystem types, traditional societies has often ensured application of a management

Box 4. Biodiversity/landscape functions linked to biodiversity, for sustainable development (from Ramakrishnan, 1992a; Ramakrishnan, et al.,1994a, 1996)

There are 'process' and 'system' level linkages between biodiversity and ecosystem function. The following are a few illustrative linkages:

Process level linkages

o The sub-specific level biodiversity in rice is effectively used by the Apatani tribal farmer of north-east India to capture and maximize production under differential nutrient status of the soil. Thus he is able to have pisciculture combined with a long duration variety of rice in a nutrient-rich environment whilst restricting short-duration variety to nutrient-poor sites of the landscape.

o As part of ecosystem function, the way shifting agricultural farmers manipulate biodiversity for determining ecosystem function within the landscape and for societal benefits is interesting. The shift effected, (a) towards nutrient-use efficient tuber and vegetable crops under shorter Jhum cycles from less efficient cereals emphasised under longer cycles, and (b) a similar shift in crop placement with nutrient-use efficient species being more concentrated on top of the hill slope with less efficient species at the bottom of the slope in order to capture the soil nutrient gradient in a Jhum plot, are elegant examples of adaptation biodiversity towards optimisation of resource use and risk coverage.

o Through mixed cropping involving a large number of species, and traditional weed management strategies (where 20% weed bio mass is left *in situ*), shifting agricultural farmer of north-east India, like other traditional farmer elsewhere, ensures not only effective check on nutrient loss but also recycle nutrients during the cropping phase.

o Ecologically important keystone species in natural and human-managed ecosystems are often socially selected. Such species often have important ecosystem level functions such as improving soil fertility, strengthening nutrient cycling processes or contributing to more effective soil moisture conservation - functions that help in biodiversity conservation and enhancement both in space and time. Such keystone species can be a basis for peoples' participation in sustainable development.

System level linkages

o Water was identified as a triggering agent for a variety of sustainable development functions, such as agroecosystem redevelopment, rehabilitation of degraded ecosystems, watershed management, etc., with peoples' participation in the Himalayan belt and elsewhere in India. By redeveloping traditional water harvesting technologies and providing water during the dry season outside the monsoon season when there is severe water shortage, ecosystem redevelopment possibilities were demonstrated.

o The concept of the 'sacred groves' enshrined in the culture of many forest dwelling traditional societies may be viewed as 'refugia' in an ecological sense, that could have value for sustainable forestry management practices. Apart from the rich biodiversity that these harbour, they also harbour keystone species of value for forest rehabilitation.

o The diversity of cropping and resource systems that form part of the landscape amongst many traditional societies serve not only as a major means of protecting ecological integrity at the landscape level, but also acts as the knowledge and resource base that makes adaptivity possible.

Understanding the these complex interactions in a landscape and the way traditional knowledge and technology is used for meeting livelihood needs in the past, would give us a better handle to cope up with land use management under future 'global change' threat.

strategy, that is now being rediscovered in the context of a decentralised network of protected patches of biodiversity in a heterogeneous landscape, (Box 5).

Box 5. Loosely coupled landscape management strategy (Ehrenfeld, 1991)
* Management plan should stimulate loose coupling within the system, rather than maximise tight coupling. Such a strategy would reduce the system's vulnerability to accidents. * Avoid manipulative management plan based on large, prioritised, single-protocol, single theory, generalised scheme. Instead for large protected areas have only a low key management plan, if required. Have individualised management for fragmented landscapes to accommodate singularities and uniqueness of each piece and ensure loose coupling. * In the ultimate analysis, management strategy should be flexible, capable of small-scale operations, information-sensitive, and composed of elements that are integrated and yet independent.

POLICY ISSUES

With rapid and continued decline occurring in the quality and the number of sacred groves, because of changes in value system and pressure on land due to increasing population and dwindling natural resources, there is an urgent need to document and monitor the existing groves, analyse the scientific basis of these relict ecosystem functional units, and evaluate their value for biodiversity conservation. The research efforts should particularly be directed towards evaluating keystone species in these ecosystems, the range of sub-specific variability met within them, and to exploit them for nature reserve management and for rehabilitation of degraded landscape. The efforts in this direction should be well coordinated between Governmental agencies such as the Ministry of Environment and Forests, the Wasteland Development Board, other governmental and many non-governmental agencies concerned and the scientific community at large. The research and management of these groves should be part of a larger effort, extending beyond the sacred sites themselves, to develop a meaningful action plan for natural resource management at the landscape level, based on traditional knowledge and technology in regions where traditional societies live. In doing this ecological perceptions should be closely knit with socio-economic and cultural concerns of local communities so as to ensure their participation both in conserving and indeed enhancing the quality of these groves. In this context, the locally recognized and accepted land tenure system is a case in point. Joint management committees, that may be appropriate for improving the quality of these groves and to use them for local benefit to the extent feasible, should operate within local traditional set up.

Identifying and conserving selected large groves that are well preserved and which are ecologically significant systems should be taken up on a priority basis. Indeed, the governments concerned should initiate steps to declare a few of selected groves representing different ecological zones in the country as 'National Heritage' sites, for effective conservation and use wherever feasible (e.g., eco-tourism). The Mawsmai forest in Cherrapunji is an example because of its geographical significance and ecological fragility. The economically valuable germ plasm, such as medicinal plants, could then be exploited through cultivation outside the sacred grove, as part of a larger rural development plan.

All identifiable sacred landscapes, such as the Ganga system, the 'Demojong' of the West Sikkim, Sabarimalai in the Western Ghat region of Kerala identified from India, Adam's Peak in Sri Lanka could be fit candidates for declaration as 'National Heritage' sites, with a view on their conservation and sustainable management on the basis of a well developed plan of action. These cultural landscapes with traditional societies living within, the natural and managed biodiversity within them forming an interconnected and interacting ecological landscape should be rigorously researched upon with a view to link conservation with sustainable livelihood/ development.

Thus, the Sikkimese sacred landscape, for example, is a unique case where ecological considerations cannot be separated from historical, social, cultural and religious dimensions of the problem. Here is a sacred landscape where the people living in the region are truly integrated within the landscape unit itself, in a socio-economic sense. Therefore, one has to consider sustainable development of the region as an integrated issue with vernacular conservation. Declaring the sacred landscapes as 'National Heritage sites' and their eventual recognition as 'World Heritage sites' of UNESCO, would be a step in the right direction not only for conservation but also for evolving and implementing a meaningful sustainable development action plan, with peoples' participation.

REFERENCES

Bhadauria, T. and Ramakrishnan, P.S. 1989. Earthworm population dynamics and contribution to nutrient cycling during cropping and fallow phases of shifting agriculture (Jhum) in north-east India. *Journal of Applied Ecology,* 26: 505-520.

Bhadauria, T. and Ramakrishnan, P.S. 1990. Population dynamics of earthworms and their activity in forest ecosystems of north-east India. *Journal of Tropical Ecology,* 7: 305-318.

Ehrenfeld, D. 1991. The management of biodiversity: A conservation paradox. In: F.H Bormann and S.R. Kellert (Eds.) *Ecology, Economics, Ethics: The Broken Circle.* Yale University Press, New Haven. pp. 26-39.

Gokhale, A.M., Zeliang, D.K. Kevichusa, R. and Angami, T. 1985. *Nagaland: The Use of Alder*

Trees. Education Dept., Kohima, Nagaland.

Khiewtam, R.S. and Ramakrishnan, P.S. 1993. Litter and fine root dynamics of a relict sacred grove forest at Cherrapunji in north-eastern India. *Forest Ecology and Management,* 60: 327-344.

Mishra, B.K. and Ramakrishnan, P.S. 1984. Nitrogen budget under rotational bush fallow agriculture (Jhum) at higher elevations of Meghalaya in north-eastern India. *Plant and Soil,* 81: 37-46.

Ram, S.C. and Ramakrishnan,P.S. 1992. Fire and nutrient cycling in seral grasslands of Cherrapunji. *International Journal of Wildland Fire,* 2: 131-138.

Ramakrishnan, P.S. 1992a. *Shifting Agriculture and Sustainable Development of North-Eastern India*. UNESCO-MAB Series, Paris, Parthenon Publ., Carnforth, Lancs. U.K. (republished by Oxford University Press, New Delhi, 1993).

Ramakrishnan, P.S. 1992b. Tropical forests: Exploitation, conservation and management. In: special issue on *Environment and Development Impact (UNESCO)*. 42 (No. 166), 149-162.

Ramakrishnan, P.S. 1993. Evaluating Sustainable Development with Peoples' Participation. In: F. Franz Moser (Ed.). *Proceedings of the International Symposium. Sustainability - Where Do We Stand ?* Technische Universitat Graz, Austria, pp.165-182.

Ramakrishnan, P.S. 1994a Participatory development in managing population pressure on natural resources. In: R. Krishnan (Ed.), *Growing Numbers and Dwindling Resources*. Tata Energy Research Inst., New Delhi. pp. 86-96.

Ramakrishnan, P.S. 1994b. The jhum agroecosystem in north-eastern India: A case study of the biological management of soils in a shifting agricultural system. In: P.L. Woomer and M.J. Swift (Eds.), *The Management of Tropical Soil Biology and Fertility*. TSBF, Nairobi and Wiley-Sayce Publ., Exeter, U.K. pp. 189-207.

Ramakrishnan, P.S. 1995a. Ecological functions of biodiversity: The human dimension. In: F. di Castri and T. Younes (Eds.), *Biodiversity, Science and Development*. CAB International, Oxford. pp.114-129.

Ramakrishnan, P.S. 1995b. Sustainable rural development with people's participation in the Asian context. In: S. Samad, T. Watanabe and S. Kim (Eds.), *People's Initiatives for Sustainable Development: Lessons of Experience* Asia & Pacific Development Centre, Kuala Lumpur, pp. 421-439.

Ramakrishnan, P.S. 1996. Conserving the sacred: From species to landscapes. *Nature and Resources,* 32: 11-19.

Ramakrishnan, P.S. and Ram, S.C. 1988. Vegetation, biomass and productivity of seral grasslands of Cherrapunji in north-east India. *Vegetatio,* 74: 47-53.

Ramakrishnan, P.S., Purohit, A.N., Saxena, K.G. and Rao, K.S. 1994a. *Himalayan Environment and Sustainable Development*. Diamond Jubilee Publication of Indian National Science Academy, New Delhi.

Ramakrishnan, P.S., Campbell, J., Demierre, L., Gyi, A., Malhotra, K.C., Mehndiratta, S., Rai, S.N., & Sashidharan, E.M. 1994b. *Ecosystem Rehabilitation of the Rural Landscape in South and Central Asia: An Analysis of Issues*. Special Publication, UNESCO (ROSTCA), New Delhi.

Ramakrishnan, P.S. Saxena,K.G. Swift, M.J. and Seward, P.A. (Eds.) 1994c. *Tropical Soil Biology and Fertility Research: South Asian Context*. G.B. Pant Inst. of Himalayan Environment and Development, Kosi and Oriental Enterprises. Dehra Dun.

Ramakrishnan, P.S., Das, A.K. and Saxena, K.G. 1996. *Managing Biodiversity for Sustainable Development*. Indian National Science Academy, New Delhi.

Sharma, E. and Ambasht, R.S. 1988. Nitrogen accretion and its energetics in the Himalayan alder. *Functional Ecololgy,* 2: 229-235.

Swamy, P.S. and Ramakrishnan, P.S. 1988. Nutrient budget under slash and burn agriculture (jhum) with different weeding regimes in north-eastern India. *Acta Oecologica-Oecol. Applic.*, 9: 85-102.

Woomer, P.L. and Swift, M.J. (Eds.) 1994. *The Biological Management of Tropical Soil Fertility*. John Wiley, Chichester.

Conserving the Sacred: Epilogue

P.S. Ramakrishnan

School of Environmental Sciences, Jawaharlal Nehru University,
New Delhi 110067, India

When it was decided to title this volume as 'Conserving the Sacred: For Managing Biodiversity', the intention was not to look at the issues involved here in the traditional sense of religious parochialism. The intention was to bring out the 'Unity in Diversity' of all religious faiths, as is seen from the wide range of articles from diverse situations and from different parts of the world. This objective has been amply realised as seen from the preceding pages. However, one would like to see revivalism and indeed a greater focus on traditional value systems of relevance for natural resource management, geared to meet the needs of an industrialised and/ or industrialising society like ours, in a sustainable manner. At the same time, we need to reassure ourselves against narrow opportunistic tendencies overtaking us in the name of 'development' (Burrows, this Volume).

The exaltation of the World as the body of 'God' as enshrined in the ancient Indian tradition-the 'Vedas' of antiquity, is conceptualised by this author elsewhere (Ramakrishnan, this Volume), through the sacred landscape linked through the Ganga river system right from the highest mountain reaches of the 'Goumukh' in the Garhwal Himalaya extending up to the Gangetic delta merging into the Bay of Bengal, a few hundred kilometers away. Indeed, this concept of sacred landscape is concertised more precisely by Buddhism and Jainism (Ramakrishnan, this volume); The concept of 'Ahimsa'-non-violence, and 'Sarvabhutadaya'-compassion towards all living beings enshrined in Buddhism, further sets the tone for incorporating ethics into eco-philosophy.

"Tat Twam Asi', literally translated is "That Thou Art", is a very profound Vedantic statement, from the Hindu scriptures of antiquity. Here the individual (Twam) is identified as part of the creation-the 'Brahman' (Tat). This statement impregnated with a variety of meanings, in an ecological sense projects the individual as part of, 'Prakriti' or the world as we perceive it, the 'Nature'. Saraswati (this Volume) talks of the 'Cosmic Tree', embracing the entire Universe, and rooted in 'Brahman'. Summarised as a world-view of ecology linked to ethics, one could derive the following from this profound Vedantic statement.

Evolution, in which all creation plays its part, and in which human beings, though still at the cutting edge of the evolutionary process are always a part of the ecosystem function (Ramakrishnan, this Volume) and not apart from it. This would then imply compassion and fellowship, not just between humans but extended to the entire biosphere. Such a relationship alone will give meaning and value to our world, lest we fall into the kind of anthropo-centred thinking that has been the bane of traditional ecological thought.

Index